INTRODUCTION TO MODERN DYNAMICS

Introduction to Modern Dynamics

Chaos, Networks, Space and Time

David D. Nolte

Purdue University

OXFORD

UNIVERSITY PRESS

OXFORD
UNIVERSITY PRESS

Great Clarendon Street, Oxford, OX2 6DP,
United Kingdom

Oxford University Press is a department of the University of Oxford.
It furthers the University's objective of excellence in research, scholarship,
and education by publishing worldwide. Oxford is a registered trade mark of
Oxford University Press in the UK and in certain other countries

Published in the United States of America by Oxford University Press
198 Madison Avenue, New York, NY 10016, United States of America

British Library Cataloguing in Publication Data

Data available

Library of Congress Control Number: 2014939331

ISBN 978–0–19–965703–2 (hbk.)
ISBN 978–0–19–965704–9 (pbk.)

Printed and bound by
CPI Group (UK) Ltd, Croydon, CR0 4YY

Preface
The Best Parts of Physics

The best parts of physics are the last topics that our students ever see. These are the exciting new frontiers of nonlinear and complex systems that are at the forefront of university research and are the basis of many of our high-tech businesses. Topics such as traffic on the World Wide Web, the spread of epidemics through globally-mobile populations, or the synchronization of global economies are governed by universal principles just as profound as Newton's laws. Nonetheless, the conventional university physics curriculum reserves most of these topics for advanced graduate study. Two justifications are given for this situation: first, that the mathematical tools needed to understand these topics are beyond the skill set of undergraduate students, and second, that these are specialty topics with no common theme and little overlap.

Introduction to Modern Dynamics: Chaos, Networks, Space and Time dispels these myths. The structure of this book combines the three main topics of modern dynamics—chaos theory, dynamics on complex networks, and the geometry of dynamical spaces—into a coherent framework. By taking a geometric view of physics, concentrating on the time evolution of physical systems as trajectories through abstract spaces, these topics share a common and simple mathematical language with which any student can gain a unified physical intuition. Given the growing importance of complex dynamical systems in many areas of science and technology, this text provides students with an up-to-date foundation for their future careers.

While pursuing this aim, *Introduction to Modern Dynamics* embeds the topics of modern dynamics—chaos, synchronization, network theory, neural networks, evolutionary change, econophysics, and relativity—within the context of traditional approaches to physics founded on the stationarity principles of variational calculus and Lagrangian and Hamiltonian physics. As the physics student explores the wide range of modern dynamics in this text, the fundamental tools that are needed for a physicist's career in quantitative science are provided as well, including topics the student needs to know for the graduate record examination (GRE). The goal of this textbook is to modernize the teaching of junior-level dynamics, responsive to a changing employment landscape, while retaining the core traditions and common language of dynamics texts.

A unifying concept: geometry and dynamics

Instructors or students may wonder how an introductory textbook can contain topics, under the same book cover, on econophysics and evolution as well as the physics of black holes. However, it is not the physics of black holes that matters, rather it is the description of general dynamical spaces that is important, and the understanding that can be gained of the geometric aspects of trajectories governed by the properties of these spaces. All changing systems, whether in biology or economics or computer science or photons in orbit around a black hole, are understood as trajectories in abstract dynamical spaces.

Newton takes a back seat in this text. He will always be at the heart of dynamics, but the modern emphasis has shifted away from $F = ma$ to a newer perspective where Newton's laws are special cases of broader concepts. There are economic forces and forces of natural selection that are just as real as the force of gravity on point particles. For that matter, even the force of gravity recedes into the background as force-free motions in curved space–time takes the fore.

Unlike Newton, Hamilton and Lagrange retain their positions here. The variational principle and the extrema of dynamical quantities are core concepts in dynamics. Maxima or minima of action integrals and metric distances provide trajectories—geodesics—in dynamical spaces. Conservation laws arise naturally from Lagrangians, and energy conservation enables simplifications using Hamiltonian dynamics. Space and geometry are almost synonymous in this context. Defining the space of a dynamical system takes first importance, and the geometry of the dynamical space then determines the set of all trajectories that can exist in it.

A common tool: dynamical flows and the ODE solver

A mathematical flow is a set of first-order differential equations that are solved using as many initial values as there are variables, which defines the dimensionality of the dynamical space. Mathematical flows are one of the foundation stones that appears continually throughout this textbook. Nearly all of the subjects explored here—from evolving viruses to orbital dynamics—can be captured as a flow. Therefore, a common tool used throughout this text is the numerical solution to the ordinary differential equation (ODE). Computers can be both a boon and a bane to the modern physics student. On the one hand, the easy availability of ODE solvers makes even the most obscure equations easy to simulate numerically, enabling any student to plot a phase plane portrait that contains all manner of behavior. On the other hand, physical insight and analytical understanding of complex behavior tend to suffer from the computer-game nature of simulators. Therefore, this textbook places a strong emphasis on analysis, and behavior in limits, with the goal to reduce a problem to a few simple principles, while making use of computer simulations to capture both the whole picture as well as the details of system behavior.

Traditional junior-level physics: how to use this book

All the traditional topics of junior-level physics are here. From the simplest description of the harmonic oscillator, through Lagrangian and Hamiltonian physics, to rigid body motion and orbital dynamics—the core topics of advanced undergraduate physics are retained and are found interspersed through this textbook. The teacher and student can plan a path through these topics here:

- Newton's laws (Section 1.1)
- Harmonic oscillators (Section 1.1)
- Coordinate systems and inertial frames (Sections 1.2 and 1.3)
- Rotating frames and fictitious forces (Section 1.5)
- Rigid body motion (Section 1.6)
- Lagrangian physics (Section 2.1)
- Conservation laws and collisions (Section 2.2)
- The Hamiltonian function (Section 2.3)
- Orbital dynamics and planetary motion (Section 2.4)
- Phase space (Section 2.5)
- Nonlinear oscillators (Section 3.4)
- Coupled oscillators (Section 4.1)
- Special relativity: covered in comprehensive detail (Chapter 10)

What's simple in complex systems?

The traditional topics of mechanics are integrated into the broader view of modern dynamics that draws from the theory of complex systems. The range of subject matter encompassed by complex systems is immense, and a comprehensive coverage of this topic is outside the scope of this book. However, there is still a surprisingly wide range of complex behavior that can be captured using the simple concept that the geometry of a dynamic space dictates the set of all possible trajectories in that space. Therefore, simple analysis of the associated flows provides many intuitive insights into the origins of complex behavior. The special topics covered in this textbook are:

- Chaos theory (Chapter 3)

Much of nonlinear dynamics can be understood through *linearization* of the *flow* equations (equations of motion) around special *fixed points*. Visualizing the dynamics of multi-parameter systems within multidimensional spaces is made simpler by concepts such as the *Poincaré section*, *strange attractors* that have *fractal geometry*, and *iterative maps*.

- Synchronization (Chapter 4)

The nonlinear *synchronization* of two or more oscillators is a starting point for understanding more complex systems. As the whole can be greater than the sum of the parts, the global properties often emerge from the local interactions of the parts. Synchronization of oscillators is surprisingly common and robust, leading to *frequency-entrainment*, *phase-locking*, and *fractional resonance* that allow small perturbations to control large networks of interacting systems.

- Network theory (Chapter 5)

Everywhere we look today we see networks. The ones we interact with daily are social networks and related networks on the World Wide Web. In this chapter, individual nodes are joined into networks of various geometries, such as *small-world networks* and *scale-free networks*. The *diffusion* of disease across these networks is explored, and the synchronization of *Poincaré phase oscillators* can induce a *Kuramoto transition* to complete synchronicity.

- Neural networks (Chapter 6)

Perhaps the most complex of all networks is the brain. This chapter starts with the single neuron, which is a *limit-cycle oscillator* that can show interesting *bistability* and *bifurcations*. When neurons are placed into simple neural networks, such as *perceptrons* or *feed-forward networks*, they can do simple tasks after training by *error back-propagation*. The complexity of the tasks increases with the complexity of the networks, and *recurrent networks*, like the *Hopfield neural net*, can perform associated memory operations that challenge even the human mind.

- Evolutionary dynamics (Chapter 7)

Some of the earliest explorations into nonlinear dynamics came from studies of *population dynamics*. In a modern context, populations are governed by evolutionary pressures and by genetics. Topics such as viral mutation and spread, as well as the evolution of species within a *fitness landscape*, are understood as simple balances within *quasi-species* equations.

- Econophysics (Chapter 8)

A most baffling complex system that influences our daily activities, as well as the trajectory of our careers, is the economy in the large and the small. The dynamics of *microeconomics* determines what and why we buy, while the dynamics of *macroeconomics* drives entire nations up and down economic swings. These forces can be (partially) understood in terms of nonlinear dynamics and flows in economic spaces. *Business cycles* and the diffusion of prices on the *stock market* are no less understandable than evolutionary dynamics (Chapter 7) or network dynamics (Chapter 5), and indeed draw closely from those topics.

- Geodesic motion (Chapter 9)

This chapter is the bridge between the preceding chapters on complex systems, and the succeeding chapters on relativity theory (both special and general). This is where the geometry of space is first fully defined in terms of a *metric tensor*, and where trajectories through a *dynamical space* are discovered to be paths of *force-free motion*. The *geodesic equation* (a geodesic flow) supersedes Newton's second law as the fundamental equation of motion that can be used to define the path of masses through potential landscapes and of light through space–time.

- Special relativity (Chapter 10)

In addition to traditional topics of *Lorentz transformations* and *mass–energy* equivalence, this chapter presents the broader view of trajectories through Minkowski *space–time* whose geometric properties are defined by the *Minkowski metric*. Relativistic forces and non-inertial (accelerating) frames connect to the next chapter that generalizes all relativistic behavior.

- General relativity (Chapter 11)

The physics of *gravitation*, more than any other topic, benefits from the over-arching theme developed throughout this book—that the geometry of a space defines the properties of all trajectories within that space. Indeed, in this geometric view of physics, Newton's force of gravity disappears and is replaced by force-free geodesics through *warped* space–time. Mercury's orbit around the Sun, and trajectories of light past *black holes*, are simply geodesic flows whose properties are easily understood using the tools developed in Chapter 3 and expanded upon throughout this textbook.

- Book Website:

The website for this book is www.works.bepress.com/ddnolte where students and instructors can find the Appendix and additional support materials. The Appendix provides details for some of the mathematical techniques used throughout the textbook, and provides a list of Matlab programming examples for specific Homework problems among the computational projects. The Matlab programming examples are located in a single pdf file on the website.

Acknowledgments

I gratefully acknowledge the many helpful discussions with my colleagues Ephraim Fischbach, Andrew Hirsch, Sherwin Love, and Hisao Nakanishi during the preparation of this book, and for the help of Bradley C. Abel and Su-Ju Wang who carefully checked all of the equations. Special thanks to my family, Laura and Nicholas, for putting up with my "hobby" for so many years, and also for their encouragement and moral support. I also thank the editors at Oxford, Jessica White for her help in preparing the manuscript, and especially Sonke Adlung for helping me realize my vision.

Contents

Part 1

Geometric Mechanics

A unifying viewpoint of physics has emerged, over the past century, that studying the geometric properties of special points and curves within dynamical spaces makes it possible to gain a global view of the dynamical behavior, rather than focusing on individual trajectories. Dynamical spaces can have many different dimensions and many different symmetries. They carry with them names like *configuration space*, *state space*, *phase space*, and *space–time*. This section introduces the mathematical tools necessary to study the geometry of dynamical spaces and the resulting dynamical behavior within those spaces. Central to geometric mechanics is Hamilton's principle, which states that the dynamical path taken through a space, among all possible paths, has the special property that the action integrated along that path is independent of small deviations. The *principle of stationary action* is the origin of all extrema principles that lie at the heart of dynamics, ultimately leading to the geodesic equation of general relativity, in which matter warps space, and trajectories execute force-free motion through that space.

Physics and Geometry

Foucault's pendulum in the Pantheon in Paris.
<http://www.flickr.com/photos/woto/6153399105/in/photosof-woto/>

Modern dynamics, like classical dynamics, is concerned with trajectories through space—the descriptions of trajectories (kinematics) and the causes of trajectories (dynamics). However, unlike classical mechanics that emphasizes motions of physical masses and the forces acting on them, modern dynamics generalizes the notion of trajectories to encompass a broad range of time-varying behavior that goes beyond material particles to include animal species in ecosystems, market prices in economies, and virus spread on connected networks. The spaces that these trajectories inhabit are abstract, and can have a high number of dimensions. These generalized spaces may not have Euclidean geometry, and may be curved like the surface of a sphere or warped space–time in gravitation.

The central object of interest in dynamics is the evolving state of a system. The state description of a system must be unambiguous, meaning that the next state to develop in time is uniquely determined by the current state. This is called deterministic dynamics. Much of this book deals with nonlinear dynamics where chaotic trajectories may have an apparent randomness to their character. However, deterministic chaos shares the same character as the (nearly) timeless orbits of the planets.

This chapter lays the foundation for the description of dynamic systems that move continuously from state to state. Families of trajectories, called mathematical flows, are the fundamental elements of interest; they are the field lines of dynamics. These field lines are to deterministic dynamics what electric and magnetic field lines are to electromagnetism. Where Maxwell's equations are the foundation upon which electricity and magnetism is erected, so the flow equations of a dynamical system govern all the possible behavior of a system. One key difference is that there is only one set of Maxwell's equations, while every nonlinear dynamical system has its own set of equations, providing a nearly limitless number of possibilities for us to study.

This chapter begins by describing trajectories in Euclidean (flat) space and within specific coordinate systems. To describe trajectories, we will establish the notation that helps us describe high-dimensional, abstract, and possibly curved spaces. This is accomplished through the use of matrix (actually tensor) indices that look strange at first to a student familiar only with vectors, but which are convenient devices for doing simple accounting. The next step is a coordinate transformation from one coordinate system to another. The main question is how two observers, one in each system, describe the common phenomena that they observe. If the systems are inertial, then the *physics* must be invariant to the choice of coordinate frame, but the descriptions can differ widely. However, if the transformation is non-inertial, as for a rotating frame, then fictitious forces like the Coriolis force emerge, that are not true forces, but are mere consequences of the non-inertial frame.

1.1   Newton and geometry

Newton's laws consist of the law of inertia, the law of acceleration, and the law of action–reaction. These are the cornerstones of classical mechanics from which so much else is derived. The Newtonian emphasis is on the material particle possessing a mass with a force acting upon it. The particle moves in ordinary space, and the tools of analytic (coordinate) geometry are well suited to describing its trajectory. But as the system acquires more particles, and as these particles are constrained in how they may move, then analytic geometry gives way to differential geometry, which is also the language of general relativity that we will encounter in the final chapters of this book. Despite its central role in classical dynamics, Newton's second law is actually a special case of more general equations of motion (mathematical flows) that go beyond masses and forces.

1.1.1   Newton's second law of motion

Consider the case of a single massive particle moving in one dimension subject to a force. Newton's second law is expressed as

$$\ddot{x} = \frac{F}{m},\qquad(1.1)$$

where each dot over the x variable denotes a time derivative. The two dots over the variable x make this a second-order time derivative (the acceleration), and the integral of this equation has two constants of integration, which often are taken as the initial position $x(0)$ and the initial speed $v(0)$. Once these initial values are set, the trajectory $x(t)$ of the particle is defined uniquely as an initial value problem (IVP) of an ordinary differential equation (ODE).

Example 1.1 Damped one-dimensional harmonic oscillator

A damped harmonic oscillator in one coordinate has the single second-order ODE

$$m\ddot{x} + \gamma\dot{x} + kx = 0,\qquad(1.2)$$

where m is the mass of the particle, γ is the drag coefficient, and k is the spring constant. Consider the initial values $x(0) = A$ and $\dot{x}(0) = 0$. To solve this equation, assume a solution in the form of a complex exponential in time evolving with an angular frequency ω as (see Appendix A.1 online)

$$x(t) = Xe^{i\omega t}.\qquad(1.3)$$

Insert this expression into Eq. (1.2) to yield

$$-m\omega^2 Xe^{i\omega t} + i\omega\gamma Xe^{i\omega t} + kXe^{i\omega t} = 0\qquad(1.4)$$

with the characteristic equation

$$\begin{aligned}0 &= m\omega^2 - i\omega\gamma - k\\ &= \omega^2 - i2\omega\beta - \omega_0^2,\end{aligned}\qquad(1.5)$$

where the damping parameter is $\beta = \gamma/2m$ and the resonant angular frequency is given by $\omega_0^2 = k/m$. The solution of the quadratic equation (1.5) is

$$\omega = i\beta \pm \sqrt{\omega_0^2 - \beta^2}.\qquad(1.6)$$

Using this expression for the angular frequency in the assumed solution, Eq. (1.3) gives

$$x(t) = X_1 \exp(-\beta t)\exp\left(i\sqrt{\omega_0^2 - \beta^2}\,t\right) + X_2 \exp(-\beta t)\exp\left(-i\sqrt{\omega_0^2 - \beta^2}\,t\right)\qquad(1.7)$$

and the two initial conditions impose the values

$$X_1 = \frac{A}{2}\left(\frac{\sqrt{\omega_0^2 - \beta^2} - i\beta}{\sqrt{\omega_0^2 - \beta^2}}\right)$$

$$X_2 = \frac{A}{2}\left(\frac{\sqrt{\omega_0^2 - \beta^2} + i\beta}{\sqrt{\omega_0^2 - \beta^2}}\right) = X_1^*.\qquad(1.8)$$

continued

Example 1.1 *continued*

The final solution is

$$x(t) = A \exp\left(-\beta t\right) \left[\cos\left(\sqrt{\omega_0^2 - \beta^2}\, t\right) + \frac{\beta}{\sqrt{\omega_0^2 - \beta^2}} \sin\left(\sqrt{\omega_0^2 - \beta^2}\, t\right) \right], \tag{1.9}$$

which is plotted in Figure 1.1 for the case where the initial displacement is a maximum and the initial speed is zero. The oscillator "rings down" with the exponential decay constant β. The angular frequency of the ring-down is not equal to ω_0, but is reduced to the value $\sqrt{\omega_0^2 - \beta^2}$. Hence, the damping decreases the frequency of the oscillator from its natural resonant frequency.

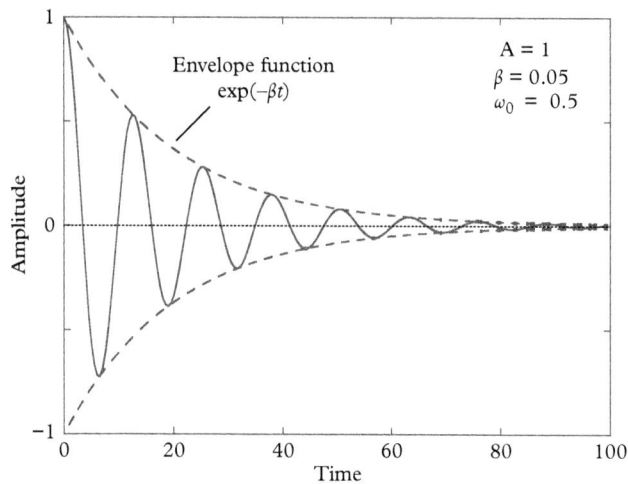

Figure 1.1 *A damped harmonic oscillator "rings down" with an exponential decay rate given by the damping coefficient β.*

1.1.2 Coordinate notation and configuration space

The position of a free particle in three-dimensional ($D = 3$) space is specified by three values that conventionally can be assigned the Cartesian coordinate values $x(t)$, $y(t)$, and $z(t)$. These coordinates define the instantaneous configuration of the system. If a second particle is added, then there are three additional coordinates and the *configuration space* of the system is now six dimensional. Rather than specifying three new coordinate names, such as $u(t)$, $v(t)$, or $w(t)$, it is more convenient to use a simplified notation that is extended easily to any number of dimensions. Index notion accomplishes this by having the index span across all the coordinate values. For instance, a system composed of N particles in three dimensions requires a total of $3N$ indices.

The indexed quantity that is adopted to describe a physical system is the position vector that points to a location in the configuration space that uniquely defines the instantaneous configuration of the system. Rather than using subscripts, vector components throughout this text will be denoted with a superscript. For instance, the position vector of a free particle in three-dimensional Euclidean space is a three-tuple of values

$$\vec{x} = \begin{pmatrix} x^1 \\ x^2 \\ x^3 \end{pmatrix}. \tag{1.10}$$

Vectors throughout this book are represented by column matrices (which is the meaning of the superscript).[1] It is important to remember that these superscripts are not "powers." A coordinate component raised to an nth power will always be expressed as $(x^a)^n$. For N free particles, a single $3N$-dimensional position vector defines the instantaneous configuration of the system. To abbreviate the coordinate description, one can use the notation

$$\vec{x} = \{x^a\} \qquad a = 1{:}3N, \tag{1.11}$$

where the brackets denote the full set of coordinates. An even shorter, and more common, notation for a vector is simply

$$x^a, \tag{1.12}$$

where the full set $a = 1{:}3N$ is implied. Cases where only a single coordinate is intended will be clear from the context. The position coordinates develop in time as

$$x^a(t), \tag{1.13}$$

which describes a trajectory of the system in its $3N$-dimensional configuration space.

1.1.3 Trajectories in three-dimensional (3D) configuration space

A trajectory is a set of position coordinate values that vary continuously with a single parameter and define a smooth curve in the configuration space. For instance,

$$x^a = x^a(t) \quad \text{or} \quad x^a = x^a(s), \tag{1.14}$$

where t is the time and s is the path length along the trajectory. Once the trajectory of a particle has been defined within its configuration space, it is helpful to define properties of the trajectory, like the tangent to the curve and the normal. The

[1] The reason for the superscript will become clear in Chapter 9, which discusses tensors and manifolds and introduces a different type of component called a covector that is denoted by a subscript. In Cartesian coordinates a superscript denotes a column vector and a subscript denotes a row vector (see Appendix A.3 online).

tangent to a curve is related to the velocity of the particle because the velocity vector is tangent to the path. For a single particle in 3D this would be

$$\vec{v}(s) = \frac{ds}{dt}\left(\frac{dx^1(s)}{ds}, \frac{dx^2(s)}{ds}, \frac{dx^3(s)}{ds}\right)^T,$$ (1.15)

where the ds/dt term in front is simply the speed of the particle. In the simplified index notation, this is

$$v^a(s) = \frac{ds}{dt}\frac{dx^a(s)}{ds}$$
$$= \frac{ds}{dt}T^a,$$ (1.16)

where T^a is a unit tangent vector in the direction of the velocity,

$$T^a = \frac{dx^a}{ds}.$$ (1.17)

Each point on the trajectory (see Fig. 1.2) has an associated tangent vector. In addition to the tangent vector, another important vector property of a trajectory is the normal to the trajectory, defined by

$$\frac{dT^a}{ds} = \kappa N^a,$$ (1.18)

Serret-Frenet vectors

$$\frac{d}{ds}\begin{pmatrix} \vec{T} \\ \vec{N} \\ \vec{B} \end{pmatrix} = \begin{pmatrix} 0 & \kappa & 0 \\ -\kappa & 0 & \tau \\ 0 & -\tau & 0 \end{pmatrix}\begin{pmatrix} \vec{T} \\ \vec{N} \\ \vec{B} \end{pmatrix}$$

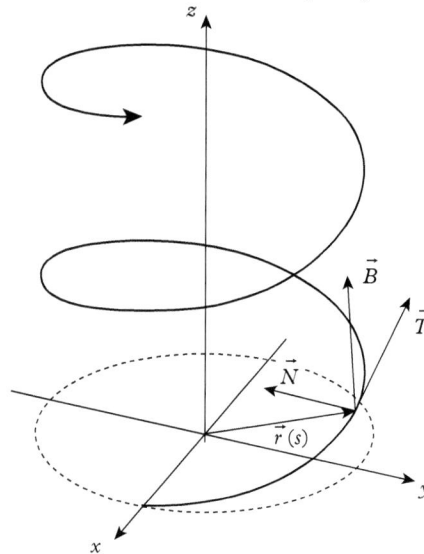

Figure 1.2 *A trajectory in three dimensions, parameterized by time $\vec{r}(t)$, or equivalently by path length $\vec{r}(s)$, showing the tangent, normal, and binormal vectors.*

where N^a is the unit vector normal to the curve and the curvature of the trajectory is

$$\kappa = \frac{1}{R},\tag{1.19}$$

where R is the radius of curvature at the specified point on the trajectory. To complete the set of unit vectors associated with a trajectory, there is the binormal unit vector B^a defined by

$$\frac{dN^a}{ds} = -\kappa\,T^a + \tau B^a,\tag{1.20}$$

where τ is called the torsion of the trajectory. The three unit vectors T^a, N^a, and B^a are mutually orthogonal. The relationships among these three unit vectors are collectively known as the Serret–Frenet formulas

Serret – Frenet formulas:

$$\frac{dx^a}{ds} = T^a$$
$$\frac{dT^a}{ds} = \kappa N^a$$
$$\vec{T} \times \vec{N} = \vec{B} \tag{1.21}$$
$$\frac{dN^a}{ds} = -\kappa\,T^a + \tau B^a$$
$$\frac{dB^a}{ds} = -\tau N^a$$

The parameterization of a trajectory in terms of its path length s is often a more "natural" way of describing the trajectory, especially under coordinate transformations. For instance, in special relativity, time is no longer an absolute parameter, because it is transformed in a manner similar to the positions. Then it is possible to define a path length interval ds^2 in space–time that remains invariant under Lorentz transformation (see Chapter 10) and hence can be used to specify the path through space–time.[2]

Example 1.2 Parabolic trajectory in a gravitational field

This is a familiar problem that goes back to freshman physics. However, it is seen here in a slightly different light. Consider a particle in a constant gravitational field thrown with initial velocity v_0 in the x-direction. The mathematical description of this motion is

$$\frac{dx}{dt} = v_0$$
$$\frac{dy}{dt} = -gt \tag{1.22}$$

continued

[2] More generally, the invariant path length interval ds^2 is an essential part of the metric description of the geometry of space–time and other dynamical spaces, and is a key aspect of geodesic motion for bodies moving through those spaces (see Chapter 9).

Example 1.2 *continued*

with the solution, for initial conditions $x = 0$, $y = 0$, $\dot{x} = v_0$, $\dot{y} = 0$,

$$x = v_0 t$$
$$y = -\frac{1}{2}gt^2 \tag{1.23}$$

giving the spatial trajectory

$$y = -\frac{1}{2}\frac{g}{v_0^2}x^2. \tag{1.24}$$

The speed of the particle is

$$\left(\frac{ds}{dt}\right)^2 = \left(\frac{dx}{dt}\right)^2 + \left(\frac{dy}{dt}\right)^2$$
$$= v_0^2 - 2gy \tag{1.25}$$
$$= v_0^2 + g^2 t^2 = v^2$$

with the arc-length element

$$ds = \sqrt{v_0^2 + g^2 t^2}\, dt$$
$$= \sqrt{1 + \frac{g^2 x^2}{v_o^4}}\, dx \tag{1.26}$$

and the tangent vector components

$$T^1 = \frac{dx}{ds} = \frac{1}{\sqrt{1 + \frac{g^2 x^2}{v_0^4}}}$$

$$T^2 = \frac{dy}{ds} = \frac{-\frac{gx}{v_0^2}}{\sqrt{1 + \frac{g^2 x^2}{v_0^4}}}. \tag{1.27}$$

The trajectory and its tangent vector are described as functions of position—a geometric curve rather than an explicit function of time. While the results for this familiar problem may look unfamiliar, it is similar to the description of trajectories in special relativity, or geodesic trajectories near gravitating bodies in space–time that will be treated in later chapters.

1.1.4   Generalized coordinates

The configuration coordinates considered so far have been Cartesian coordinates (x, y, z). However, there are abstract coordinates, called *generalized coordinates*, that may be more easily employed to solve dynamical problems. Generalized co-ordinates arise in different ways. They may be dictated by the symmetry of the problem, like polar coordinates for circular motion. They may be defined by constraints on the physical system, like a particle constrained to move on a surface. Or they may be defined by coupling (functional dependence) between the coordinates of a multicomponent system, leading to generalized coordinates known as normal modes. Generalized coordinates are often denoted by q's. They may be described in terms of other coordinates, for instance Cartesian coordinates, as

$$q^a = q^a \left(x^b, t \right)$$
$$x^b = x^b \left(q^a, t \right),$$

(1.28)

where the transformations associated with each index may have different functional forms, and do not need to be linear functions of their arguments. The generalized coordinates do not need to have the dimension of length, and each can have different units. However, it is required that the transformation is invertible (one-to-one).

Generalized coordinates can be used to significantly simplify the description of the motions of complex systems composed of large numbers of particles. If there are N particles, each with three coordinates, then the total dimension of the configuration space is $3N$ and there is a single system trajectory that threads its way through this configuration space. However, often there are constraints on the physical system, such as the requirement that particles are constrained to reside on a physical surface such as the surface of a sphere. In this case there are equations that connect two or more of the coordinates. If there are K equations of constraints, then the number of independent generalized coordinates is $3N - K$ and the motion occurs on a $(3N - K)$-dimensional hypersurface within the configuration space. This hypersurface is called a *manifold*. In principle, it is possible to find the $3N - K$ generalized coordinates that span this manifold, and the manifold becomes the new configuration space spanned by the $3N - K$ generalized coordinates. Furthermore, some of the generalized coordinates may not participate in the dynamics. These are called *ignorable coordinates* (also known as *cyclic coordinates*), and they arise due to symmetries in the configuration space plus constraints, and are associated with conserved quantities. The dimensionality of the dynamical manifold on which the system trajectory resides is further reduced by each of these conserved quantities. Ultimately, after all the conserved quantities and all the constraints have been accounted for, the manifold that contains the system trajectory may have a dimension much smaller than the dimension of the original Cartesian configuration space.

Example 1.3 Bead sliding on a frictionless helical wire

Consider a bead sliding without friction on a helical wire with no gravity. The trajectory is defined in 3D Cartesian coordinates by

$$x(t) = R\cos\omega t$$
$$y(t) = R\sin\omega t \qquad (1.29)$$
$$z(t) = v_z t$$

parameterized by time t. There are two constraints

$$x^2 + y^2 = R^2$$
$$z = a\theta, \qquad (1.30)$$

where a is the pitch of the helix, and $\theta = \omega t$. These constraints reduce the three-dimensional dynamics to one-dimensional motion $(3 - 2 = 1)$, and the one-dimensional trajectory has a single generalized coordinate

$$q(t) = t\sqrt{R^2\omega^2 + v_z^2}, \qquad (1.31)$$

which is also equal to the path length s. The speed of the particle is a constant and is

$$\dot{s} = \sqrt{R^2\omega^2 + v_z^2}. \qquad (1.32)$$

For a system composed of massive particles within three-dimensional space, as both the number N of particles increases and the number K of constraints increases, it becomes less helpful to think in terms of N trajectories of the individual particles in 3-space, and instead to think in terms of a single trajectory for the system through a $D = 3N - K$ dimensional configuration space. We do not need to think of this D-dimensional subspace as a hypersurface within $3N$-dimensional space, but can view the D generalized coordinates as a complete description of the configuration of the system. When this happens, the complex system behavior is no longer referenced to the behavior of the individual particles. While it is still possible to track how each individual particle moves in 3-space, it is the system behavior, and the properties of the single trajectory through the D-dimensional configuration space, that carries meaning.

1.2 State space and flows

Consider the single system trajectory of a number of massive particles in a D-dimensional generalized configuration space that visits the same position at two different times along its trajectory. For instance, this could happen in a loop as the system trajectory folds back on itself to go another way, and the trajectory literally crosses itself. Although the system acquires the same configuration at two different times, it is not in the same state. The difference is in the velocity vector that pointed one direction at the earlier time and pointed in a different direction at a later time. Therefore, the trajectory in configuration space does not uniquely define the state of the system—only its configuration. What is needed is a complete and unique description of the dynamical system in which the current state uniquely defines all subsequent states.[3] This complete description is accomplished by expanding the configuration space into state space.

1.2.1 State space

Any set of second-order time-dependent ODEs (Newton's second law) can be written as a larger set of first-order equations. For instance, the single second-order equation (1.2) can be rewritten as two first-order equations

$$m\dot{v} + \gamma v + kx = 0$$
$$\dot{x} = v. \tag{1.33}$$

It is conventional to write these with a single time derivative on the left as

$$\dot{x} = v$$
$$\dot{v} = -2\beta v - \omega_0^2 x \tag{1.34}$$

in the two variables (x, v) with $\beta = \gamma/2m$ and $\omega_0^2 = k/m$. *State space* for this system of equations consists of two coordinate axes in the two variables (x, v). A system trajectory starts at an initial condition, and traces the time evolution of the system. The solution with initial conditions $\dot{x} = -\beta A$ and $x = A$ is

$$x = Ae^{-\beta t} \cos\left(\sqrt{\omega_0^2 - \beta^2}\, t\right) \tag{1.35}$$

with the speed as the time derivative

$$v = -Ae^{-\beta t}\left[\sqrt{\omega_0^2 - \beta^2} \sin\left(\sqrt{\omega_0^2 - \beta^2}\cdot t\right) + \beta \cos\left(\sqrt{\omega_0^2 - \beta^2}\cdot t\right)\right]. \tag{1.36}$$

The graph in Figure 1.3 shows one particular solution in the state space for the given initial condition. However, any point on the plane can be an initial condition to a trajectory.

The system described by Eq. (1.33) is an inertial system, which means that the time derivative of the momentum $m\dot{v}$ appears explicitly in the system dynamics, which is a consequence of Newton's second law. However, there are broader classes of dynamical systems for which inertial effects are negligible, or which don't involve masses at all. For instance, most biology occurs under the conditions

[3] See A. E. Jackson, *Perspectives of Non linear Dynamics* (Cambridge University Press, 1989).

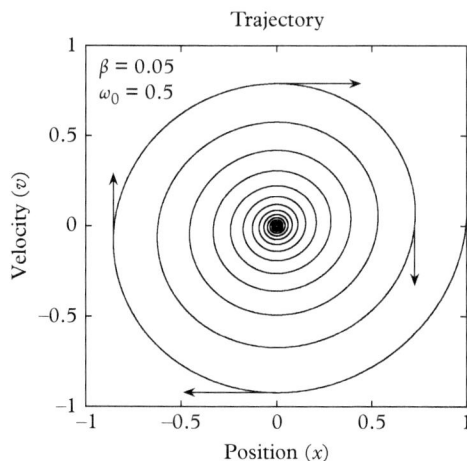

Figure 1.3 *State-space diagram of a damped one-dimensional (1D) harmonic oscillator trajectory in a space spanned by the two variables $(x, \dot{x}) = (x, v)$ of the 1D damped harmonic oscillator for $\omega_0^2 = 1$ and $\beta = 0.05$.*

of "low Reynolds number," which is another word for saying that inertial effects are unimportant for transport and dynamics in biology.[4] While particles are subject to spatially and temporally varying forces, every particle travels at its instantaneous terminal velocity.

Example 1.4 Force on a non-inertial particle

Consider a single non-inertial mass (a particle at low Reynolds number) constrained to move in two dimensions attached to a spring and subject to spatially varying forces, such as

$$\dot{x} = ay - \frac{k}{\gamma}x$$
$$\dot{y} = -ax - \frac{k}{\gamma}y, \tag{1.37}$$

where the first terms on the right define the spatially varying force on the particles, and the second terms are related to the spring force and damping. This is a coupled ODE that is solved by adding the two equations together and making the substitution to the complex variable (see Appendix A.2 online):

$$q = x + iy. \tag{1.38}$$

The resulting one-dimensional ODE is

$$\dot{q} = -\left(\frac{k}{\gamma} + ia\right)q \tag{1.39}$$

---

[4] For insight into biological dynamics, see E. M. Purcell, Life at low Reynolds-number, *Am. J. Physics*, 45, pp. 3–11 (1977).

with the solution

$$q(t) = q_0 \exp\left(-\frac{k}{\gamma}t\right) \exp\left(iat - i\phi\right). \tag{1.40}$$

The expressions for $x(t)$ and $y(t)$ are regained by taking the real and imaginary parts

$$x(t) = \mathrm{Re}\{q\} = q_0 \exp\left(-\frac{k}{\gamma}t\right) \cos\left(at - \phi\right)$$

$$y(t) = \mathrm{Im}\{q\} = q_0 \exp\left(-\frac{k}{\gamma}t\right) \sin\left(at - \phi\right). \tag{1.41}$$

The initial conditions can be specified by giving q_0 and φ, or equivalently by giving the initial (x, y) position. This system has two degrees of freedom (DOF) and the state space is two-dimensional. The two variables that span this state space are simply the x-position and the y-position of the particle. For this dynamical system, by specifying the location of the particle, the trajectory of the particle is uniquely defined. Therefore, *the configuration space of this system is equal to the state space.* Several trajectories in this two-dimensional state space are shown in Figure 1.4 as tightly decaying spirals centered on the origin.[5]

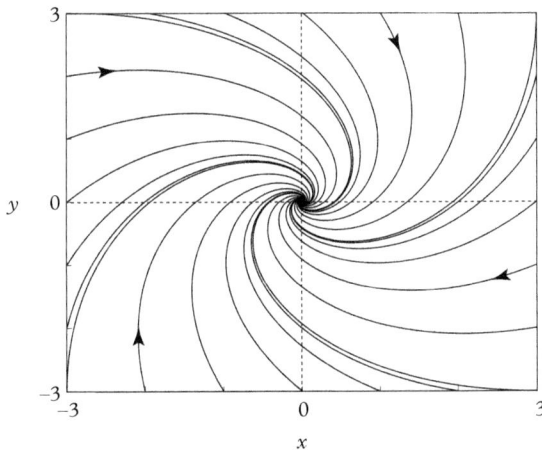

Figure 1.4 *Selected trajectories in the state space for the low Reynold's number system of Eq. (1.37) with a = 1 and k/γ = 1. The particles travel at their terminal velocity, subject to a central spring constant and an additional force that has a finite curl.*

[5] If this particle, constrained to move in two space dimensions, had inertia (meaning that velocity terms would appear on the right of Eq. (1.37)) then the state space would be four dimensional, and the configuration space would *not* be sufficient to define unique trajectories.

1.2.2   Flow, velocity fields, and field lines

The inertial term in Newton's second law has a second-order time derivative that was converted to two first-order time derivatives in Eq. (1.34). Much of classical dynamics treats "inertial" systems such as these. However, modern dynamics treats general and abstract systems that are as diverse as biological rhythms and the highs and lows of the stock market. Therefore, this book works with a more general form of dynamical equations called a mathematical flow. The flow for a system of N variables is defined by

$$
\frac{dq^1}{dt} = F^1\left(q^1, q^2, \ldots, q^N; t\right)
$$
$$
\frac{dq^2}{dt} = F^2\left(q^1, q^2, \ldots, q^N; t\right)
$$
$$
\vdots
$$
$$
\frac{dq^N}{dt} = F^N\left(q^1, q^2, \ldots, q^N; t\right)
$$

(1.42)

or more succinctly

$$
\frac{dq^a}{dt} = F^a\left(q^a; t\right),
$$

(1.43)

which is a system of N simultaneous equations, where the vector function F^a is a function of the time-varying coordinates of the position vector. If F^a is not an explicit function of time, then the system is *autonomous* with an N-dimensional state space. On the other hand, if F^a is an explicit function of time, then the system is *non-autonomous* with an $(N+1)$-dimensional state space (space plus time). The solution of the system of equations in Eq. (1.43) is a trajectory $q^i(t)$ through the state space.[6]

The damped harmonic oscillator equations in Eq. (1.34) describe a flow. Each point on the (x–v) plane has a tangent vector associated with it (given by Eq. (1.34)). The vector field is plotted as the field of arrows in Fig. 1.5. Therefore, the flow equations define a vector field on the state space. Each point on the state space has exactly one trajectory passing through it, which is drawn as a stream line. In Figure 1.5, only one trajectory (stream line) is drawn. Stream lines are the field lines of the vector field. Much of the study of modern dynamics is the study of the geometric properties of the vector field and field lines associated with a defined set of flow equations.

[6] In this book, the phrase *configuration space* is reserved for the dynamics of systems of massive particles (second-order time derivatives). For mathematical flows (first-order time derivatives), state space and configuration space are the same thing and the phrase *state space* will be used.

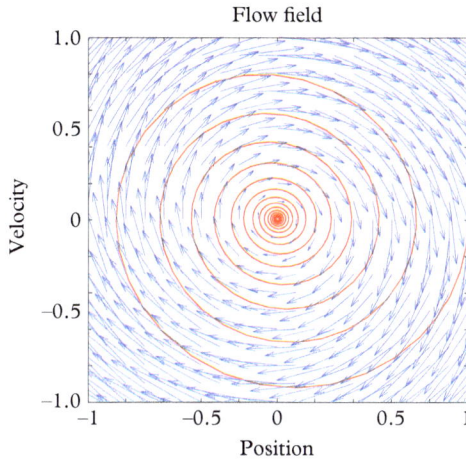

Figure 1.5 *State-space vector field of the damped harmonic oscillator. Every point in state space has a tangent vector associated with it. Stream lines are the field lines of the vector field. In this figure, only a single stream line is shown.*

Example 1.5 Velocity field and flow lines

Consider the two-dimensional (2D) flow

$$\dot{x} = y - \alpha x$$
$$\dot{y} = \text{atan}\,(\mu x) - y \tag{1.44}$$

with parameters α and μ. The 2D vector field and selected stream lines are shown in Figure 1.6. There are several special locations in this flow. The solid curves are sets of points where one of the time derivatives vanishes. These curves are called *nullclines*. There is one nullcline for all points where $\dot{x}$ is zero, and a second nullcline for all points where $\dot{y}$ is zero. The point where the two nullclines cross is called a *fixed point*. Both time derivatives vanish at this point, and initial conditions with this value do not change—they are fixed at that point in state space. The dashed line is called a *separatrix*. It is a line that separates the flow lines into distinct domains in state space— flow lines never cross a separatrix. There are many different types of fixed points, defined by their differing stability properties (the one here is called an attractor node because it attracts the field lines to it from all directions). The study of nonlinear dynamical systems pays close attention to the study of fixed points and nullclines. The mathematical details of flows will be covered in Chapter 3, and these principles will be applied in many of the following chapters in this book.

continued

Example 1.5 *continued*

2D Flow

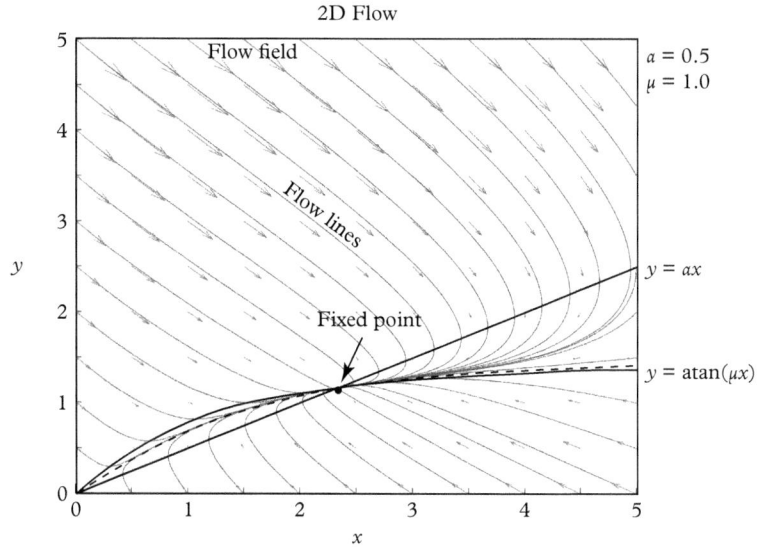

Figure 1.6 *Example of a flow diagram in state space showing the vector field and selected stream lines (also called field lines or flow lines). The solid dark curves are nullclines. The dashed curve (that closely follows the ẏ = 0 nullcline) is called a separatrix.*

1.3 Coordinate transformations

Although *physics* must be independent of any coordinate frame, the description of what we see *does* depend on which frame we are viewing it from. Therefore, we will often need to transform from one frame to another to describe what observers in each frame see. For this reason, we need to find the transformation laws that take the description from one frame to another.

1.3.1 Transformations

For a general coordinate transformation, the original Cartesian coordinates x, y, and z are related to coordinates q^1, q^2, and q^3 by the function

$$x = x\left(q^1, q^2, q^3\right)$$
$$y = y\left(q^1, q^2, q^3\right)$$
$$z = z\left(q^1, q^2, q^3\right).$$

(1.45)

These equations can be inverted to yield

$$q^1 = q^1(x, y, z)$$
$$q^2 = q^2(x, y, z) \tag{1.46}$$
$$q^3 = q^3(x, y, z),$$

which may be generalized coordinates that are chosen to simplify the equations of motion of a dynamical system. For the transformation functions to exist, it is required that the functions are smooth and that the Jacobian[7] determinant is nonzero. The Jacobian matrix of the transformation is

$$\mathcal{J} = \begin{pmatrix} \dfrac{\partial x}{\partial q^1} & \dfrac{\partial x}{\partial q^2} & \dfrac{\partial x}{\partial q^3} \\ \dfrac{\partial y}{\partial q^1} & \dfrac{\partial y}{\partial q^2} & \dfrac{\partial y}{\partial q^3} \\ \dfrac{\partial z}{\partial q^1} & \dfrac{\partial z}{\partial q^2} & \dfrac{\partial z}{\partial q^3} \end{pmatrix} \qquad \mathcal{J}^{-1} = \begin{pmatrix} \dfrac{\partial q^1}{\partial x} & \dfrac{\partial q^1}{\partial y} & \dfrac{\partial q^1}{\partial z} \\ \dfrac{\partial q^2}{\partial x} & \dfrac{\partial q^2}{\partial y} & \dfrac{\partial q^2}{\partial z} \\ \dfrac{\partial q^3}{\partial x} & \dfrac{\partial q^3}{\partial y} & \dfrac{\partial q^3}{\partial z} \end{pmatrix} \tag{1.47}$$

and the determinant $|\mathcal{J}| = \det(\mathcal{J})$ is called the Jacobian.

The Jacobian matrix requires two indices to define its individual elements, just as a vector required one index. Because the Jacobian matrix is generated using derivatives, an index notation that distinguishes between the differential vector in the numerator relative to the differential vector in the denominator is

$$\text{row index} \searrow$$
$$\mathcal{J}^a_b = \frac{\partial x^a}{\partial q^b} \tag{1.48}$$
$$\text{column index} \nearrow$$

where the superscript and subscript relate to x^a and q^b, respectively. The superscript is called a contravariant index, and the subscript is called a covariant index. One way to remember this nomenclature is that "co" goes "below." The covariant index relates to the columns of the matrix, and the contravariant index relates to the rows. Column vectors have contravariant indices because they have multiple rows, while row vectors have covariant indices because they have multiple columns. Row vectors are also known as covariant vectors, or covectors (see Appendix A.3 online for the connection between indices and matrices).

When transforming between Cartesian and generalized coordinates, the Jacobian matrix describes an infinitesimal transformation

$$dx^a = \sum_b \frac{\partial x^a}{\partial q^b} dq^b = \sum_b \mathcal{J}^a_b dq^b, \tag{1.49}$$

where $\mathcal{J}^a_b$ can depend on position. If the transformation is linear, then the Jacobian matrix is constant. Equation (1.49) is a matrix equation where the generalized

[7] Jacobian matrices are names after the German mathematician Carl Gustav Jacob Jacobi (1804–1851) who helped develop multidimensional analysis.

coordinates dq^b form a column vector, and $\mathcal{J}_b^a$ is the Jacobian matrix. The operation of the Jacobian matrix on the generalized coordinates generates a new column vector dx^a in the new coordinate system.

Rather than always expressing the summation explicitly, there is a common convention, known as the *Einstein summation convention*, in which the summation symbol is dropped and a repeated index—one above and one below—implies summation,

$$\text{Einstein summation convention:} \qquad x^a = \sum_b \Lambda_b^a q^b \equiv \Lambda_b^a q^b, \qquad (1.50)$$

where the "surviving" index—a—is the non-repeated index. For example, in three dimensions this is

$$
\begin{aligned}
x^1 &= \Lambda_1^1 q^1 + \Lambda_2^1 q^2 + \Lambda_3^1 q^3 \\
x^2 &= \Lambda_1^2 q^1 + \Lambda_2^2 q^2 + \Lambda_3^2 q^3 \\
x^3 &= \Lambda_1^3 q^1 + \Lambda_2^3 q^2 + \Lambda_3^3 q^3,
\end{aligned}
\qquad (1.51)
$$

which is recognized in matrix multiplication as

$$
\begin{pmatrix} x^1 \\ x^2 \\ x^3 \end{pmatrix} =
\begin{pmatrix} \Lambda_1^1 & \Lambda_2^1 & \Lambda_3^1 \\ \Lambda_1^2 & \Lambda_2^2 & \Lambda_3^2 \\ \Lambda_1^3 & \Lambda_2^3 & \Lambda_3^3 \end{pmatrix}
\begin{pmatrix} q^1 \\ q^2 \\ q^3 \end{pmatrix}
\qquad (1.52)
$$

that is simplified to

$$x^a = \Lambda_b^a q^b \qquad (1.53)$$

with the Einstein repeated-index summation. The Einstein summation convention also is convenient when defining the inner (or "dot") product between two vectors. For instance

$$
\begin{aligned}
\vec{A} \cdot \vec{B} &= A_x B^x + A_y B^y + A_z B^z \\
&= A_a B^a
\end{aligned}
\qquad (1.54)
$$

with implicit summation over the repeated indices produces a scalar quantity from the two vector quantities. The inner product in matrix notation multiplies a column vector from the left by a row vector.

The Jacobian matrix and its uses are recurring themes in modern dynamics. Its use goes beyond simple coordinate transformations, and it appears any time a nonlinear system is "linearized" around fixed points to perform stability analysis (Chapter 3). The eigenvalues of the Jacobian matrix define how rapidly nearby initial conditions diverge (called Lyapunov exponents (Chapters 3 and 7)). The determinant of the Jacobian matrix arises as the coefficient between area and volume elements (Chapter 9), and can be used to prove which processes conserve volumes in phase space or Minkowski space (Chapters 2 and 10).

The path length element ds^2 plays a central role in many of the topics of this book. In orthogonal curvilinear coordinates the squared length element is

$$ds^2 = g_{11}\left(dq^1\right)^2 + g_{22}\left(dq^2\right)^2 + g_{33}\left(dq^3\right)^2, \tag{1.55}$$

where the coefficients are related to the elements in the rows of the Jacobian matrix through[8]

$$g_{aa} = \left(\frac{\partial x^1}{\partial q^a}\right)^2 + \left(\frac{\partial x^2}{\partial q^a}\right)^2 + \left(\frac{\partial x^3}{\partial q^a}\right)^2. \tag{1.56}$$

The path length element takes on different forms for the different curvilinear coordinate systems, like spherical coordinates, and for warped space–time around gravitating bodies (Chapter 11). It also plays an essential role in the definition of geodesic curves, or shortest paths between two points. In Cartesian coordinates the coefficients are unity, and the path length element is given by

$$ds^2 = dx^2 + dy^2 + dz^2. \tag{1.57}$$

Expressions for the squared line element in cylindrical and spherical coordinates are given in Figure 1.7.

1.3.2 Rotations

A common class of coordinate transformation consists of rotations. The rotation matrix is an operator that operates on the components of a vector to express them in terms of a new rotated set of coordinate axes. In two dimensions the components of a vector A^a are transformed as

$$A^{\bar{b}} = R^{\bar{b}}_a A^a, \tag{1.58}$$

where $R^{\bar{b}}_a$ is the rotation matrix and $A^{\bar{b}}$ are the components of the vector as viewed in the primed (bar) frame.[9] For a 2D primed coordinate frame $\bar{O}$ that has been rotated counterclockwise by an angle θ relative to the unprimed frame O, the rotation matrix that transforms the vector components (describes them with respect to the new frame) is

$$2D\ rotation\ matrix: \quad R^{\bar{b}}_a = \begin{pmatrix} \cos\theta & \sin\theta \\ -\sin\theta & \cos\theta \end{pmatrix} \tag{1.59}$$

[8] In Chapter 9, we will see that the coefficients are components of the metric tensor that defines distances within a generalized space. The metric tensor is also used to convert between vectors and covectors.

[9] The prime and bar notations are used interchangeably in this chapter.

Cylindrical coordinates (r, θ, z)

$$x^a = \begin{pmatrix} x \\ y \\ z \end{pmatrix} = \begin{pmatrix} r\cos\theta \\ r\sin\theta \\ z \end{pmatrix}$$

$$ds^2 = dr^2 + r^2 d\theta^2 + dz^2$$

z-axis

r

$P(r, \theta, z)$

z

O

y-axis

θ

x-axis

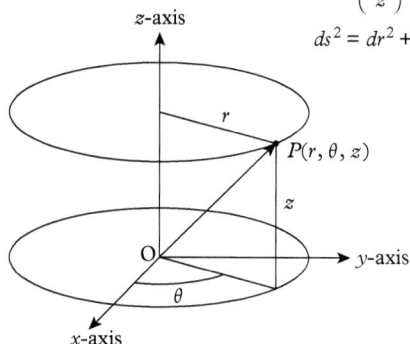

Spherical coordinates (r, θ, ϕ)

z-axis

$P(r, \theta, \phi)$

$$x^a = \begin{pmatrix} x \\ y \\ z \end{pmatrix} = \begin{pmatrix} r\sin\theta\,\cos\phi \\ r\sin\theta\,\sin\phi \\ r\cos\theta \end{pmatrix}$$

$$ds^2 = dr^2 + r^2 d\theta^2 + r^2\sin^2\theta\, d\phi^2$$

θ

r z

O

y-axis

ϕ

x-axis

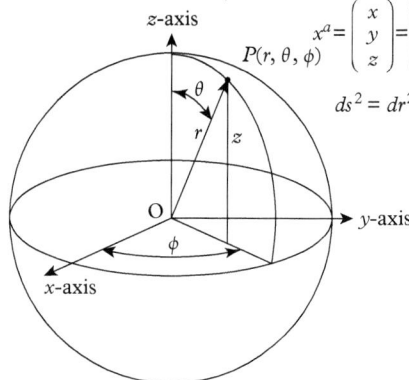

Figure 1.7 *Cylindrical and spherical coordinate systems with line elements* ds^2.

with the inverse transformation

$$\left(R_a^{\bar{b}}\right)^{-1} = R_{\bar{b}}^a = \begin{pmatrix} \cos\theta & -\sin\theta \\ \sin\theta & \cos\theta \end{pmatrix}. \tag{1.60}$$

The inverse transformation $R_{\bar{b}}^a$ has the primed index below, while in the forward transformation of vector components, $R_a^{\bar{b}}$, the primed index is above.

It is important to keep a clear distinction between basis vectors (like $\hat{x}$ and $\hat{y}$) that point along the coordinate axes and the components of the vector $\vec{A}$ projected onto these axes. If the basis vectors, and hence the coordinate frame, is rotated counterclockwise, then the vector components, as seen from the transformed coordinate frame, appear to have rotated clockwise. This is shown in Figure 1.8.

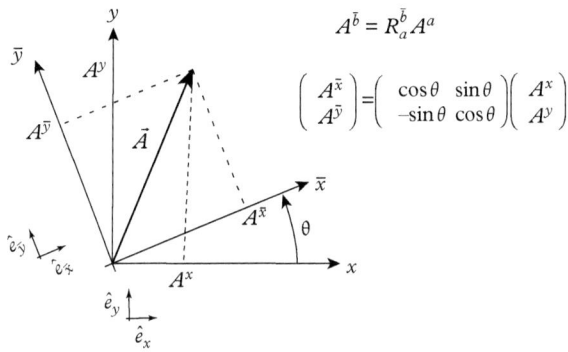

$$A^{\bar{b}} = R^{\bar{b}}_a A^a$$

$$\begin{pmatrix} A^{\bar{x}} \\ A^{\bar{y}} \end{pmatrix} = \begin{pmatrix} \cos\theta & \sin\theta \\ -\sin\theta & \cos\theta \end{pmatrix} \begin{pmatrix} A^x \\ A^y \end{pmatrix}$$

Figure 1.8 *Rotated coordinate axes through the transformation $R^{\bar{b}}_a$. The vector $\vec{A}$ remains the same—only the description of the vector (vector components projected onto the axes) changes.*

Therefore, basis vectors are rotated through the inverse rotation matrix. The basis vectors of general coordinate systems are not necessarily unit vectors and are denoted as $\vec{e}_a$ with an "arrow" and the subscript below. The transformation of the basis vectors is therefore

$$\vec{e}_{\bar{b}} = R^a_{\bar{b}}\vec{e}_a, \tag{1.61}$$

which transforms inversely to Eq. (1.58). A vector quantity is expressed in terms of the basis vectors as

$$\vec{A} = A^a\vec{e}_a = A^{\bar{b}}\vec{e}_{\bar{b}}, \tag{1.62}$$

which is independent of the coordinate system—vectors are invariant quantities (elements of reality) that exist independently of their coordinate description.

Three-dimensional rotations can be constructed as successive two-dimensional rotations applied around different axes. Three angles are required to express an arbitrary three-dimensional rotation, and the general rotation matrix can be expressed as

$$R^{d'''}_a = R^{d'''}_{c''}(\psi)R^{c''}_{b'}(\theta)R^{b'}_a(\phi), \tag{1.63}$$

where each rotation is applied around a different axis. When applied to a vector x^a this produces the successive transformations

$$\begin{aligned} x^{b'} &= R^{b'}_a(\phi)x^a \\ x^{c''} &= R^{c''}_{b'}(\theta)x^{b'} \\ x^{d'''} &= R^{d'''}_{c''}(\psi)x^{c''}, \end{aligned} \tag{1.64}$$

where the original unprimed frame is rotated into the primed frame, then the primed frame is rotated into the double-primed frame, which is rotated into the

Euler angles

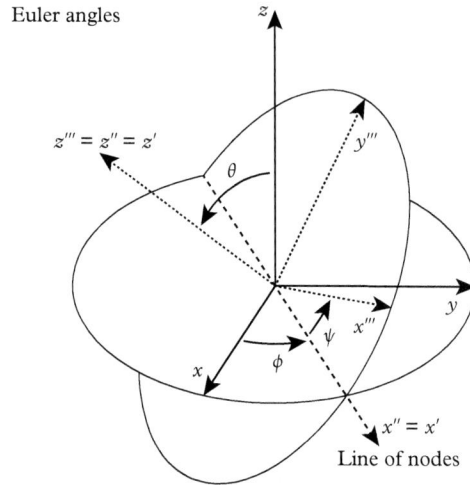

Figure 1.9 *Euler angles for a 3D rotation. The original unprimed axes are first rotated by φ around the z-axis, then by θ around the x′-axis (also known as the line of nodes), and finally by ψ around the z″-axis.*

triple-primed frame which is the resultant frame of the 3D rotation. Although there is no unique choice for the rotation axes, one conventional choice known as Euler angles[10] (Fig. 1.9) uses a rotation by φ around the z-axis, by θ around the x'-axis, and then by ψ around the z''-axis. The rotation matrices for this choice are

$$R_z(\phi) = Z_a^{b'} = \begin{pmatrix} \cos\phi & \sin\phi & 0 \\ -\sin\phi & \cos\phi & 0 \\ 0 & 0 & 1 \end{pmatrix} \tag{1.65}$$

$$R_{x'}(\theta) = X_{b'}^{c''} = \begin{pmatrix} 1 & 0 & 0 \\ 0 & \cos\theta & \sin\theta \\ 0 & -\sin\theta & \cos\theta \end{pmatrix} \tag{1.66}$$

$$R_{z''}(\psi) = Z_{c''}^{d'''} = \begin{pmatrix} \cos\psi & \sin\psi & 0 \\ -\sin\psi & \cos\psi & 0 \\ 0 & 0 & 1 \end{pmatrix} \tag{1.67}$$

It is often convenient to construct a single rotation matrix in 3D by selecting a rotation axis defined by a unit vector $\hat{u}_a$ and a rotation angle θ. The rotation matrix that accomplishes this is

[10] Leonhard Euler (1707–1783) introduced these angles in 1765 in his paper on rotational mechanics.

3D rotation matrix: $R_a^{\bar{b}} = I_a^{\bar{b}}\cos\theta + S_a^{\bar{b}}\sin\theta + T_a^{\bar{b}}(1-\cos\theta),$ \tag{1.68}

where $I_a^{\bar{b}} = \delta_a^{\bar{b}}$ is the identity matrix, and the other matrices are

$$
\begin{aligned}
S_a^{\bar{b}} &= \begin{pmatrix} 0 & -u_z & u_y \\ u_z & 0 & -u_x \\ -u_y & u_x & 0 \end{pmatrix} \\
T_a^{\bar{b}} &= \begin{pmatrix} u_x u_x & u_x u_y & u_x u_z \\ u_y u_x & u_y u_y & u_y u_z \\ u_z u_x & u_z u_y & u_z u_z \end{pmatrix}
\end{aligned}
\tag{1.69}
$$

and where u_a are the Cartesian components of the unit vector. The matrix $T_a^{\bar{b}}$ is the tensor product of the unit vector with itself, denoted in vector notation as $(\hat{u} \otimes \hat{u})$. The matrix $S_a^{\bar{b}}$ is a skew-symmetric matrix constructed from the unit vector and is denoted in vector notation as the operator $(\hat{u} \times)$ with the cross product. The structure of the skew-symmetric matrix reflects the geometry of rotations in three-dimensional space. Ultimately, it is this intrinsic property of 3-space that is the origin of physics equations containing cross products, such as definitions of angular momentum and torque as well as equations that depend on the moments of inertia, which are encountered later in this chapter.

1.3.3 Translating frames

Galilean relativity[11] concerns the description of kinematics when the same trajectory is viewed from different inertial frames. An inertial frame (Fig. 1.10) is one that moves at constant velocity (no acceleration). The transformation between the two inertial frames is a linear transformation.

Consider a general trajectory in frame O defined by $\vec{x}(t) = x^a$ which might be a mass subjected to time-varying forces. A second "primed" frame $\bar{O}$ is moving relative to the first with a velocity $\vec{u}(t) = u^a$. The same trajectory seen in the primed frame $\bar{O}$ is given by

$$
\begin{aligned}
\bar{x}^1 &= x^1 - u^1 t \\
\bar{x}^2 &= x^2 - u^2 t \\
\bar{x}^3 &= x^3 - u^3 t
\end{aligned}
\tag{1.70}
$$

with the inverse transformation given by

$$
\begin{aligned}
x^1 &= \bar{x}^1 + u^1 t \\
x^2 &= \bar{x}^2 + u^2 t \\
x^3 &= \bar{x}^3 + u^3 t.
\end{aligned}
\tag{1.71}
$$

Linear transformations can be expressed in matrix form. In Galilean relativity, the linear transformation is a simple translation. The forward transformation of a vector component is

$$
x^{\bar{b}} = \delta_a^{\bar{b}} x^a + u^{\bar{b}} t
\tag{1.72}
$$

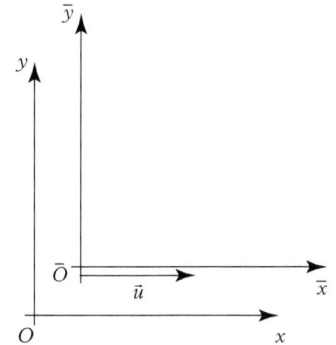

Figure 1.10 *Galilean frames in uniform relative motion. The primed frame is observed from the unprimed frame to move with velocity u.*

[11] The first systematic application of classical relativity was made by Christiaan Huygens (1669) and might more accurately be called Galilean–Huygens relativity.

with the inverse transformation

$$x^a = \delta^a_{\bar{b}} x^{\bar{b}} + u^a t, \tag{1.73}$$

where $u^{\bar{b}} = -u^a$ and where the Kronecker delta is

$$\delta^{\bar{b}}_a = \begin{cases} 1 & \bar{a} = b \\ 0 & \bar{a} \neq b \end{cases}, \tag{1.74}$$

which is also the identity matrix. The primed velocity is the relative velocity observed by the primed observer and is the negative of the unprimed velocity.

For a linear transformation (between inertial frames) straight trajectories in one frame remain straight trajectories in the other. The velocity transformation is expressed as

$$v^{\bar{b}} = \delta^{\bar{b}}_a v^a + u^{\bar{b}}, \tag{1.75}$$

which is the classic Galilean relativity equation, and the transformation of accelerations is

$$a^{\bar{b}} = \delta^{\bar{b}}_a a^a, \tag{1.76}$$

which guarantees that velocity-independent forces in the one frame are equal to the forces in the other. In other words, "physics" is observed to be the same in both inertial frames.

1.4 Non-inertial transformations

Linear transformations are only a subset of more general transformations that may not be linear. An especially important class of transformations is between non-inertial frames, such as a frame that experiences constant acceleration relative to a fixed frame.

1.4.1 Uniformly accelerating frame (non-relativistic)

Accelerating frames are quite different from inertial frames. Observers in the different frames interpret trajectories differently, invoking forces in one frame that are not needed or felt in the other. These forces arise from coordinate transformations and are called fictitious forces.

The simplest non-inertial frame is one that is accelerating with constant acceleration. Consider a primed coordinate frame that is accelerating with acceleration α along the x^3 direction relative to the original unprimed frame. An event at (x^1, x^2, x^3) at time t is observed in the primed frame at

$$x^{\bar{1}} = x^1$$
$$x^{\bar{2}} = x^2$$
$$x^{\bar{3}} = x^3 - \frac{1}{2}\alpha t^2 \qquad (1.77)$$

or

$$x^{\bar{b}} = \delta^{\bar{b}}_a x^a - \frac{1}{2}\delta^{\bar{b}}_3 \alpha t^2 \qquad (1.78)$$

(Einstein summation convention assumed), and a linear trajectory of a particle in the unprimed frame

$$x^a = x^a_0 + v^a t \qquad (1.79)$$

becomes a parabolic trajectory in the primed frame

$$x^{\bar{b}} = \delta^{\bar{b}}_a \left(x^a_0 + v^a t \right) - \frac{1}{2}\delta^{\bar{b}}_3 \alpha t^2. \qquad (1.80)$$

Therefore an observer in the primed frame $\bar{O}$ would describe the dynamics of the trajectory in terms of a fictitious force along the $-z$-axis with a magnitude $F_z = -m\alpha$, while the observer in O assumes no force. However, there is an important difference between the observers. While the observer in the unprimed frame O experiences no force on *himself or herself*, the observer in the primed frame $\bar{O}$ *does*. This force could be experienced, for instance, as the force of a space-ship floor on the feet of the observer as it accelerates forward. If the primed observer does not know that the space-ship is the cause of the force, he or she might assume that the force is a gravitational force, and that the same gravitational force was the cause of the observed parabolic trajectory. This description of the non-inertial force leads to an important principle of physics:

Principle of equivalence:

An observer in a uniformly accelerating frame is equivalent to an observer in a uniform gravitational field.

According to this principle, physics in a uniform gravitational field is equivalent to physics in a frame that experiences constant linear acceleration. This principle is an important motivation for general relativity.

1.4.2 The deflection of light by gravity

The equivalence principle makes it easy to prove that gravity must bend the path of a light ray (see Fig. 1.11). Consider the cases of an Earth-bound elevator (elevator stationary in a gravitational field) compared to a constantly accelerating outer-space elevator (no gravitation). Each has a pinhole at the top through which a photon enters at time $t = 0$, and the photon hits the far wall at a time $t = L/c$. Where does the photon hit the far wall?

Accelerating elevator (no gravity)

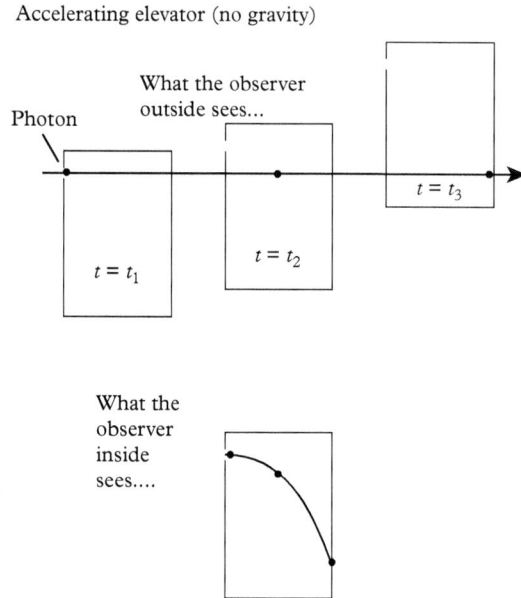

Figure 1.11 *Einstein thought experiment showing that gravity (equivalent to an accelerating elevator) deflects light.*

Example 1.6 Light in an elevator

Consider the case of the constantly accelerating elevator in space far from any gravitating body. The light path in the fixed frame is

$$x^1 = ct$$
$$x^3 = 0. \tag{1.81}$$

However, the observer in the elevator sees

$$x^{\bar{1}} = ct$$
$$x^{\bar{3}} = -\frac{1}{2}gt^2 = -\frac{1}{2}g\left(\frac{L}{c}\right)^2. \tag{1.82}$$

By the principle of equivalence, an Earth-bound observer in a stationary elevator must see the same thing. The photon enters the pinhole at the top of the elevator, and hits the far wall a distance $gL^2/2c^2$ from the ceiling. The inescapable conclusion is that gravity bends light! Admittedly, the effect is small. The gravitational deflection for a 1 m wide elevator accelerating with g is

$$\Delta x^{\bar{3}} = -\frac{1}{2}g\left(\frac{L}{c}\right)^2 = -6 \times 10^{-17}\text{m}, \tag{1.83}$$

which is about a million times smaller than the radius of an atom. To see the effect of gravity on light, it is usually necessary to make observations over scales of solar systems or larger, and to use large gravitating bodies such as the Sun or even whole galaxies. On the other hand, exquisitely sensitive solid state detectors, called Mössbauer resonance detectors, can measure effects of gravity on light over the length of a tall tower. But this is measured as a nuclear absorption effect rather than as a deflection.

The equivalence principle is a central motivation for general relativity, which is the topic of Chapter 11. However, an important caveat must be mentioned when applying the equivalence principle. In the elevator thought experiment, the "equivalent" elevator on Earth was assumed to be in a *uniform* gravitational field. By *uniform* is meant that it is constant in both space and time. Such uniform gravitational fields are not realistic, and may only apply as approximations over very local ranges. For instance, the equivalence principle was first used by Einstein to calculate the deflection of a light ray by the Sun at a time when he was exploring the consequences of relativity in non-inertial frames, but before he had developed the full tensor theory of general relativity. The equivalence principle, when applied to uniform fields, only modifies the time component of the space–time metric, but ignores any contribution from the curvature of space–time. In fact, the deflection calculated using the equivalence principle alone leads to an answer that is exactly half of the correct deflection obtained using the full theory of general relativity.

1.5 Uniformly rotating frames

A uniformly rotating frame is an important example of a non-inertial frame. In this case, acceleration is not constant, which leads to fictitious forces such as the centrifugal and Coriolis forces.

Consider two frames: one fixed and one rotating. These could be, for instance, a laboratory frame and a rotating body in the laboratory, as in Figure 1.12. The fixed frame has primed coordinates, and the rotating frame has unprimed coordinates. The position vector in the fixed lab frame is

$$\vec{\bar{r}} = x^{\bar{a}}\hat{e}_{\bar{a}} = \bar{x}\hat{e}_{\bar{x}} + \bar{y}\hat{e}_{\bar{y}} + \bar{z}\hat{e}_{\bar{z}}. \tag{1.84}$$

These basis vectors $\hat{e}_a$ have a subscript instead of a superscript.[12] The position vector in the rotating frame is

$$\vec{r} = x^a\hat{e}_a = x\hat{e}_x + y\hat{e}_y + z\hat{e}_z \tag{1.85}$$

relative to the origin of the rotating frame. The primed position vector is then

$$\vec{\bar{r}} = \vec{R} + \vec{r}. \tag{1.86}$$

Taking the time derivative gives

$$\begin{aligned} \dot{\vec{\bar{r}}} &= \dot{\vec{R}} + \frac{d}{dt}x^a\hat{e}_a \\ &= \dot{\vec{R}} + \left(\dot{x}^a\hat{e}_a + x^a\dot{\hat{e}}_a\right) \\ &= \dot{\vec{R}} + \dot{\vec{r}} + x^a\dot{\hat{e}}_a \end{aligned} \tag{1.87}$$

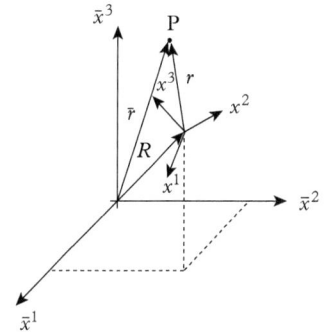

Figure 1.12 *Coordinates for a rotating frame. The body frame is the unprimed frame. The lab, or fixed, frame is primed (with bars).*

[12] Basis vectors in Cartesian coordinates are unit vectors denoted with a "hat." However, in general curvilinear coordinates, basis vectors might not be unit vectors. The general properties of basis vectors are described in Chapter 9.

(Einstein summation convention on the repeated index is assumed), where the last term is a non-inertial term because the basis vectors of the rotating frame are changing in time.

To obtain the time derivative of the basis vectors, consider an infinitesimal rotation transformation that operates on the basis vectors

$$\hat{e}_{\bar{b}} = R_{\bar{b}}^a \hat{e}_a = \delta_{\bar{b}}^a \hat{e}_a + \left(\frac{d\hat{e}_a}{dt}\right) dt, \tag{1.88}$$

where the infinitesimal rotation matrix $R_a^{\bar{b}}$ from Eq. (1.68) is expressed to lowest order in $d\theta = \omega dt$ as

$$R_{\bar{b}}^a \approx \left[\begin{pmatrix} 1 & 0 & 0 \\ 0 & 1 & 0 \\ 0 & 0 & 1 \end{pmatrix} - \begin{pmatrix} 0 & -\omega_z & \omega_y \\ \omega_z & 0 & -\omega_x \\ -\omega_y & \omega_x & 0 \end{pmatrix} \right] \begin{pmatrix} \hat{e}_x \\ \hat{e}_y \\ \hat{e}_z \end{pmatrix} dt, \tag{1.89}$$

where ω_a are the Cartesian components of the angular velocity vector $\vec{\omega}$ along the axes.[13] Therefore, the time derivatives of the basis vectors are

$$\frac{d\hat{e}_x}{dt} = \omega_z \hat{e}_y - \omega_y \hat{e}_z$$
$$\frac{d\hat{e}_y}{dt} = -\omega_z \hat{e}_x + \omega_x \hat{e}_z \tag{1.90}$$
$$\frac{d\hat{e}_z}{dt} = \omega_y \hat{e}_x - \omega_x \hat{e}_y.$$

The rotation of the basis vectors by the different components ω_a is shown in Figure 1.13.

Using Eq. (1.90) to express the non-inertial term in Eq. (1.87) gives

$$x^a \dot{\hat{e}}_a = x\omega_z \hat{e}_y - x\omega_y \hat{e}_z - y\omega_z \hat{e}_x + y\omega_x \hat{e}_z + z\omega_y \hat{e}_x - z\omega_x \hat{e}_y. \tag{1.91}$$

Combining terms gives

$$x^a \dot{\hat{e}}_a = \hat{e}_x \left(\omega_y z - \omega_z y\right) - \hat{e}_y \left(\omega_x z - \omega_z x\right) + \hat{e}_z \left(\omega_x y - \omega_y x\right), \tag{1.92}$$

where the result is recognized as the cross product

$$x^a \dot{\hat{e}}_a = \vec{\omega} \times \vec{r}. \tag{1.93}$$

Cross products occur routinely in the physics of rotating frames and rotating bodies, and are efficiently expressed in vector notion, which will be used through most of the remainder of the chapter instead of the index notation.[14] By using Eq. (1.93) in Eq. (1.87), the fixed and rotating velocities are related by

$$\vec{v}_f = \vec{V} + \vec{v}_r + \vec{\omega} \times \vec{r}. \tag{1.94}$$

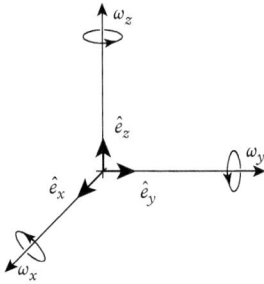

Figure 1.13 *Angular velocities related to changes in the basis vectors.*

[13] Cartesian vector components can be denoted with subscripts, but they are not to be confused with general covectors that are defined in Chapter 9.

[14] Vector cross products arise from the wedge product $A \wedge B$ of Hermann Grassmann (1844), introduced in *The Theory of Linear Extension, a New Branch of Mathematics.*

This result is general, and for any vector $\vec{Q}$, the time rate of change in the fixed frame is related to the time rate of change in the rotating frame as

$$\left.\frac{d\vec{Q}}{dt}\right|_{fixed} = \left.\frac{d\vec{Q}}{dt}\right|_{rotating} + \vec{\omega} \times \vec{Q}. \tag{1.95}$$

As an example, consider the case $\vec{Q} = \vec{\omega}$

$$\left.\frac{d\vec{\omega}}{dt}\right|_{fixed} = \left.\frac{d\vec{\omega}}{dt}\right|_{rotating} + \vec{\omega} \times \vec{\omega}, \tag{1.96}$$

where the last term is clearly zero. Therefore

$$\dot{\vec{\omega}}_f = \dot{\vec{\omega}}_r, \tag{1.97}$$

proving that angular accelerations are observed to be the same, just as linear accelerations were the same when transforming between inertial frames. This equality is because the rotating frame is in constant angular motion.

As a second, and more important example, take the time derivative of Eq. (1.94). This is

$$\left.\frac{d\vec{v}_f}{dt}\right|_{fixed} = \left.\frac{d\vec{V}}{dt}\right|_{fixed} + \left.\frac{d\vec{v}_r}{dt}\right|_{fixed} + \dot{\vec{\omega}} \times \vec{r} + \vec{\omega} \times \left.\frac{d\vec{r}}{dt}\right|_{fixed}. \tag{1.98}$$

The second term on the right is expanded using Eq. (1.95) as

$$\left.\frac{d\vec{v}_r}{dt}\right|_{fixed} = \left.\frac{d\vec{v}_r}{dt}\right|_{rotating} + \vec{\omega} \times \vec{v}_r. \tag{1.99}$$

The fourth term in Eq. (1.98) becomes

$$\begin{aligned} \vec{\omega} \times \left.\frac{d\vec{r}}{dt}\right|_{fixed} &= \vec{\omega} \times \left.\frac{d\vec{r}}{dt}\right|_{rotating} + \vec{\omega} \times (\vec{\omega} \times \vec{r}) \\ &= \vec{\omega} \times \vec{v}_r + \vec{\omega} \times (\vec{\omega} \times \vec{r}) \end{aligned} \tag{1.100}$$

The acceleration in the fixed frame is then

$$\vec{a}_f = \ddot{\vec{R}} + \vec{a}_r + \dot{\vec{\omega}} \times \vec{r} + \vec{\omega} \times (\vec{\omega} \times \vec{r}) + 2\vec{\omega} \times \vec{v}_r. \tag{1.101}$$

For a particle of mass m, Newton's second law is

$$\vec{F}_f = m\ddot{\vec{R}} + m\vec{a}_r + m\dot{\vec{\omega}} \times \vec{r} + m\vec{\omega} \times (\vec{\omega} \times \vec{r}) + 2m\vec{\omega} \times \vec{v}_r, \tag{1.102}$$

which is the force in the fixed frame.

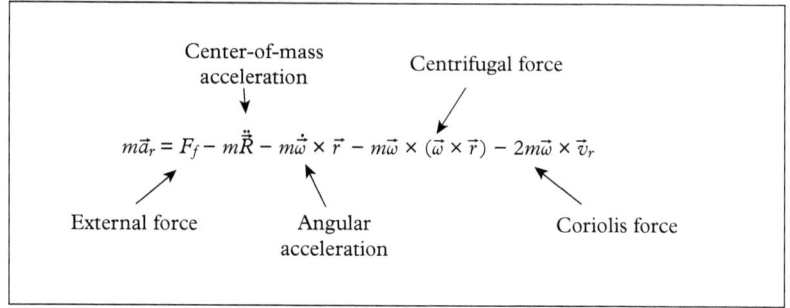

$$m\vec{a}_r = F_f - m\ddot{\vec{R}} - m\dot{\vec{\omega}} \times \vec{r} - m\vec{\omega} \times (\vec{\omega} \times \vec{r}) - 2m\vec{\omega} \times \vec{v}_r$$

Center-of-mass acceleration — Centrifugal force — External force — Angular acceleration — Coriolis force

Figure 1.14 *Effective force in a frame rotating with an angular velocity ω.*

Therefore, in the rotating frame there is an effective force (see Fig. 1.14):

$$\vec{F}_{eff} = m\vec{a}_r = F_f - m\ddot{\vec{R}} - m\dot{\vec{\omega}} \times \vec{r} - m\vec{\omega} \times (\vec{\omega} \times \vec{r}) - 2m\vec{\omega} \times \vec{v}_r. \quad (1.103)$$

The first two terms on the right are the fixed-frame forces. The third term is the angular acceleration of the spinning frame. The fourth term is the centrifugal force, and the last term is the Coriolis force.[15] The centrifugal and the Coriolis forces are fictitious forces. They are only apparent in the rotating frame because the rotating frame is not inertial.

1.5.1 Motion relative to the Earth

For a particle subject to the Earth's gravitational field, the effective force (Fig. 1.15) experienced by the particle is

$$\vec{F}_{eff} = \vec{F}_{ext} + m\vec{g}_0 - m\ddot{\vec{R}} - m\dot{\vec{\omega}} \times \vec{r} - m\vec{\omega} \times (\vec{\omega} \times \vec{r}) - 2m\vec{\omega} \times \vec{v}_r. \quad (1.104)$$

The fourth term is related to the deceleration of the Earth and is negligible. The centrifugal term is re-expressed as

$$\begin{aligned}\ddot{\vec{R}} &= \vec{\omega} \times (\vec{\omega} \times \vec{R}) \\ &= \vec{\omega} \times \dot{\vec{R}}\end{aligned} \quad (1.105)$$

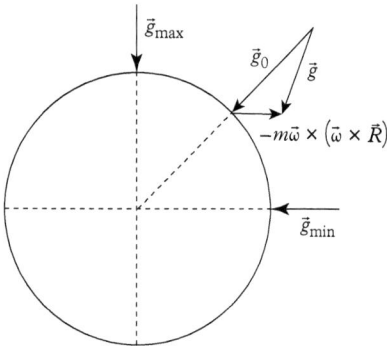

Figure 1.15 *Geometry for motion relative to the Earth.*

The effective force is then

$$\vec{F}_{eff} = \vec{F}_{ext} + m\vec{g}_0 - m\vec{\omega} \times (\vec{\omega} \times (\vec{r} + \vec{R})) - 2m\vec{\omega} \times \vec{v}_r \quad (1.106)$$

and redefining the effective gravitational acceleration through

$$\vec{g}_{eff} = \vec{g}_0 - \vec{\omega} \times [\vec{\omega} \times (\vec{r} + \vec{R})] \quad (1.107)$$

[15] Gaspard-Gustave de Coriolis (1792–1843) was a French mathematician, mechanical engineer and scientist. He was the first to coin the term "work" for the product of force and distance, but is best known for his mathematical work on the motion of rotating bodies.

Figure 1.16 *Dramatic example of cyclone motion in the Northern Hemisphere for a low-pressure center between Greenland and Iceland.*
<http://visibleearth.nasa.gov/view.php?id=68992> Image credit: Jacques Descloitres, MODIS Rapid Response Team, NASA/GSFC

gives

$$\vec{F}_{eff} = \vec{F}_{ext} + m\vec{g}_{eff} - 2m\vec{\omega} \times \vec{v}_r. \tag{1.108}$$

This last equation adds the centrifugal contribution to the measured gravitational acceleration. The last term $-2m\vec{\omega} \times \vec{v}_r$ in Eq. (1.108) is the Coriolis force that has important consequences for weather patterns (Fig. 1.16) on Earth, and hence has a powerful effect on the Earth's climate. It is also a sizeable effect for artillery projectiles. On the other hand, it plays a negligible role in the motion of whirlpools in bathtubs.

[16] Jean Bernard Lèeon Foucault (1819–1868) was a French physicist who performed one of the earliest measurements of the speed of light. In 1851 he constructed a long and heavy pendulum that was installed in the Pantheon in Paris.

1.5.2 Foucault's pendulum

Foucault's pendulum[16] is one of the most dramatic demonstrations of the rotation of the Earth. It also provides an accurate measure of latitude.

A simple pendulum (Figure 1.17) swinging at a latitude λ will precess during the day as a consequence of the Coriolis force. The acceleration in the rotating Earth frame is

$$\vec{a}_r = \vec{g} + \frac{1}{m}\vec{T} - 2\vec{\omega} \times \vec{v}_r, \tag{1.109}$$

where the components of the tension and angular velocity are

$$
\begin{aligned}
T_x &= -T\frac{x}{\ell} & \omega_x &= -\omega\cos\lambda \\
T_y &= -T\frac{y}{\ell} & \omega_y &= 0 \\
T_z &= T\frac{z}{\ell} & \omega_z &= \omega\sin\lambda
\end{aligned}
\tag{1.110}
$$

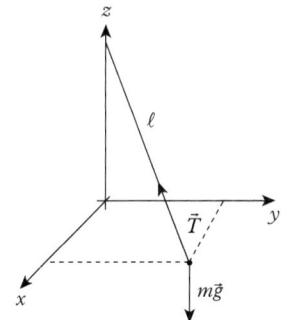

Figure 1.17 *Geometry of a Foucault pendulum of mass m attached to a massless string of length ℓ supporting a tension T.*

The cross product is

$$\vec{\omega} \times \vec{v}_r = \begin{pmatrix} 0 & -\omega_z & \omega_y \\ \omega_z & 0 & -\omega_x \\ -\omega_y & \omega_x & 0 \end{pmatrix} \begin{pmatrix} v_x \\ v_y \\ v_z \end{pmatrix}$$

$$= \begin{pmatrix} 0 & -\omega\sin\lambda & 0 \\ \omega\sin\lambda & 0 & \omega\cos\lambda \\ 0 & -\omega\cos\lambda & 0 \end{pmatrix} \begin{pmatrix} v_x \\ v_y \\ 0 \end{pmatrix} \qquad (1.111)$$

$$= \begin{pmatrix} -\omega\dot{y}\sin\lambda \\ \omega\dot{x}\sin\lambda \\ -\omega\dot{y}\cos\lambda \end{pmatrix}$$

and the acceleration in the rotating frame is

$$a_{rx} = \ddot{x} = -\frac{T}{m}\frac{x}{\ell} + 2\dot{y}\omega\sin\lambda$$

$$a_{ry} = \ddot{y} = -\frac{T}{m}\frac{y}{\ell} - 2\dot{x}\omega\sin\lambda \qquad (1.112)$$

leading to the coupled equations

$$\begin{matrix} \ddot{x} + \omega_0^2 x = 2\omega_z\dot{y} \\ \ddot{y} + \omega_0^2 y = -2\omega_z\dot{x} \end{matrix} \qquad \text{where} \quad \omega_0^2 = \frac{T}{m\ell} \approx \frac{g}{\ell}. \qquad (1.113)$$

The coupled equations are added in quadrature (see Appendix A.2 online) to yield

$$(\ddot{x} + i\ddot{y}) + \omega_0^2 (x + iy) = -2i\omega_z (\dot{x} + i\dot{y}). \qquad (1.114)$$

This is converted into a single second-order equation through the substitution of variable $q = x + iy$ to give

$$\ddot{q} + 2i\omega_z\dot{q} + \omega_0^2 q = 0, \qquad (1.115)$$

which is the equation of a harmonic oscillator with imaginary damping. The solution is

$$q(t) = e^{-i\omega_z t} \left[Ae^{i\sqrt{\omega_z^2 + \omega_0^2}\, t} + Be^{-i\sqrt{\omega_z^2 + \omega_0^2}\, t} \right]. \qquad (1.116)$$

For a typical pendulum $\omega_0 \gg \omega_z$ and the solution simplifies to

$$q(t) = e^{-i\omega_z t} \left[Ae^{i\omega_0 t} + Be^{-i\omega_0 t} \right], \qquad (1.117)$$

where the term in the brackets is the solution of a conventional pendulum that is not rotating. Expressing this solution as

$$q(t) = e^{-i\omega_z t} [x_0 + iy_0] \qquad (1.118)$$

it can also be written as (see Appendix A.2 online)

$$\begin{pmatrix} x(t) \\ y(t) \end{pmatrix} = \begin{pmatrix} \cos\omega_z t & \sin\omega_z t \\ -\sin\omega_z t & \cos\omega_z t \end{pmatrix} \begin{pmatrix} x_0 \\ y_0 \end{pmatrix}. \qquad (1.119)$$

The matrix is the rotation matrix that rotates the pendulum through an angle

$$\theta = \omega_z t. \qquad (1.120)$$

To find the precession time T_p of the rotation, use

$$\omega_z T_p = 2\pi$$
$$\frac{2\pi}{T_E} \sin\lambda T_p = 2\pi \qquad (1.121)$$

or

$$T_p = T_E / \sin\lambda. \qquad (1.122)$$

At a latitude of 45°, the precession period is 34 hours. At the North (or South) Pole, the precession time is equal to the Earth's rotation period. At the equator, the precession time is infinite—it does not precess.

1.6 Rigid-body motion

A rigid body is a multiparticle system with a very large number of constituents. In principle, the dynamical equations of the system would include the positions and velocities of each of the particles. In a rigid body, the particle coordinates are subject to a similarly large number of equations of constraints. Therefore, the dynamical equations are greatly simplified in practice, consisting of only a few DOF. Nonetheless, the dynamics of rotating bodies is a topic full of surprises and challenges, with a wide variety of phenomena, many of which are not immediately intuitive.

1.6.1 Inertia tensor

The dynamics of rotating bodies follow closely from the principles of rotating frames.[17] Consider a collection of N individual point masses whose relative positions are constant in the body frame. The masses are m_α for $\alpha = 1:N$, and their

[17] The analysis in this section is carried out relative to the inertial fixed frame, which is unprimed in this discussion.

locations are $\vec{r}_\alpha$. Because the body is rigid, there are no internal motions and the velocities in the fixed frame are

$$\vec{v}_\alpha = \vec{\omega} \times \vec{r}_\alpha \qquad (1.123)$$

for angular velocity $\vec{\omega}$, where the fixed frame is traveling with the center-of-mass motion of the rigid body. The only kinetic energy in this frame is rotational. By using the identity

$$\left(\vec{A} \times \vec{B}\right)^2 = A^2 B^2 - \left(\vec{A} \cdot \vec{B}\right)^2 \qquad (1.124)$$

the kinetic energy is obtained as

$$T_{rot} = \frac{1}{2} \sum_\alpha m_\alpha \left(\omega^2 r_\alpha^2 - (\vec{\omega} \cdot \vec{r}_\alpha)^2 \right). \qquad (1.125)$$

The rotational energy can be expressed in terms of the components of the position vectors and the components of the angular velocity as[18]

$$T_{rot} = \frac{1}{2} \sum_\alpha m_\alpha \left[\left(\sum_a \omega_a^2 \right) \left(\sum_b (x_\alpha^b)^2 \right) - \left(\sum_a \omega_a x_\alpha^a \right) \left(\sum_b \omega_b x_\alpha^b \right) \right], \qquad (1.126)$$

where the sum over α is over point masses and the sums over a and b are over coordinates. This expression can be simplified using $\omega_b = \sum_a \omega_a \delta_{ab}$ to give

$$\begin{aligned} T_{rot} &= \frac{1}{2} \sum_\alpha \sum_{a,b} m_\alpha \left(\omega_a \omega_b \delta_{ab} \left(\sum_c (x_\alpha^c)^2 \right) - \omega_a \omega_b x_\alpha^a x_\alpha^b \right) \\ &= \frac{1}{2} \sum_{a,b} \omega_a \omega_b \sum_\alpha m_\alpha \left[\delta_{ab} \left(\sum_c (x_\alpha^c)^2 \right) - x_\alpha^a x_\alpha^b \right] \end{aligned} \qquad (1.127)$$

This procedure has separated out a term summed over the masses α that makes the rotational kinetic energy particularly simple:

$$T_{rot} = \frac{1}{2} \sum_{ab} I_{ab} \omega_a \omega_b, \qquad (1.128)$$

where the expression

$$\textit{Inertia tensor:} \qquad I_{ab} = \sum_\alpha m_\alpha \left[\delta_{ab} \left(\sum_c (x_\alpha^c)^2 \right) - x_\alpha^a x_\alpha^b \right] \qquad (1.129)$$

is the moment of inertia tensor for a rigid body. In matrix form this is

$$
I_{ab} =
\begin{pmatrix}
\sum_\alpha m_\alpha \left((y_\alpha)^2 + (z_\alpha)^2 \right) & -\sum_\alpha m_\alpha x_\alpha y_\alpha & -\sum_\alpha m_\alpha x_\alpha z_\alpha \\
-\sum_\alpha m_\alpha y_\alpha x_\alpha & \sum_\alpha m_\alpha \left((x_\alpha)^2 + (z_\alpha)^2 \right) & -\sum_\alpha m_\alpha y_\alpha z_\alpha \\
-\sum_\alpha m_\alpha z_\alpha x_\alpha & -\sum_\alpha m_\alpha z_\alpha y_\alpha & \sum_\alpha m_\alpha \left((x_\alpha)^2 + (y_\alpha)^2 \right)
\end{pmatrix}
$$

$$(1.130)$$

The moment of inertia tensor is a rank two symmetric tensor that is quadratic in the coordinate positions of the constituent masses. There are only six independent tensor components that capture all possible mass configurations. If the rigid body has symmetries, then the number of independent components is reduced. For instance, if the rigid body has spherical symmetry, then all off-diagonal terms are zero, and the diagonal terms are all equal.

The summations over the individual particles can be replaced (in the limit of a continuous distribution of mass) by integrals. The moment of inertia tensor for a continuous mass distribution is given by

$$
I_{ab} = \int_V \rho(\vec{r}) \left[\delta_{ab} \left(\sum_c (x_\alpha^c)^2 \right) - x_\alpha^a x_\alpha^b \right] d^3x,
\qquad (1.131)
$$

which is expanded as

$$
I_{ab} =
\begin{pmatrix}
\int_V \left(y^2 + z^2 \right) \rho(\vec{r})\, dxdydz & -\int_V xy\rho(\vec{r})\, dxdydz & -\int_V xz\rho(\vec{r})\, dxdydz \\
-\int_V xy\rho(\vec{r})\, dxdydz & \int_V \left(x^2 + z^2 \right) \rho(\vec{r})\, dxdydz & -\int_V yz\rho(\vec{r})\, dxdydz \\
-\int_V xz\rho(\vec{r})\, dxdydz & -\int_V yz\rho(\vec{r})\, dxdydz & \int_V \left(x^2 + y^2 \right) \rho(\vec{r})\, dxdydz
\end{pmatrix}
$$

$$(1.132)$$

The integral is carried out over the limits of the mass distribution. Several examples of moments of inertia of symmetric solids with constant mass densities are shown in Table 1.1.

The solids in Table 1.1 all have high symmetry, the coordinate axes are directed along the primary symmetry axes, and the coordinate origin is located at the center of mass. Under these conditions all of the inertia tensors are diagonal with zero off-diagonal elements, the coordinate axes are called principal axes, and the diagonal elements of the inertia tensor are called the principal moments of inertia. In general, the coordinate axes may not be chosen along the main symmetry axes, and the origin may not be at the center of mass. In this general case, all elements of the tensor may be nonzero. However, principal moments and principal

Table 1.1 *Inertia tensors.*

Solid sphere	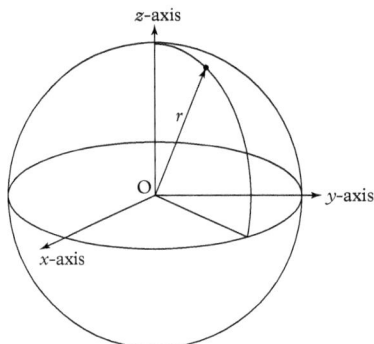	$I = \begin{pmatrix} 2/5 & 0 & 0 \\ 0 & 2/5 & 0 \\ 0 & 0 & 2/5 \end{pmatrix} mr^2$
Spherical shell	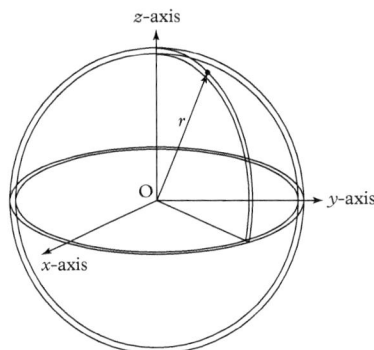	$I = \begin{pmatrix} 2/3 & 0 & 0 \\ 0 & 2/3 & 0 \\ 0 & 0 & 2/3 \end{pmatrix} mr^2$
Rectangular prism	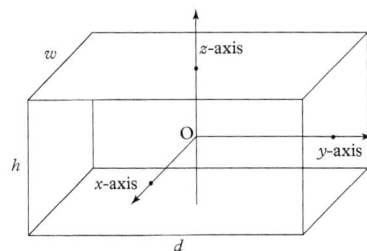	$I = \begin{pmatrix} d^2 + h^2 & 0 & 0 \\ 0 & w^2 + h^2 & 0 \\ 0 & 0 & w^2 + d^2 \end{pmatrix} \dfrac{m}{12}$
Right solid cylinder	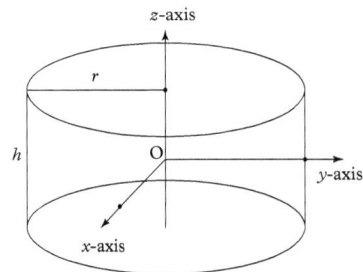	$I = \begin{pmatrix} (3r^2 + h^2)/12 & 0 & 0 \\ 0 & (3r^2 + h^2)/12 & 0 \\ 0 & 0 & r^2/2 \end{pmatrix} m$

axes can always be regained by finding the eigenvalues and the eigenvectors of the inertia matrix.

1.6.2   Angular momentum

The total angular momentum for N discrete particles is obtained by summing the angular momentum of the individual particles as

$$\vec{L} = \sum_{\alpha=1}^{N} \vec{r}_\alpha \times \vec{p}_\alpha. \tag{1.133}$$

The individual momenta are

$$\vec{p}_\alpha = m_\alpha \vec{v}_\alpha = m_\alpha \vec{\omega} \times \vec{r}_\alpha \tag{1.134}$$

and hence the angular momentum is

$$\begin{aligned} \vec{L} &= \sum_\alpha \vec{r}_\alpha \times (m_\alpha \vec{\omega} \times \vec{r}_\alpha) \\ &= \sum_a m_\alpha \left[r_\alpha^2 \vec{\omega} - \vec{r}_\alpha (\vec{r}_\alpha \cdot \vec{\omega}) \right] \end{aligned} \tag{1.135}$$

By again using the identity Eq. (1.124) the angular momentum can be broken into terms that are the components of the position vectors as well as the components of the angular velocity, just as was done for the moment of inertia in Eq. (1.126). The angular momentum is

$$\begin{aligned} L^a &= \sum_\alpha m_\alpha \left(\omega_a \sum_c \left(x_\alpha^c\right)^2 - x_\alpha^a \sum_b x_\alpha^b \omega_b \right) \\ &= \sum_\alpha m_\alpha \sum_b \left(\omega_a \delta_a^b \sum_c \left(x_\alpha^c\right)^2 - x_\alpha^a x_\alpha^b \omega_b \right) \\ &= \sum_b \omega_b \sum_\alpha m_\alpha \left(\delta_a^b \sum_c \left(x_\alpha^c\right)^2 - x_\alpha^a x_\alpha^b \right), \end{aligned} \tag{1.136}$$

which is re-expressed as

$$L_a = \sum_b I_{ab} \omega_b \tag{1.137}$$

by using the definition in Eq. (1.129) for the moment of inertia tensor. This is a tensor relation that connects the angular velocity components to the angular momentum. The rotational kinetic energy Eq. (1.128) is expressed in terms of the angular momentum as

$$T_{rot} = \frac{1}{2} \vec{L} \cdot \vec{\omega}, \tag{1.138}$$

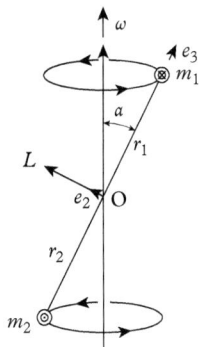

Figure 1.18 *Non-parallel angular momentum and angular velocity for a rigid rotating dumbell. The rotation axis is at an angle to the rigid (massless) rod connecting the masses.*

where $\vec{L}$ and $\vec{\omega}$ are not necessarily parallel because of off-diagonal elements in the inertia tensor.

A simple example when $\vec{L}$ and $\vec{\omega}$ are not parallel is a tilted rigid dumbbell, as shown in Figure 1.18. Two masses are attached at the ends of a rigid (but massless) rod that is tilted relative to the rotation axis. The angular velocity $\vec{\omega}$ is directed along the central rotation axis, but the angular momentum $\vec{L}$ is directed perpendicular to the rod connecting the two masses. As the dumbbell spins, the angular velocity remains constant, but the angular momentum continuously changes direction. This represents a steady time rate of change of the angular momentum, which requires a steady external torque, which can be found using Euler's equations.

1.6.3   Euler's equations

The time rate of change of the angular momentum in Eq. (1.133) is

$$\dot{\vec{L}} = \sum_\alpha \dot{\vec{r}}_\alpha \times \vec{p}_\alpha + \sum_\alpha \vec{r}_\alpha \times \dot{\vec{p}}_\alpha$$
$$= \sum_\alpha m_\alpha \left(\dot{\vec{r}}_\alpha \times \dot{\vec{r}}_\alpha \right) + \sum_\alpha \vec{r}_\alpha \times \dot{\vec{p}}_\alpha \qquad (1.139)$$
$$= \sum_\alpha \vec{r}_\alpha \times \dot{\vec{p}}_\alpha = \sum_\alpha \vec{r}_\alpha \times \vec{F}_\alpha,$$

where $\left(\dot{\vec{r}}_\alpha \times \dot{\vec{r}}_\alpha \right)$ is identically zero. The force $\vec{F}_\alpha$ acting on the αth particle consists of the force $\vec{f}_{\beta\alpha}$ of particle β acting on particle α plus any external force $\vec{F}_\alpha^{ext}$ acting on α. The time derivative of the angular momentum is then

$$\dot{\vec{L}} = \sum_\alpha \vec{r}_\alpha \times \vec{F}_\alpha$$
$$= \sum_\alpha \vec{r}_\alpha \times \vec{F}_\alpha^{ext} + \sum_\alpha \sum_{\beta \neq \alpha} \vec{r}_\alpha \times \vec{f}_{\alpha\beta} \qquad (1.140)$$
$$= \sum_\alpha \vec{r}_\alpha \times \vec{F}_\alpha^{ext}$$
$$= \vec{N}^{ext},$$

where the double sum vanishes because of action–reaction pairs and the time derivative of the angular momentum is equal to the externally applied torque,

$$\left(\frac{d\vec{L}}{dt} \right)_{fixed} = \vec{N}^{ext}. \qquad (1.141)$$

In the body frame, from Eq. (1.95), this is

$$\left(\frac{d\vec{L}}{dt} \right)_{body} + \vec{\omega} \times \vec{L} = \vec{N}. \qquad (1.142)$$

There is a considerable simplification by choosing the body axes along the principal axes of the inertia tensor (defined by the directions of the eigenvalues of the inertia tensor). Then the component along the body axis is

$$\dot{L}_3 + \omega_1 L_2 - \omega_2 L_1 = N_3 \tag{1.143}$$

and

$$L_i = I_i \omega_i \tag{1.144}$$

to give

$$I_3 \dot{\omega}_3 - (I_1 - I_2)\, \omega_1 \omega_2 = N_3. \tag{1.145}$$

By taking all three permutations this yields

Euler's equations:
$$
\begin{aligned}
I_1 \dot{\omega}_1 - (I_2 - I_3)\, \omega_2 \omega_3 &= N_1 \\
I_2 \dot{\omega}_2 - (I_3 - I_1)\, \omega_3 \omega_1 &= N_2 \\
I_3 \dot{\omega}_3 - (I_1 - I_2)\, \omega_1 \omega_2 &= N_3
\end{aligned}
\tag{1.146}
$$

These are Euler's equations that relate the time rate of change of the angular velocity to the applied torque.

Example 1.7 Torque on a dumbbell

The torque required to rotate the dumbbell in Figure 1.18 is obtained by finding the components of Euler's equations as

$$
\begin{aligned}
\omega_1 &= 0 \\
\omega_2 &= \omega \sin \alpha \\
\omega_3 &= \omega \cos \alpha
\end{aligned}
\tag{1.147}
$$

and

$$
\begin{aligned}
I_1 &= (m_1 + m_2)\, d^2 \\
I_2 &= (m_1 + m_2)\, d^2 \\
I_3 &= 0
\end{aligned}
\tag{1.148}
$$

giving the angular momentum

$$
\begin{aligned}
L_1 &= I_1 \omega_1 = 0 \\
L_2 &= I_2 \omega_2 = (m_1 + m_2)\, d^2 \omega \sin \alpha \\
L_3 &= I_3 \omega_3 = 0.
\end{aligned}
\tag{1.149}
$$

continued

Example 1.7 *continued*

Using these equations in Euler's equations leads to the required externally applied torque

$$N_1 = -(m_1 + m_2)\, d^2\omega^2 \sin\alpha \cos\alpha$$
$$N_2 = 0 \tag{1.150}$$
$$N_3 = 0$$

to sustain a constant angular velocity. The torque is directed perpendicular to the angular momentum, and causes the angular momentum to precess around the rotation axis as a function of time, as shown in Fig. 1.18.

1.6.4 Force-free top

Euler's equations also apply to a force-free top (no gravity and no external constraints), but now the angular velocity is no longer constant. If the top is symmetric with $I_1 = I_2$, and with no torques acting, Euler's equations are

$$(I_1 - I_3)\, \omega_2\omega_3 - I_1\dot\omega_1 = 0$$
$$(I_3 - I_1)\, \omega_3\omega_1 - I_1\dot\omega_2 = 0 \tag{1.151}$$
$$I_3\dot\omega_3 = 0$$

and the angular velocity along the body axis is a constant, i.e., ω_3 = const. The equations for the time rate of change of the other two components are

$$\dot\omega_1 = -\left(\frac{I_3 - I_1}{I_1}\omega_3\right)\omega_2$$
$$\dot\omega_2 = \left(\frac{I_3 - I_1}{I_1}\omega_3\right)\omega_1. \tag{1.152}$$

By substituting

$$\Omega = \frac{I_3 - I_1}{I_1}\omega_3 \tag{1.153}$$

these become

$$\dot\omega_1 + \Omega\omega_2 = 0$$
$$\dot\omega_2 - \Omega\omega_1 = 0. \tag{1.154}$$

This set of coupled equations is combined in the expression

$$\dot\eta - i\Omega\eta = 0 \tag{1.155}$$

with the substitution

$$\eta = \omega_1 + i\omega_2. \tag{1.156}$$

This gives the solution

$$\omega_1(t) = A\cos\Omega t$$
$$\omega_2(t) = A\sin\Omega t, \tag{1.157}$$

where

$$A = \sqrt{\omega^2 - \omega_3^2}. \tag{1.158}$$

The solution to the dynamics of a force-free prolate spheroid is illustrated in Figure 1.19 for a prolate top $(I_1 > I_3)$. The symmetry axis of the rigid body is the x_3-axis. Because both A and ω_3 are constants, the magnitude of the angular velocity is a constant

$$\omega = \sqrt{A^2 + \omega_3^2}, \tag{1.159}$$

where A is the projection of the angular velocity vector on the body x_1–x_2 plane. Equation (1.157) describes the precession of the angular velocity vector around the x_3 body symmetry axis with a precession angular speed of Ω. The angular momentum $\boldsymbol{L}$ is also a constant of the motion, is steady in space, and points along the fixed x_3'-axis. Another constant of the motion is the kinetic energy which is defined by the projection of the angular velocity onto the angular momentum vector as

$$T_{rot} = \frac{1}{2}\vec{\omega}\cdot\vec{L}, \tag{1.160}$$

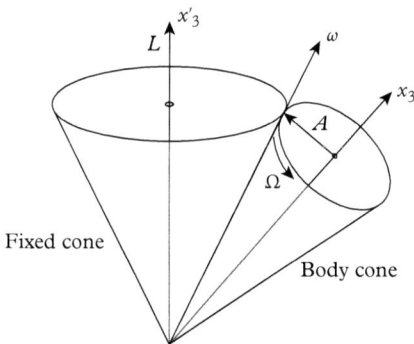

Figure 1.19 *Precession of a force-free prolate $(I_1 > I_3)$ top visualized as the body cone rolling around the surface of the fixed cone. The angular velocity ω precesses around the body symmetry axis with a rate Ω. The vectors $\vec{L}$, $\vec{\omega}$, and $\vec{x}_3$ are coplanar.*

which means that the projection of $\vec{\omega}$ onto $\vec{L}$ is a constant. Therefore, $\vec{\omega}$ also precesses around $\vec{L}$ and the fixed x_3'-axis. The result is visualized in Fig. 1.19 for a prolate top $(I_1 > I_3)$ as a body cone that rolls without slipping on a fixed cone. The three vectors $\vec{L}$, $\vec{\omega}$, and $\vec{x}_3$ are coplanar, with the angular velocity vector oriented along the contact between the cones. For an oblate symmetric top $(I_1 < I_3)$, the fixed cone is inside the body cone, but in both cases it is the body cone that rolls, and the space cone is fixed because $\vec{L}$ is constant.

For a general body, with $I_1 \neq I_2 \neq I_3$, Euler's equations yield three eigenfrequency solutions that may be real or imaginary valued. Real eigenfrequencies correspond to stable configurations, while imaginary eigenfrequencies correspond to unstable configurations. For $I_1 < I_2 < I_3$ motion around the x_2-axis is unstable. This is why a book tossed into the air, originally rotating about its x_2-axis, will begin to tumble no matter how carefully you launch it.

The force-free top has no external torque acting on it—it is a spinning or tumbling object in free fall. On the other hand, a gyroscope on Earth has a point on its rotation axis fixed in the fixed frame, and it experiences an external torque caused by gravity acting on the center of mass. The behavior of a spinning top tilting in gravity, in contact with a supporting plane but not constrained to a fixed point, becomes even more complicated.[19] There are even flipping tops that can flip over, and banana-shaped tops that can reverse their direction of rotation. The physics of these objects is highly complex and subtle.[20]

1.7   Summary

This chapter has emphasized the importance of a geometric approach to dynamics. The central objects of interest are trajectories of a dynamical system through multidimensional spaces composed of generalized coordinates. Trajectories through *configuration space* are often parameterized by the path length element ds, which becomes important in problems in special relativity when time transforms between frames, and in general relativity when space–time is warped by mass and energy. Trajectories through *state space* uniquely define the dynamical properties of a system, and *flow fields* and *flow lines* become the fields and field lines of *mathematical flows*. *Coordinate transformations* and *Jacobian matrices* are among the central tools that are used throughout this text, and the transformation to non-inertial frames introduces fictitious forces like the Coriolis force that are experienced by an observer in the non-inertial frame. Uniformly rotating frames are the reference frames for rigid-body motion and contribute concepts of the *moment of inertia* and *angular momentum*.

Degrees of freedom

The number of degrees of freedom (DOF) of a system is the number of conditions needed to uniquely specify a trajectory of the system in state space. If the

[19] See Thornton and Marion, Chap. 12.
[20] For example, see C. M. Braams, On the influence of friction on the motion of a top, *Physica*, 18, pp. 503–514 (1952) and S. Ebenfeld and F. Scheck, A new analysis of the Tippe top, *Annals of Physics*, 243, pp. 195–217 (1995).

system is composed of N coordinates, and there are K additional equations of constraint among the coordinates, then DOF $= N - K$ and the trajectories are constrained to lie on a $D = (N - K)$-dimensional manifold in the configuration space.[21] There can be D generalized coordinates that span the manifold.

Serret–Frenet formulas

A position on a general 3D trajectory is defined by three mutually orthogonal vectors: tangent T^a, normal N^a, and binormal B^a.

$$\frac{dx^a}{ds} = T^a$$
$$\frac{dT^a}{ds} = \kappa N^a$$
$$\frac{dN^a}{ds} = -\kappa T^a + \tau B^a \qquad N^a \times T^a = B^a \qquad (1.21)$$
$$\frac{dB^a}{ds} = -\tau N^a$$

Equations of a mathematical flow

A mathematical flow is an ODE:

$$\frac{dq^a}{dt} = F^a (q^a; t). \qquad (1.43)$$

This is perhaps the most important equation of this textbook. Most of the chapters in this book explore aspects of mathematical flows.

Jacobian matrix of a linear transformation

The Jacobian matrix plays important roles in coordinate transformations and stability analysis.

$$\text{row index} \searrow$$
$$\mathcal{J}^a_b = \frac{\partial x^a}{\partial q^b} \qquad (1.48)$$
$$\text{column index} \nearrow$$

The upper index (contravariant) is a row index, and the lower index (covariant) is a column index.

[21] The phrase "degrees of freedom" may be used in different ways. For instance, in a molecular gas, degrees of freedom are related to the partition of internal energy in the equipartition theorem. In mechanics, "degrees of freedom" sometimes refer to the dimensions of the state space rather than of the configuration space.

Metric distance

The arc-length differential ds,

$$ds^2 = g_{11}(dq^1)^2 + g_{22}(dq^2)^2 + g_{33}(dq^3)^2, \tag{1.55}$$

plays a central role in the mathematical treatment of trajectories through complex, and possible curved, spaces.

Time derivatives in a rotating frame

The relationship between vectors observed in the rotating and fixed frames is given by

$$\left.\frac{d\vec{Q}}{dt}\right|_{fixed} = \left.\frac{d\vec{Q}}{dt}\right|_{rotating} + \vec{\omega} \times \vec{Q}, \tag{1.95}$$

where the second term gives rise to fictitious forces.

Effective force in a rotating frame

The effective force on a mass moving in a rotating frame contains several terms that are called fictitious forces:

$$\vec{F}_{eff} = m\vec{a}_r = F - m\ddot{\vec{R}} - m\dot{\vec{\omega}} \times \vec{r} - m\vec{\omega} \times (\vec{\omega} \times \vec{r}) - 2m\vec{\omega} \times \vec{v}_r. \tag{1.103}$$

The third term relates to the angular acceleration of the rotating frame. The fourth term is the centrifugal term. The fifth term is the Coriolis term.

Moment of inertia

The physics of rotating solid bodies uses the concept of moments of inertia. The inertia tensor is

$$I_{ab} = \sum_{\alpha} m_{\alpha}\left[\delta_{ab}\sum_{b}(x_{\alpha}^b)^2 - x_{\alpha}^a x_{\alpha}^b\right]. \tag{1.129}$$

Euler's equations

The equations of motion of a rotating body subject to torques are

$$\begin{aligned}
I_1\dot{\omega}_1 - (I_2 - I_3)\,\omega_2\omega_3 &= N_1 \\
I_2\dot{\omega}_2 - (I_3 - I_1)\,\omega_3\omega_1 &= N_2 \\
I_3\dot{\omega}_3 - (I_1 - I_2)\,\omega_1\omega_2 &= N_3
\end{aligned} \tag{1.146}$$

1.8   Bibliography

G. Arfkin, *Mathematical Methods for Physicists* (Academic Press, 1985). 3rd Ed.
   A standard mathematical reference text for all physicists. Has a solid introduction on coordinate transformations and the Jacobian matrix.

T. Frankel, *The Geometry of Physics: An Introduction* (Cambridge University Press, 2003).
   This is a good graduate introduction to the modern geometric view of mechanics.

A. E. Jackson, *Perspectives of Nonlinear Dynamics* (Cambridge University Press, 1989).
   This is an early classic that delves deeper than most into the mathematical structure of the solutions of dynamical equations.

R. C. Hilborn, *Chaos and Nonlinear Dynamics: An Introduction for Scientists and Engineers* (Oxford University Press, 2000).
   Full of treasures and clear explanations of the details of mathematical flows.

D. D. Holm, *Geometric Mechanics: Part I Dynamics and Symmetry* (World Scientific/Imperial College Press, 2008).
   This book has a second part on rotation and uses a strongly geometric approach for undergrdauates.

B. F. Schutz, *A First Course in General Relativity* (Cambridge University Press, 1985).
   Gives a clear description of contravariant and covariant vector notations, as well as an excellent introduction to general relativity.

S. H. Strogatz, *Nonlinear Dynamics and Chaos* (Westview Press, 1994).
   This should be the first read for the beginner on the topic of chaos and nonlinear dynamics.

R. Talman, *Geometric Mechanics: Toward a Unification of Classical Physics* (Wiley, 2007).
   Is a good undergraduate introduction to the modern geometric view of mechanics.

S. T. Thornton and J. B. Marion, *Classical Dynamics of Particles and Systems* (Thomson, 2004). 5th Ed.
   The classic undergraduate text on the topic of mechanics. Very clear exposition on rotating frames and fictitious forces.

1.9   Homework exercises

1. **Damped harmonic oscillator:** Derive the response function $X(\omega)$ for a driven damped harmonic oscillator

$$m\ddot{x} + \gamma\dot{x} + kx = Fe^{i\omega t}.$$

(a) Find the real and imaginary parts of $X(\omega)$.

(b) Solve for the motion if the right-hand side equals $F \cos \omega t$, and the initial conditions are $x(0) = 0$ and $\dot{x}(0) = \mathrm{B}$.

2. **Terminal velocity:** Solve for the terminal velocity of a massive particle falling under gravity

$$m\ddot{z} + \gamma\dot{z} = mg.$$

   (a) Choose the initial condition $\dot{z}(0) = 0$.

   (b) Choose the initial condition $\dot{z}(0) = \dfrac{2mg}{\gamma}$.

3. **Arc length:** Calculate the total path length as a function of time of a general parabolic trajectory with arbitrary initial velocity (v_x, v_y) in a constant gravitational field. Evaluate (expand) the short-time and long-time behavior and interpret as simple limits.

4. **Path length and tangent vector:** A pendulum is constructed as a mass on a rigid massless rod. The mass is dropped from rest at a height h above the center of rotation. Find the path length as a function of time. Find the unit tangent vector as a function of time.

5. **Generalized coordinate:** Consider a bead on a helical wire subject to a z-dependent potential $V(z)$. Show that $q = z$ can be the single generalized coordinate, and the helix can be described as

$$x(q) = R \cos (\omega^* q)$$
$$y(q) = R \sin (\omega^* q)$$
$$z(q) = q,$$

   where the parameter ω^* is the pitch of the helical wire.

6. **Serret–Frenet equations of a geometric spiral:** Find T, N, and B of a geometric spiral $r = r_0 e^{b\theta}$ and $z = a\theta$.

7. **Curvature:** Find the curvature κ of a general parabolic trajectory of a mass m in gravity g.

8. **Affine transformation:** Find the affine parameters in Eq. (1.53) for the variables (x,t) that match the Lorentz coordinate transformations.

9. **Path element:** Derive, using Eq. (1.56), the expressions for ds^2 for cylindrical and spherical curvilinear coordinates.

10. **3D rotations:** Show that Eq. (1.63) and Eq. (1.68) are equivalent representations of a 3D rotation.

11. **Jacobian matrix:** Consider the coordinate system sometimes used in electrostatics and hydrodynamics

$$xy = u$$
$$x^2 - y^2 = v$$
$$z = z.$$

Find the Jacobian matrix and the Jacobian of the transformation.

12. **Deflection of light by gravity:** Do a back-of-the-envelope calculation (within an order of magnitude) of the deflection of a light ray passing just above the surface of the Sun using the elevator analogy. Express you answer in terms of a deflection angle. How does it compare to the correct answer $\Delta\phi = \dfrac{4MG}{c^2R}$?

13. **Coriolis force:** If a projectile is fired due east from a point on the surface of Earth at a northern latitude λ with a velocity of magnitude v_0 and at an angle of inclination to the horizontal of α, show that the lateral deflection when the projectile strikes Earth is

$$d = \frac{4v_0^3}{g^2}\omega \sin\lambda \sin^2\alpha \cos\alpha,$$

where ω is the angular rotation velocity of Earth.

14. **Falling mass in rotating frame:** Consider a particle falling in Earth's gravitational field. Take g to be defined at ground level and use the zeroth-order result for the time of fall, $T = \sqrt{2h/g}$. Perform a calculation in second-order approximation (retain terms in ω^2) and calculate the southerly deflection. There are three components to consider: (a) Coriolis force to second order (C_1), (b) variation of centrifugal force with height (C_2), and (c) variation of gravitational force with height (C_3). Show that each of these components gives a result equal to

$$C_i \frac{h^2}{g}\omega^2 \sin\lambda \cos\lambda,$$

where $C_1 = 2/3$, $C_2 = 5/6$, and $C_3 = 5/2$. The total southerly deflection is therefore

$$4\frac{h^2}{g}\omega^2 \sin\lambda \cos\lambda$$

15. **Coriolis effect:** A warship fires a projectile due south at latitude $50°$S. If the shells are fired at $37°$ elevation with a speed of 800 m/s, by how much do the shells miss their target and in what direction? Ignore air resistance.

16. **Effective gravity:** Calculate the effective gravitational field vector g at Earth's surface at the poles and the equator. Take account of the difference in the equatorial (6378 km) and polar (6357 km) radius as well as the centrifugal force. How well does the result agree with the difference calculated with the result $g = 9.780356(1 + 0.0052885 \sin^2\lambda - 0.0000059 \sin^2(2\lambda))$ m/s^2, where λ is the latitude?

17. **Bathtub:** Estimate the Coriolis deflection for water in a bathtub whirlpool going down the drain. Then estimate the Coriolis deflection for a hurricane. What can you conclude about bathtub whirlpools?

18. **Inertia tensor:** Derive the inertia tensor for a cube of side L with the origin at the center. Then derive it again with the origin at an apex. Construct a parallel axis theorem to transform from one expression to the other.

19. **Stability:** Using Euler's equations, show that for an object with $I_1 < I_2 < I_3$ there are two stable rotation axes and one unstable axis.

20. **Jet stream:** High pressure at the equator drives a northward air speed of nearly 100 mph in the upper atmosphere. This leads to the jet stream. Estimate the air speed of the jet stream as a function of latitude.

Hamiltonian Dynamics and Phase Space

2

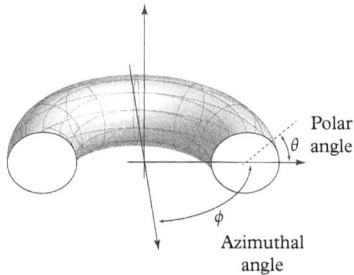

"*Hamiltonian Mechanics is geometry in phase space.*" V. I. Arnold (1978)

Newton's second law, which relates forces to changes in momenta, is the cornerstone upon which classical physics was built. It provides the causative explanation of why things move. There is cause—force—and effect—change in motion—analyzed using the mathematical tools of dynamics. However, motion obeys deeper principles that are more general and more powerful. These deeper principles have a strongly geometric character, and a central concept in these geometric principles is the extremum (usually a minimum) principle, in which a trajectory minimizes or maximizes some quantity calculated upon it. For instance, geodesics minimize metric distance between points. In this chapter, that connection is made more formal, by introducing the principle of stationary action and its extension to the more general Hamilton's principle that is the foundation of Lagrangian and Hamiltonian dynamics. Hamiltonian dynamics is described most naturally in *phase space*, which is a state space with new types of properties described by symplectic geometry. A fundamental property of Hamiltonian systems is the conservation of phase-space volume along a trajectory, known as Liouville's theorem. One of the most valuable features of Hamiltonian dynamics is the ability to find transformations, known as canonical transformations, of dynamical variables into new variables that simplify the description of the dynamical system. Action–angle variables on multidimensional tori are among the most important of these new variables, and are the starting point for understanding nonlinear Hamiltonian dynamics that will be studied in Chapter 3.

2.1 Hamilton's principle

The trajectory of a particle in configuration space $x^a(t)$ is a single-parameter geometric curve that connects two points: the initial position and the final position. The simplest question one can ask about such a trajectory is: what property of that curve distinguishes it from all neighboring ones? To answer this question, one introduces the "action integral" of a dynamical system, and applies the calculus of variations to find the trajectory that yields the minimum or maximum action. Hamilton's principle generalizes these results to derive Lagrange's (and Hamilton's) equations of motion.

2.1.1 Calculus of variations

Consider a trajectory $y(x)$ and a function $f(x, y, y')$, where the prime denotes the derivative with respect to x. We seek a trajectory such that the integral of $f(x, y, y')$, defined as

$$I = \int_{x_1}^{x_2} f(x, y, y') dx, \tag{2.1}$$

between the points x_1 and x_2 is independent of small changes in the shape of the trajectory.[1] This condition of independence is called *stationarity* and is expressed as $\delta I = 0$. The calculus of variations is used to find the stationarity conditions by defining the related integral

$$I = \int_{x_1}^{x_2} f(x, Y, Y') dx, \tag{2.2}$$

where the general (twice-differentiable) functions $Y(x)$ are constructed as

$$Y(x) = y(x) + \varepsilon \eta(x), \tag{2.3}$$

where $\varepsilon = 0$ yields the sought-after stationary behavior. The function $\eta(x)$ is arbitrary, except that it vanishes at the end points, as shown in Fig. 2.1.

The integral becomes

$$I(\varepsilon) = \int_{x_1}^{x_2} f(x, y + \varepsilon \eta, y' + \varepsilon \eta') dx. \tag{2.4}$$

Differentiating Eq. (2.4) with respect to ε yields

$$I'(\varepsilon) = \int_{x_1}^{x_2} \left[\left(\frac{\partial f}{\partial Y} \right) \eta(x) + \left(\frac{\partial f}{\partial Y'} \right) \eta'(x) \right] dx \tag{2.5}$$

---

[1] For instance, the function $f(x, y, y')$ will be a dynamical function of coordinates y and velocities y', the variable x will be the time t, and the integral I will be a property of the system known as an action integral.

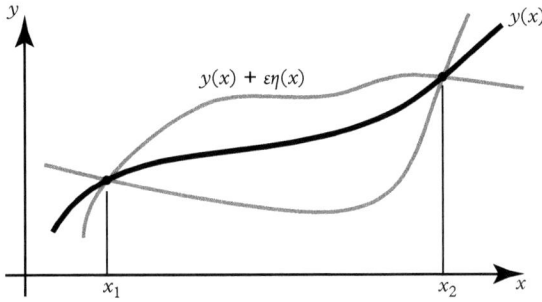

Figure 2.1 *A function defined on the trajectory y(x) between two points in configuration space is stationary with respect to variation by an arbitrary function η(x).*

and integrating Eq. (2.5) by parts yields

$$I'(\varepsilon) = \int_{x_1}^{x_2} \eta(x) \left[\left(\frac{\partial f}{\partial Y} \right) - \frac{d}{dx} \left(\frac{\partial f}{\partial Y'} \right) \right] dx$$
$$- \frac{\partial f}{\partial Y'} \, \eta|_{x_1} + \frac{\partial f}{\partial Y'} \, \eta|_{x_2}. \tag{2.6}$$

The last terms vanish at the endpoints because $\eta(x_1) = 0$ and $\eta(x_2) = 0$ to give

$$I'(\varepsilon) = \int_{x_1}^{x_2} \eta(x) \left[\left(\frac{\partial f}{\partial Y} \right) - \frac{d}{dx} \left(\frac{\partial f}{\partial Y'} \right) \right] dx. \tag{2.7}$$

The integral vanishes when $\varepsilon = 0$, and therefore

$$\int_{x_1}^{x_2} \eta(x) \left[\left(\frac{\partial f}{\partial y} \right) - \frac{d}{dx} \left(\frac{\partial f}{\partial y'} \right) \right] dx = 0. \tag{2.8}$$

Because $\eta(x)$ is an arbitrary function, the only way the integral can be guaranteed zero is when

Euler equation: $\left(\dfrac{\partial f}{\partial y} \right) - \dfrac{d}{dx} \left(\dfrac{\partial f}{\partial y'} \right) = 0.$ (2.9)

Equation (2.9) is known as the *Euler equation* and is the fundamental lemma of the calculus of variations. It takes on central importance in dynamics when the function f is an appropriately-defined dynamical scalar quantity that depends on positions and velocities $f(t, x, \dot{x})$. The Euler equation generalizes to multiple variables as

$$\left(\frac{\partial f}{\partial q^a} \right) - \frac{d}{dx} \left(\frac{\partial f}{\partial q'^a} \right) = 0 \tag{2.10}$$

for $a = 1, 2, 3, \ldots, N$ over all of the generalized coordinates.

2.1.2   Stationary action

Just as momentum plays a central role in Newtonian dynamics, it also helps to define a dynamical function for use in the Euler equation. For instance, the integral of momentum along the path of a particle is called the *action*.[2] The simplest form of the action integral, due to Euler, integrates momentum over a path between fixed endpoints

$$S = m \int_{x_1}^{x_2} \vec{v} \cdot d\vec{s} = m \int_{x_1}^{x_2} v ds. \tag{2.11}$$

If the selected path is the dynamical trajectory, then the velocity and the path element are collinear. When the system conserves energy, this integral is

$$S = \int \sqrt{2m(E - U)} ds, \tag{2.12}$$

where E is a constant, U depends on the generalized coordinates q^a, and the path element is ds^2. The path that makes Eq. (2.12) stationary is the dynamical path of the system. In most cases the stationary integral is a minimum, in which case the principle is known as Jacobi's principle of least action.

The action integral Euler can be rewritten as an integral over time by

$$S = \int_{x_1}^{x_2} mv \, ds$$

$$= \int_{t_1}^{t_2} mv \frac{ds}{dt} dt, \tag{2.13}$$

which is also the integral

$$S = \int_{t_1}^{t_2} 2T \, dt \tag{2.14}$$

over (twice) the kinetic energy as it varies as a function of time between the start and end points of a trajectory. If the system conserves energy, there is the constraint

$$\int (T + U) \, dt = E\Delta t, \tag{2.15}$$

where Δt is the duration of the trajectory. To incorporate a constraint into the stationary integral, one uses the method of Lagrange multipliers.[3] Because E is a constant, its variation (in terms of variational calculus) is zero, and we can define an augmented action integral as

$$S = \int_{t_1}^{t_2} [2T + \lambda(T + U)] dt, \tag{2.16}$$

[2] The first to recognize the importance of the quantity now known as "action" was Maupertuis in 1744, followed by enhancements by Euler (1707–1793), Lagrange (1736–1813), Jacobi (1804–1851), and Hamilton (1805–1865).

[3] The method of Lagrange multipliers is used to add equations of constraint to the dynamical equations. Any function whose variation automatically vanishes can be added to the general function $f(t, q^a, \dot{q}^a)$ with an undetermined multiplier λ. The Euler equation relative to the variable λ provides the constraint forces.

where λ is an undetermined multiplier. Applying the Euler–Lagrange equations to $f(t, q^a, \dot{q}^a) = 2T + \lambda(T + U)$ leads to

$$m(2 + \lambda)\,\ddot{q}^a = \lambda \frac{\partial U}{\partial q^a}. \tag{2.17}$$

This equation is equivalent to Newton's second law when $\lambda = -1$, which determines the multiplier. The augmented action integral is then

$$\begin{aligned} S &= \int_{t_1}^{t_2} (T - U)\,dt \\ &= \int_{t_1}^{t_2} L dt, \end{aligned} \tag{2.18}$$

where $L = T - U$, the difference between kinetic energy T and the potential energy U, is defined as the Lagrangian function.

Hamilton's principle states that the system trajectory is the path that makes the action integral S stationary,

$$\delta S = 0 = \delta \int L\, dt, \tag{2.19}$$

and the Euler–Lagrange equations immediately follow as

Euler–Lagrange equations: $\dfrac{\partial L}{\partial q^a} - \dfrac{d}{dt}\left(\dfrac{\partial L}{\partial \dot{q}^a}\right) = 0.$ (2.20)

The Euler–Lagrange equations (Fig. 2.2) are the fundamental mathematical tool used to solve dynamical problems. Once the appropriate Lagrangian is constructed with appropriate generalized coordinates q^a, the equations of motion describing the system follow immediately from Eq. (2.20). Despite the use of energy conservation in the derivation of the Euler–Lagrange equations, they are more general. They apply any time a system can be described in terms of a potential energy, even if the potential energy is changing in time. They also apply to cases in which constraints do work on the system.

2.2 Conservation laws

Although the study of dynamics is the study of change, the key insights into motion arise from looking for those things that do not change in time—the constants of the motion. The invariant properties of trajectories provide the best way to gain a deeper understanding of the physical causes behind the motion and the range of motions that are possible deriving from those causes. Chief among

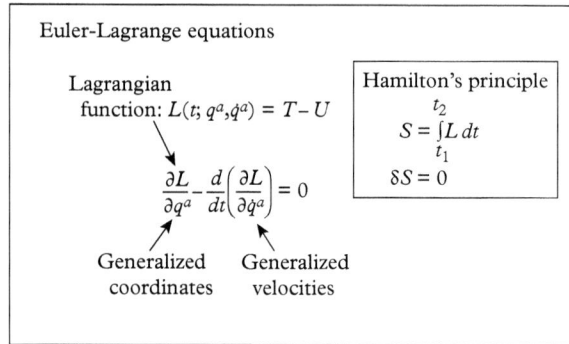

Figure 2.2 *The Euler–Lagrange equations, derived from Hamilton's principle, applied to the difference between the kinetic and the potential energies.*

the conservation principles is the conservation of energy that emerges naturally as a first integral in Lagrangian dynamics.[4] However, there is a deep connection between conserved quantities and embedded symmetries of the dynamical spaces: for every symmetry implicit in Lagrange's equations there is a conserved quantity.[5]

The Euler–Lagrange equations have two common special cases that lead directly to what are known as *first integrals*. A first integral is a differential equation that is one order less than the original set of differential equations, and retains a constant value along an orbit in state space. First integrals are conserved quantities of the motion[6] and arise in two general cases: when the Lagrangian has no explicit dependence on time, and when the Lagrangian has no explicit dependence on a coordinate.

2.2.1 Conservation of energy

When there is no explicit dependence of L on t, then

$$\frac{\partial L}{\partial t} = 0 \qquad (2.21)$$

and the total time derivative is (implicit Einstein summation over repeated indices)

$$\frac{dL}{dt} = \frac{\partial L}{\partial q^a}\dot{q}^a + \frac{\partial L}{\partial \dot{q}^a}\ddot{q}^a. \qquad (2.22)$$

Using the Euler–Lagrange equation (Eq. (2.20)) for $\dfrac{\partial L}{\partial q^a}$ in Eq. (2.22) leads to

$$\frac{dL}{dt} = \dot{q}^a\frac{d}{dt}\frac{\partial L}{\partial \dot{q}^a} + \frac{\partial L}{\partial \dot{q}^a}\ddot{q}^a. \qquad (2.23)$$

This can be rewritten as

$$\frac{dL}{dt} = \frac{d}{dt}\left(\dot{q}^a\frac{\partial L}{\partial \dot{q}^a}\right) \qquad (2.24)$$

[4] Conservation of energy in the broader context of interconversions among different types of energy emerged in the mid-1800s chiefly through the work of Helmholtz, but also through Thomson (Kelvin) and Clausius. See J. Coopersmith, *Energy, the Subtle Concept* (Oxford, 2010).

[5] This is Noether's theorem, named after Emily Noether who published it in 1918.

[6] First integrals, also known as constants of motion or integrals of motion, are constant and conserved quantities along an orbit in state space. While there can be constant functions of the motion that are not first integrals, these will not be referred to.

or

$$\frac{d}{dt}\left[\dot{q}^a\frac{\partial L}{\partial \dot{q}^q} - L\right] = 0,\tag{2.25}$$

where the first integral is

$$\dot{q}^a\frac{\partial L}{\partial \dot{q}^a} - L = const.\tag{2.26}$$

The constant can be identified with an important dynamic quantity by noting that

$$\sum_{a=1}^{N}\dot{q}^a\frac{\partial L}{\partial \dot{q}^a} = 2T.\tag{2.27}$$

Therefore, the first integral is

$$\begin{aligned}const. = 2T - L &= 2T - (T - U) = T + U\\ &= E,\end{aligned}\tag{2.28}$$

which is the total energy of the system. This system, with no explicit time dependence in the Lagrangian, is called *conservative*, and the total energy of the system holds constant.

2.2.2 Conservation of linear momentum

When a generalized coordinate does not appear explicitly in the Lagrangian, then that coordinate is called an *ignorable coordinate* or a *cyclic coordinate*, and the Lagrangian is said to be *homogeneous* in that coordinate. In this case

$$\frac{d}{dt}\left(\frac{\partial L}{\partial \dot{q}^a}\right) = \frac{\partial L}{\partial q^a} = 0\tag{2.29}$$

and the time derivative of the quantity in the parenthesis vanishes (the quantity is constant in time). Therefore, the first integral is

$$\frac{\partial L}{\partial \dot{q}^a} = const.\tag{2.30}$$

To identify this constant with a physical quantity, consider the simple case of translational motion in free space (no potential) of a single mass. Then

$$L = \frac{1}{2}m(\dot{q}^a)^2\tag{2.31}$$

and

$$p_a = \frac{\partial L}{\partial \dot{q}^a} = m\dot{q}^a,\tag{2.32}$$

which is the linear momentum (also known as the canonical momentum). Therefore, the absence of a position-dependent potential leads to the conservation of linear momentum. Translational symmetry (or homogeneity in space) leads to the conservation of linear momentum, and rotational symmetry leads to the conservation of angular momentum. Because of the wide variety of possible generalized coordinates and forms that Lagrangians can take, there are many dynamical systems in which various forms of momenta are conserved quantities. One of the classic problems is the elastic collision of two masses.

Example 2.1 Elastic collision of two masses

The conservation of linear momentum is one of the key tools used to analyze the collisions among particles. A common problem encountered in physics is the case of a mass incident on a stationary target in the lab frame. This problem is solved easily in the center-of-mass (CM) frame, and the final trajectories are obtained by transforming the velocities from the CM frame back to the lab frame. The initial conditions are compared to the final conditions for both the lab and the CM frame in Fig. 2.3. The scattering is solved in the CM frame by considering the scattered energies as functions of the CM scattering angle θ.

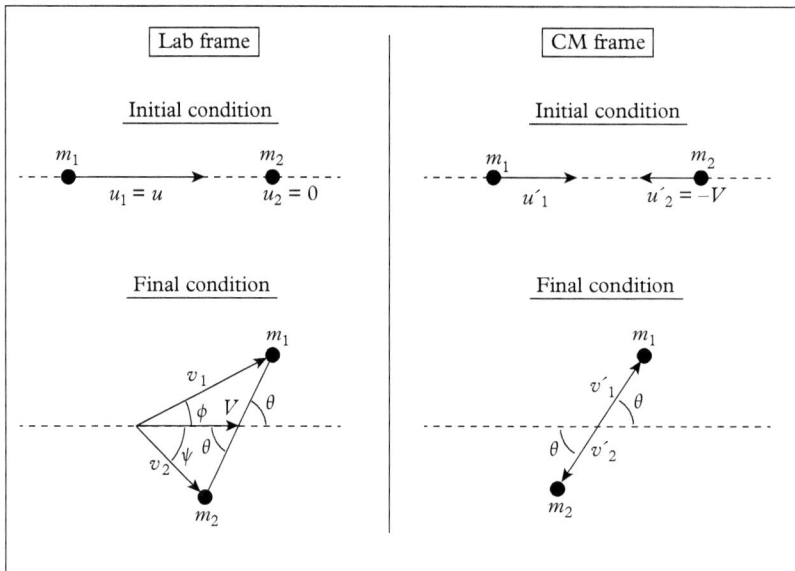

Figure 2.3 *Geometry of the elastic collision of two masses viewed from the lab frame and from the center-of-mass (CM) frame. The dynamics are analyzed in the CM frame and transformed back into the lab frame.*

The center of mass of the two particles is defined by

$$M\vec{R} = m_1\vec{r}_1 + m_2\vec{r}_2 \tag{2.33}$$

and taking the time derivative yields

$$M\vec{V} = m_1\vec{u}_1 + m_2\vec{u}_2$$
$$= m_1\vec{u}, \tag{2.34}$$

which is solved for the center-of-mass velocity as

$$\vec{V} = \frac{m_1\vec{u}}{m_1 + m_2}. \tag{2.35}$$

The initial velocities of the particles in the CM frame are

$$\vec{u}_1' = \vec{u} - \vec{V}$$
$$\vec{u}_2' = -\vec{V}. \tag{2.36}$$

Conservation of momentum holds throughout the interaction, leading to the simple balance

$$m_1\vec{u}_1' = -m_2\vec{u}_2'$$
$$m_1\vec{v}_1' = -m_2\vec{v}_2'. \tag{2.37}$$

Because the masses of the particles remain unchanged by the collision, the final speeds have the same relationship as the initial speeds

$$v_1' = u_1'$$
$$v_2' = u_2'. \tag{2.38}$$

Combining this with Eq. (2.36) and using Eq. (2.35) for the CM speed, this yields the center-of-mass speeds after the collision

$$v_2' = \frac{m_1 u}{m_1 + m_2}$$
$$v_1' = \frac{m_2 u}{m_1 + m_2} \tag{2.39}$$

in which the final speeds are independent of the scattering angle θ. These are the speeds in the CM frame. To transform back to the lab frame, the components of the final velocity in the CM frame are

$$v_{1x}' = v_1' \cos\theta$$
$$v_{1y}' = v_1' \sin\theta, \tag{2.40}$$

which transform back into the lab frame as

$$v_{1x} = v_1' \cos\theta + \frac{m_1 u}{M}$$
$$v_{1y} = v_1' \sin\theta. \tag{2.41}$$

The relationship between θ and ϕ (one of the angles in the lab frame) is obtained by comparing the velocity components in the CM and lab frames

$$v_1 \sin\phi = v_1' \sin\theta$$
$$v_1 \cos\phi = v_1' \cos\theta + \frac{m_1 u}{M}. \tag{2.42}$$

continued

Example 2.1 *continued*

Dividing these equations yields

$$\tan\phi = \frac{\sin\theta}{\cos\theta + \dfrac{m_1 u}{v_1' M}}$$

$$= \frac{\sin\theta}{\cos\theta + \left(\dfrac{m_1}{m_2}\right)}, \tag{2.43}$$

which always gives a forward boost such that $\phi < \theta$. The kinetic energy of the first mass in the lab frame is

$$K_1 = \frac{1}{2}m_1\left[\left(v_1'\cos\theta + \frac{m_1 u}{M}\right)^2 + \left(v_1'\sin\theta\right)^2\right]$$

$$= \frac{1}{2}m_1\left[v_1'^2 + \left(\frac{m_1 u}{M}\right)^2 + 2\frac{m_1 v_1' u}{M}\cos\theta\right] \tag{2.44}$$

and the ratio of the kinetic energy of the first mass to its initial kinetic energy is

$$\frac{K_1}{K_0} = \left[\left(\frac{m_2}{M}\right)^2 + \left(\frac{m_1}{M}\right)^2 + 2\frac{m_1 m_2}{M^2}\cos\theta\right]$$

$$= 1 - 2\frac{m_1 m_2}{M^2}(1 - \cos\theta), \tag{2.45}$$

which is the lab-frame ratio, but it is still expressed in terms of the CM frame scattering angle θ. The ratio equals unity for forward scattering (glancing angle on the target). The relationship Eq. (2.43) between θ and ϕ can be used to express the kinetic energy ratio in terms of the lab-frame angle.

2.3 The Hamiltonian function

The Lagrangian is a quadratic function of the velocities $\dot{q}^a$, and hence it is possible to add specific types of functions to the Lagrangian that leave the Lagrange equations of motion the same. In particular, a new Lagrangian L' can be constructed as

$$L'(q^a, \dot{q}^a, t) = L(q^a, \dot{q}^a, t) + \sum_{a=1}^{N}\frac{\partial M}{\partial q^a}\dot{q}^a + \frac{\partial M}{\partial t}, \tag{2.46}$$

which obeys the same Euler–Lagrange equations as for L. There are many possible choices for the function $M(q, t)$ (see Section 2.3.2). Of these, one particularly useful choice is

$$M(q^a, \dot{q}^a, t) = \sum_{a=1}^{N}\dot{q}^a\frac{\partial L}{\partial \dot{q}^a} - L(q^a, \dot{q}^a, t), \tag{2.47}$$

which is a conserved quantity from Eq. (2.25),

$$\frac{dM}{dt} = \sum_{a=1}^{N} \left[\ddot{q}^a \frac{\partial L}{\partial \dot{q}^a} + \dot{q}^a \frac{d}{dt} \frac{\partial L}{\partial \dot{q}^a} - \dot{q}^a \frac{d}{dt} \frac{\partial L}{\partial \dot{q}^a} - \ddot{q}^a \frac{\partial L}{\partial \dot{q}^a} \right] \equiv 0. \qquad (2.48)$$

This specific choice of M is denoted with an H and is called the Hamiltonian function

$$\text{Hamiltonian function:} \qquad H(q^a, \dot{q}^a) = \sum_{a=1}^{N} \dot{q}^a \frac{\partial L}{\partial \dot{q}^a} - L(q^a, \dot{q}^a). \qquad (2.49)$$

The Hamiltonian also is expressed as

$$\begin{aligned} H &= 2T - (T - U) \\ &= T + U, \end{aligned} \qquad (2.50)$$

which equals the total energy of the system. The Hamiltonian is used in cases when the total energy is conserved, and it also plays a central role in quantum mechanical theory.

2.3.1 Legendre transformations and Hamilton's equations

The transformation from the Lagrangian to the Hamiltonian is an example of a Legendre transformation, defined for a single-variable function $f(x)$ as

$$\mathcal{L}f(x) = x \frac{df}{dx} - f = F(p). \qquad (2.51)$$

A Legendre transform takes a function $f(x)$ in one space and converts it to function $F(p)$ in a corresponding space called a dual space. Fourier transforms or Laplace transforms take functions in time and re-express them as functions of frequency. For the Legendre transform of the Lagrangian function, it takes a function expressed in $(q, \dot{q})$ space and re-expresses it as a function in (q, p). In other words, it takes a function on velocity and converts it to a function on momentum. Dual spaces are equivalent, but provide different perspectives that can be enlisted to simplify problems.

A graphical example of a Legendre transform is shown in Fig. 2.4 for $y = f(x) = x^2$. A convex function $y = f(x)$ is compared with lines $y = px$ to create a difference function $F = px - f(x)$. This difference has a maximum at position x_p. This is found through

$$\left. \frac{\partial F}{\partial x} \right|_{x_p} = F'(x_p) = 0. \qquad (2.52)$$

The function $F(p) = F(x_p)$ is the Legendre transform of $f(x)$.

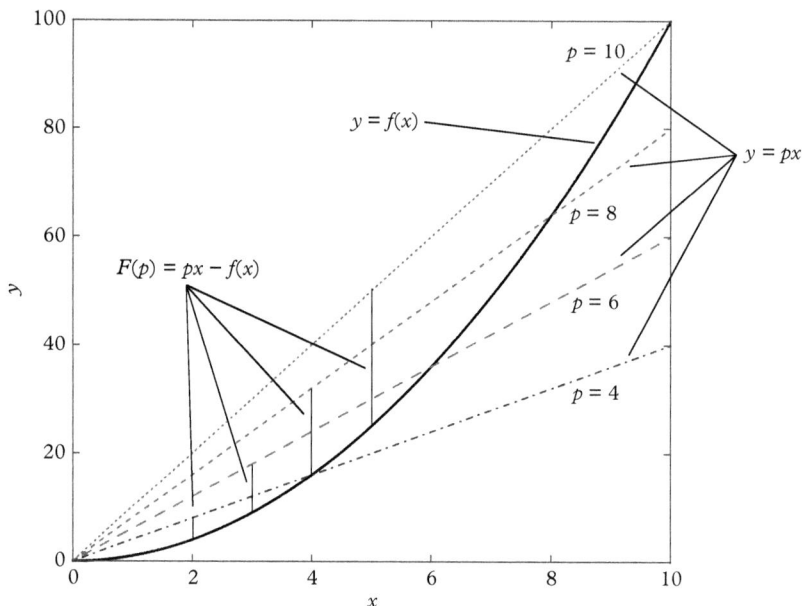

Figure 2.4 *Graphical meaning of the Legendre transform of* $y = x^2$. *The difference between* $y = px$ *and* $y = f(x)$ *is a maximum F at* x_p. *The dependence of F on p defines F(p), which is the Legendre transform of f(x).*

Example 2.2 Legendre transform

As an explicit example of a Legendre transformation of a Lagrangian, consider the one-dimensional simple harmonic oscillator. The Lagrangian is

$$L = \frac{1}{2}m\dot{x}^2 - \frac{1}{2}kx^2 \tag{2.53}$$

and the canonical momentum is $p = \dfrac{\partial L}{\partial \dot{x}} = m\dot{x}$ from Eq. (2.32). The Legendre transformation is

$$\begin{aligned}\mathcal{L}L(x, \dot{x}) &= \dot{x}\frac{dL}{d\dot{x}} - L \\ &= \frac{p}{m}p - \frac{1}{2}m\left(\frac{p}{m}\right)^2 + \frac{1}{2}kx^2 \\ &= \frac{p^2}{2m} + \frac{1}{2}kx^2 = H(x, p),\end{aligned} \tag{2.54}$$

which leads to the expected equation for the Hamiltonian. This transformation took the variable $\dot{x}$ in the Lagrangian to the new variable p in the Hamiltonian. The Legendre transform is also its own inverse, as can be seen through

$$\begin{aligned}\mathcal{L}H(x, p) &= p\frac{dH}{dp} - H \\ &= m\dot{x}^2 - \frac{1}{2}m\dot{x}^2 - \frac{1}{2}kx^2 \\ &= \frac{1}{2}m\dot{x}^2 - \frac{1}{2}kx^2 = L(x, \dot{x})\end{aligned} \tag{2.55}$$

that leads back to the Lagrangian.

The Legendre transform for a multi-variable Lagrangian can be expressed as

$$\mathcal{L}L(q^a, \dot{q}^a, t) = \dot{q}^a (q^a, p_a, t)\, p_a - L (q^a, \dot{q}^a (q^a, p_a, t), t) = H(q^a, p_a, t), \quad (2.56)$$

where

$$p_a = \frac{\partial L (q^a, \dot{q}^a, t)}{\partial \dot{q}^a} \quad (2.57)$$

is the momentum conjugate to the generalized coordinate q^a. In terms of notation, note that momentum is the derivative of a scalar quantity with respect to a vector. This is what is known as a covariant derivative, leading to a covariant quantity p_a with the subscript below (The mnemonic for remembering that the covariant quantity uses a subscript: "co goes below").[7]

The equations of motion are re-expressed in terms of the Hamiltonian and the conjugate variables q^a and p_a, defined by

$$\text{Conjugate momentum:} \quad p_a = \frac{\partial H}{\partial \dot{q}^a} \quad (2.58)$$

which holds if

$$\det \left| \frac{\partial^2 L}{\partial \dot{q}^a \partial \dot{q}^b} \right| \neq 0, \quad (2.59)$$

which is the Hessian condition for invertibility. The total differential of the Hamiltonian is

$$dH = \frac{\partial H}{\partial q} dq + \frac{\partial H}{\partial p} dp + \frac{\partial H}{\partial t} dt, \quad (2.60)$$

which is also equal to

$$dH = \dot{q} dp - \frac{\partial L}{\partial q} dq - \frac{\partial L}{\partial t} dt. \quad (2.61)$$

Equating the terms between these equations gives

$$\dot{q} = \frac{\partial H}{\partial p} \qquad \dot{p} = \frac{\partial L}{\partial q} = -\frac{\partial H}{\partial q} \qquad \frac{\partial H}{\partial t} = -\frac{\partial L}{\partial t}. \quad (2.62)$$

The Hamilton equations of motion are

$$\text{Hamilton's canonical equations:} \qquad \begin{aligned} \dot{q}^a &= \frac{\partial H}{\partial p_a} \\ \dot{p}_a &= -\frac{\partial H}{\partial q^a} \end{aligned} \qquad (2.63)$$

[7] The distinction between contravariant vectors (vectors) and covariant quantities becomes essential when comparing equations of non-Cartesian coordinates. This usage of superscripts and subscripts will be developed extensively in Chapter 9.

which are known as the *canonical equations*. They consist of $2N$ first-order equations that replace the N second-order equations of the Euler–Lagrange equations. The Hamiltonian function is clearly independent of time (energy conservation), shown by considering the total differential

$$\frac{dH}{dt} = \frac{\partial H}{\partial q^a}\dot{q}^a + \frac{\partial H}{\partial p_a}\dot{p}_a = -\dot{p}_a\dot{q}^a + \dot{q}^a\dot{p}_a = 0, \tag{2.64}$$

which is equivalent to Eq. (2.48).

2.3.2   Canonical transformations

The choice of generalized coordinates and their conjugate momentum is not unique. This raises the question of how to define new pairs of variables in terms of which to write the Hamiltonian function. Therefore, we seek a coordinate transformation that carries

$$\begin{aligned} p &\to P(q,p) \\ q &\to Q(q,p) \end{aligned} \qquad H(q,p,t) \to H'(Q,P,t) \tag{2.65}$$

and keeps the equations of motion invariant. This requires that

$$\sum_{a=1}^{D} p_a\dot{q}^a - H(q,p,t) = \sum_{a=1}^{D} P_a\dot{Q}^a - H(Q,P,t) + \frac{d}{dt}M \tag{2.66}$$

is summed over the generalized coordinates with D degrees of freedom, where M is a function that depends on the old and new variables and time, but not their time derivatives. Canonical transformations are a special type of Legendre transform and are used to help simplify the solutions to the canonical equations. The search for an appropriate function M takes some experience and skill, but we will see in a later section that, when Hamilton's equations are integrable (when there are as many constants of the motion as degrees of freedom), a special canonical transformation to action–angle variables makes the solution particularly simple.

There are four types of generating functions M that allow canonical transformations. These are outlined in Table 2.1.

As an example, consider the first type of canonical transformation that has

$$q \to q \quad p \to Q \tag{2.67}$$

with

$$M(q,Q,t) = F_1(q,Q,t) \tag{2.68}$$

and

$$\frac{dM}{dt} = \frac{\partial F_1}{\partial t} + \sum_{a=1}^{D}\left[\frac{\partial F_1}{\partial q^a}\dot{q}^a + \frac{\partial F_1}{\partial Q^a}\dot{Q}^a\right]. \tag{2.69}$$

Table 2.1 *Four canonical transformations*

Generating function	Derivatives		Special case
$M = F_1\,(q, Q, t)$	$p_a = \dfrac{\partial F_1}{\partial q^a}$	$P_a = -\dfrac{\partial F_1}{\partial Q^a}$	$F_1 = q^a Q^a \quad Q^a = p_a \quad P_a = -q^a$
$M = F_2\,(q, P, t) - Q^a P_a$	$p_a = \dfrac{\partial F_2}{\partial q^a}$	$Q^a = \dfrac{\partial F_1}{\partial P_a}$	$F_2 = q^a P_a \quad Q^a = q^a \quad P_a = p_a$
$M = F_3\,(p, Q, t) + q^a p_a$	$q^a = -\dfrac{\partial F_3}{\partial p_a}$	$P_a = -\dfrac{\partial F_3}{\partial Q^a}$	$F_3 = p_a Q^a \quad Q^a = -q^a \quad P_a = -p_a$
$M = F_4\,(p, P, t) + q^a p_a - Q^a P_a$	$q^a = -\dfrac{\partial F_4}{\partial p_a}$	$Q^a = \dfrac{\partial F_4}{\partial P_a}$	$F_4 = p_a P_a \quad Q^a = p_a \quad P_a = -q^a$

This satisfies the condition for a canonical transformation when

$$p_a = \frac{\partial F_1}{\partial q^a} \qquad P_a = -\frac{\partial F_1}{\partial Q^a} \qquad H' = H + \frac{\partial F_1}{\partial t}. \tag{2.70}$$

This transformation is a common choice for the simple harmonic oscillator.

Example 2.3 Canonical transformation for a simple harmonic oscillator

A canonical transformation for the harmonic oscillator makes the complex transformation

$$Q = \frac{m\omega q + ip}{\sqrt{2m\omega}} = \hat{a} \qquad P = i\left(\frac{m\omega q - ip}{\sqrt{2m\omega}}\right) = i\hat{a}^\dagger. \tag{2.71}$$

The choice of the notation $\hat{a}$ is motivated by correspondence with quantum mechanics. This is a transformation of Type I. The generating function F_1 can be found from

$$\begin{aligned} p = i\left(m\omega q - \sqrt{2m\omega}Q\right) = \frac{\partial F_1}{\partial q} \\ P = i\left(\sqrt{2m\omega}q - Q\right) = -\frac{\partial F_1}{\partial Q} \end{aligned} \tag{2.72}$$

with the solution

$$F_1(q, Q) = i\left(\frac{Q^2}{2} - \sqrt{2m\omega}qQ + \frac{m\omega q^2}{2}\right). \tag{2.73}$$

This generating function is time independent, hence the new Hamiltonian is just the old re-expressed in terms of the new variables

$$H = -i\omega PQ. \tag{2.74}$$

continued

Example 2.3 *continued*

Hamilton's equations of motion are

$$\dot{Q} = -i\omega Q$$
$$\dot{P} = i\omega P \tag{2.75}$$

and the time-dependent solutions, in terms of the original coordinates q and p, are

$$q(t) = \sqrt{\frac{1}{2m\omega}}\,(Q + Q^*)$$

$$p(t) = -i\sqrt{\frac{m\omega}{2}}\,(Q - Q^*). \tag{2.76}$$

These solutions are well known from quantum mechanics, in which Q and P are recognized as the classical analogs of quantum annihilation and creation operators, respectively,

$$\hat{a} = Q$$
$$\hat{a}^\dagger = -iP \tag{2.77}$$

yielding the Hamiltonian

$$H = \omega \hat{a}^\dagger \hat{a} = \hbar\omega\hat{n}, \tag{2.78}$$

where the second equality expressed with $\hat{n}$ is simply an "amplitude" invariant. This form shares much in common with a quantum mechanical Hamiltonian. Of course, the classical treatment misses the quantum zero-point motion. However, it does put the Hamiltonian into an action–angle form, as we shall see in Section 2.4.

2.4 Central force motion

A central force originates from the gradient of a potential $V(r)$ that depends only on the radial distance from the force center. A particle of mass m attracted by the force is located in three dimensions by the spherical coordinates (r, θ, ϕ). The velocity in spherical coordinates is

$$v^2 = \dot{r}^2 + r^2\dot{\phi}^2 + r^2\sin^2\phi\dot{\theta}^2, \tag{2.79}$$

yielding the Lagrangian

$$L = \frac{1}{2}m\left[\dot{r}^2 + r^2\dot{\phi}^2 + r^2\sin^2\phi\dot{\theta}^2\right] - V(r). \tag{2.80}$$

However, there is a conserved quantity in this motion which reduces the dimensionality from three to two degrees of freedom. The central force cannot exert a torque on the particle because the force is directed along the position vector. Therefore, angular momentum is conserved, and we are free to choose an initial condition $\theta = \pi/2 = const.$ that defines the equatorial plane. The Lagrangian becomes

$$L = \frac{1}{2}m\left[\dot{r}^2 + r^2\dot{\phi}^2\right] - V(r) \qquad (2.81)$$

in the two variables r and ϕ.

2.4.1 Reduced mass for the two-body problem

For the problem of two mutually attracting bodies of finite mass with no external forces, the total momentum is conserved

$$\frac{d\vec{P}}{dt} = \frac{d}{dt}(m_1\vec{r}_1 + m_2\vec{r}_2) = 0. \qquad (2.82)$$

In the center-of-mass frame this equation can be satisfied by

$$m_1\vec{r}_1 + m_2\vec{r}_2 = 0. \qquad (2.83)$$

Combining this with

$$\vec{r} = \vec{r}_1 - \vec{r}_2 \qquad (2.84)$$

gives

$$\begin{aligned} \vec{r}_1 &= \frac{m_2}{m_1 + m_2}\vec{r} \\ \vec{r}_2 &= -\frac{m_1}{m_1 + m_2}\vec{r} \end{aligned} \qquad (2.85)$$

and the Lagrangian is

$$L = \frac{1}{2}\mu\dot{\vec{r}} \cdot \dot{\vec{r}} - V(r), \qquad (2.86)$$

where μ is the reduced mass

$$\mu = \frac{m_1 m_2}{m_1 + m_2}. \qquad (2.87)$$

This has reduced the two-body problem to one of an effective one-body central potential problem.

2.4.2 Effective potentials

The Lagrange equations applied to Eq. (2.86) are

$$\begin{aligned} \frac{d}{dt}\left(\mu r^2\dot{\phi}\right) &= 0 \\ u\ddot{r} - \mu r\dot{\phi}^2 + \frac{dV}{dr} &= 0. \end{aligned} \qquad (2.88)$$

The first equation is integrated to give the conserved angular momentum l as

$$\mu r^2 \dot{\phi} = const = \ell. \tag{2.89}$$

This is put into the second equation of Eq. (2.88) to give

$$\mu \ddot{r} - \frac{\ell^2}{\mu r^3} + \frac{dV}{dr} = 0, \tag{2.90}$$

which is a one-dimensional (1D) dynamical equation. The last two terms depend only on the variable r and hence it is possible to define an effective potential as

$$\mathcal{V}(r) = V(r) + \frac{\ell^2}{2\mu r^2} \tag{2.91}$$

such that the Lagrangian is now

$$L = \frac{1}{2}\mu \dot{r}^2 - \mathcal{V}(r). \tag{2.92}$$

The effective potential for an inverse square force is shown in Fig. 2.5 for several values of angular momentum. The orbits are bound (elliptical) for $E < 0$ and are unbounded (hyperbolic) for $E > 0$. The turning points in the 1D motion are $r_{\min}$ and $r_{\max}$.

The one-dimensional Lagrangian of orbital mechanics for an inverse-square force admits a two-dimensional (2D) flow. The flow equations are

$$\dot{r} = \rho$$
$$\dot{\rho} = \frac{1}{\mu r^2}\left(\frac{\ell^2}{\mu r} - GM\right). \tag{2.93}$$

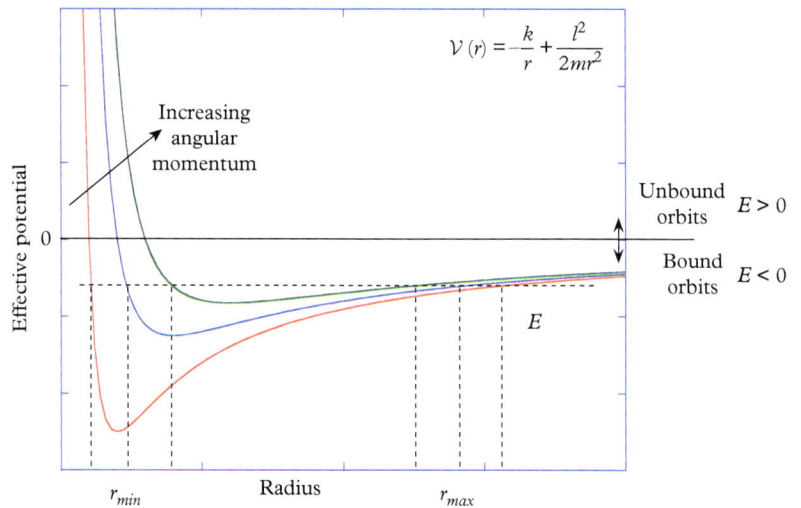

Figure 2.5 *Effective 1D potential for inverse square law for several angular momenta.*

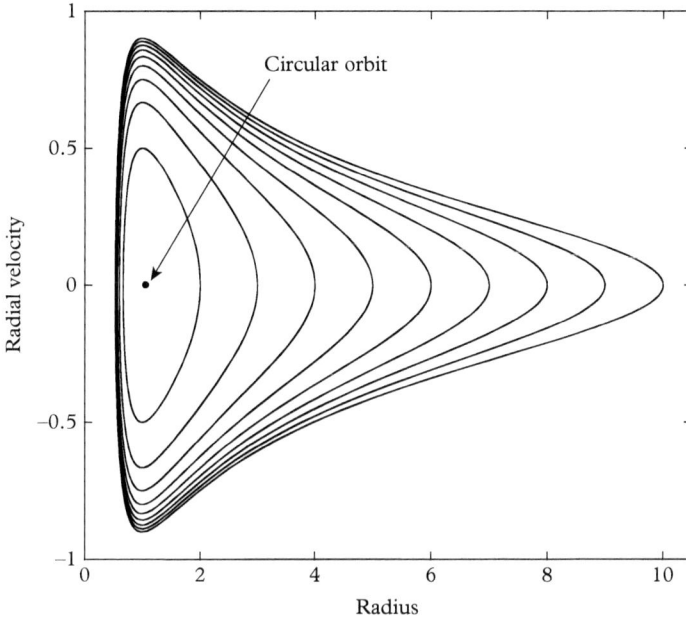

Figure 2.6 *Solved trajectories in* $(r, \dot{r})$ *state-space to the orbital dynamics equations for m = 1, ℓ = 1, and GM = 1 for initial conditions* $(r, \dot{r})$ *= (2:10, 0). These orbits are bound elliptical orbits in configuration space. The circular orbit is at* $(r, \dot{r})$ *= (1, 0).*

A set of solutions is shown in Fig. 2.6 for several different initial conditions for $m = 1$, $\ell = 1$, and $k = 1$. The solution space is radial speed ($\dot{r}$) versus radial position (r). These are bound orbits with the same angular momentum but different total energies (different initial conditions). Despite their appearance in this $(r, \dot{r})$ space, these are elliptical orbits.

The associated Hamiltonian for this central force problem is

$$H = \frac{1}{2}\mu \left[\dot{r}^2 + r^2\dot{\phi}^2 \right] + V(r) \tag{2.94}$$

with the generalized momenta

$$p_r = \mu\dot{r} \quad p_\phi = \mu r^2 \dot{\phi} = \ell, \tag{2.95}$$

where the second expression is recognized as the angular momentum. In terms of these generalized momenta, the Hamiltonian is

$$H = \frac{p_r^2}{2\mu} + \frac{p_\phi^2}{2\mu r^2} + V(r). \tag{2.96}$$

This is a particularly simple Hamiltonian that has no explicit dependence on ϕ. Because the time derivative of a conjugate momentum is related to the partial derivative of the Hamiltonian with respect to the generalized coordinate,

$$\dot{p}_\phi = \frac{\partial H}{\partial \phi} = 0, \tag{2.97}$$

then the conjugate momentum is time-invariant, i.e., it is a conserved quantity. In this simple case it is the orbital angular momentum, and the variable ϕ is called an "ignorable" coordinate also known as a "cyclic" coordinate. The reason it is called cyclic will become clear in Section 2.6 on action–angle coordinates.

2.4.3   Kepler's laws

For planetary orbits around the Sun, the Sun can be considered fixed, to good approximation, because it is much more massive than the smaller planets. In this case the Lagrangian for the motion is

$$L(r,\phi) = \frac{1}{2}\mu\left(\dot{r}^2 + r^2\dot{\phi}^2\right) - \frac{GMm}{r}, \tag{2.98}$$

where the motion is in a plane (defined by $\theta = \pi/2$) and the potential depends only on the radial position r. There is no explicit dependence on the azimuthal variable ϕ, hence ϕ is an ignorable variable and the associated conjugate momentum is conserved:

$$p_\phi = \frac{\partial L}{\partial \dot{\phi}} = \mu r^2 \dot{\phi} = \ell, \tag{2.99}$$

where ℓ is the constant angular momentum. The area swept out by a planet in a time dt is

$$dA = \frac{1}{2}r\left(v_\phi dt\right) \tag{2.100}$$

for a speed v_ϕ. From conservation of angular momentum, this becomes

$$\begin{aligned} dA &= \frac{1}{2}r^2\dot{\phi}dt \\ &= \frac{\ell}{2\mu}dt, \end{aligned} \tag{2.101}$$

where we see that "equal areas are swept out in equal times." This is Kepler's second law.

Kepler's first law, which was the most revolutionary at the time (when circular orbits were considered divine), is that all planets execute elliptical orbits. This fact can be derived by noting that

$$\frac{dr}{dt} = \sqrt{\frac{2}{\mu}(E-U) - \frac{\ell^2}{\mu^2 r^2}} = \frac{dr}{d\phi}\frac{d\phi}{dt} = \frac{dr}{d\phi}\frac{\ell}{\mu r^2}. \tag{2.102}$$

This is integrated to give angular position as a function of radius

$$\phi(r) = \int \frac{\ell/r^2}{\sqrt{2\mu\left(E + \frac{GMm}{r} - \frac{\ell^2}{2\mu r^2}\right)}}\,dr. \tag{2.103}$$

Making the substitution to $u = 1/r$, this integrates to

$$\cos\phi = \frac{\frac{\ell^2}{\mu GMm}\frac{1}{r} - 1}{\sqrt{1 + \frac{2E\ell^2}{\mu G^2 M^2 m^2}}}. \tag{2.104}$$

With the substitutions

$$\alpha = \frac{\ell^2}{\mu GMm}$$

$$\varepsilon = \sqrt{1 + \frac{2E\ell^2}{\mu G^2 M^2 m^2}} \tag{2.105}$$

the solution is

$$\frac{\alpha}{r} = 1 + \varepsilon\cos\phi. \tag{2.106}$$

This is the equation of an ellipse with an eccentricity ε and major and minor axes, respectively,

$$a = \frac{\alpha}{1-\varepsilon^2} = \frac{GMm}{2\,|E|}$$

$$b = \frac{\alpha}{\sqrt{1-\varepsilon^2}} = \frac{\ell}{\sqrt{2\mu\,|E|}} = \sqrt{\alpha a} \tag{2.107}$$

and closest and farthest approaches are, respectively,

$$r_{\min} = \frac{\alpha}{1+\varepsilon}$$

$$r_{\max} = \frac{\alpha}{1-\varepsilon}. \tag{2.108}$$

For bound motion, when $E < 0$, the planetary motion is an ellipse or a circle. It is a circle when the energy is equal to the minimum in the effective potential, $E = V_{\min}$, and $\varepsilon = 0$. For $E = 0$, the orbit is a parabola. When the orbit is unbound for $E > 0$, the trajectory is a hyperbola. Examples are shown in Fig. 2.7 with $\varepsilon = 0$ (circular), $\varepsilon = 0.2$, 0.4, 0.6, and 0.8 (elliptical), $\varepsilon = 1$ (parabolic), and $\varepsilon = 1.2$ (hyperbolic) orbits for $\alpha = 10$.

Kepler's third law, a statement about the period of the orbit and the major axis of the ellipse, begins with Kepler's second law, Eq. (2.101):

$$dt = \frac{2\mu}{\ell}dA. \tag{2.109}$$

For a full period, the area swept out is the area of the ellipse

$$T = \frac{2\mu}{\ell}A = \frac{2\mu}{\ell}\pi ab. \tag{2.110}$$

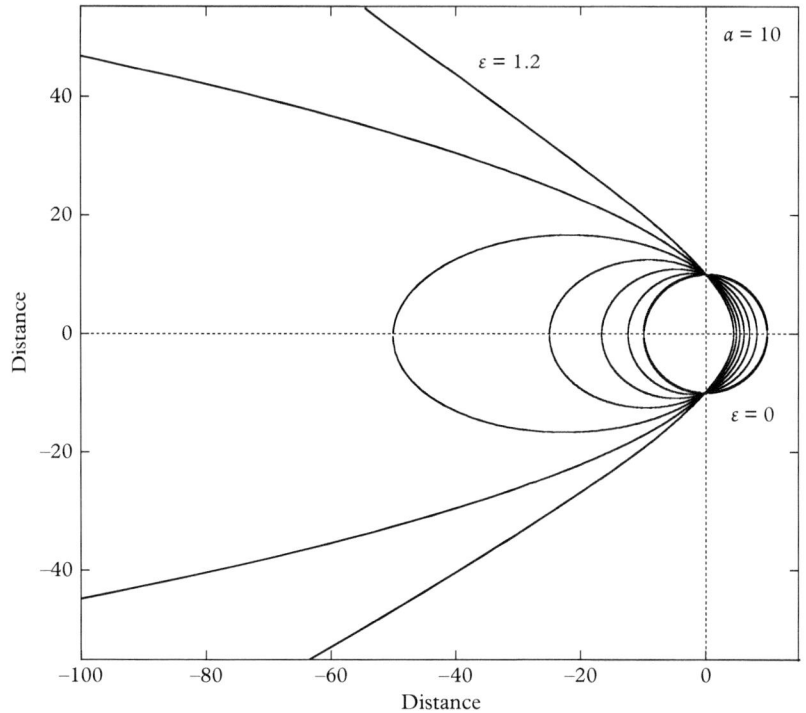

Figure 2.7 *Orbits with ε = 0, 0.2, 0.4, 0.6, 0.8, 1.0, and 1.2 for α = 10.*

The major and minor axes are from Eq. (2.107), and squaring yields

$$T^2 = \frac{4\mu^2}{\ell^2}\pi^2\alpha a^3$$

$$= \frac{4\pi^2}{G(M+m)}a^3. \tag{2.111}$$

When $M \gg m$, we have Kepler's third law: "The square of the period of the orbit varies as the cube of the major axis" with the same coefficient for every planet.

2.4.4 The restricted three-body problem

The three-body problem has a long and interesting history, and played a key role in several aspects of modern dynamics. There is no general analytical solution to the three-body problem. To find the behavior of three mutually interacting bodies requires numerical solution. However, there are subsets of the three-body problem that do yield to analytical approaches. One of these is called the restricted

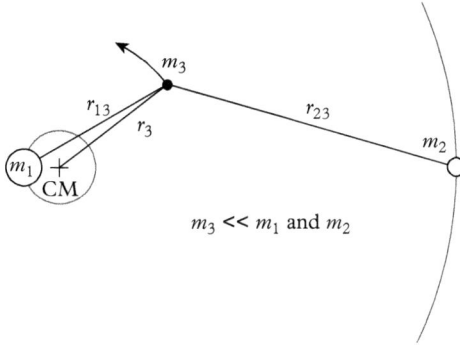

Figure 2.8 *The restricted three-body problem in a plane. The third mass is negligible relative to the first two masses that obey two-body dynamics.*

three-body problem.[8] It consists of two massive bodies plus a third massless body that all move in a plane. This restricted problem was first tackled by Euler and later by Poincaré, who discovered the existence of chaos in its solutions.

The geometry of the restricted three-body problem is shown in Fig. 2.8. The historical interest in the three-body problem is based on considerations of the stability of Earth's orbit around the Sun when considering perturbations by a second planet such as Jupiter. In this problem, take mass $m_1 = m_S$ to be the Sun's mass, $m_2 = m_J$ to be Jupiter's mass, and the third (small) mass is the Earth. The equation of motion for the Earth is

$$\ddot{\vec{r}} = -4\pi^2 \left(\frac{\vec{r} - \vec{r}_S}{\left| \vec{r} - \vec{r}_S \right|^3} + \xi \frac{\vec{r} - \vec{r}_J}{\left| \vec{r} - \vec{r}_J \right|^3} \right), \qquad (2.112)$$

where

$$\xi = \frac{M_J}{M_\odot} \qquad r_S = \frac{R_{SJ}}{1 + \xi} \qquad r_J = \frac{\xi R_{SJ}}{1 + \xi}$$

and the parameter ξ characterizes the strength of the perturbation of the Earth's orbit around the Sun. The parameters for the Jupiter–Sun system are

$$r_{SJ} = 5.203$$

with

$$\omega_J = \frac{2\pi}{11.86}$$

for the 11.86 year journey of Jupiter around the Sun. Eq. (2.112) is a four-dimensional non-autonomous flow

[8] This problem was first studied by Euler in 1760.

$$\dot{v} = -4\pi^2 \left(\frac{\left(x + r_S \cos \omega_{\jmath} t\right)}{\left(\left(x + r_S \cos \omega_{\jmath} t\right)^2 + \left(y + r_S \sin \omega_{\jmath} t\right)^2\right)^{3/2}} \right.$$

$$\left. + \xi \frac{\left(x - r_{\jmath} \cos \omega_{\jmath} t\right)}{\left|\left(x - r_{\jmath} \cos \omega_{\jmath} t\right)^2 + \left(y - r_{\jmath} \sin \omega_{\jmath} t\right)^2\right|^{3/2}} \right)$$

$$\dot{x} = v$$

$$\dot{u} = -4\pi^2 \left(\frac{\left(y + r_S \sin \omega_{\jmath} t\right)}{\left(\left(x + r_S \cos \omega_{\jmath} t\right)^2 + \left(y + r_S \sin \omega_{\jmath} t\right)^2\right)^{3/2}} \right.$$

$$\left. + \xi \frac{\left(y - r_{\jmath} \sin \omega_{\jmath} t\right)}{\left|\left(x - r_{\jmath} \cos \omega_{\jmath} t\right)^2 + \left(y - r_{\jmath} \sin \omega_{\jmath} t\right)^2\right|^{3/2}} \right)$$

$$\dot{y} = u$$

The solutions of an Earth orbit are shown in Fig. 2.9. The natural Earth–Sun–Jupiter system has a mass ratio $m_{\jmath}/m_S = 0.001$. Even in this case, Jupiter causes perturbations of the Earth's orbit by about one percent. If the mass of Jupiter increases, the perturbations would grow larger until around 0.06 when the perturbations become severe and the orbit grows unstable. The Earth gains energy from the momentum of the Sun–Jupiter system and can reach escape velocity. The simulation for a mass ratio of 0.07 shows the Earth ejected from the solar system.

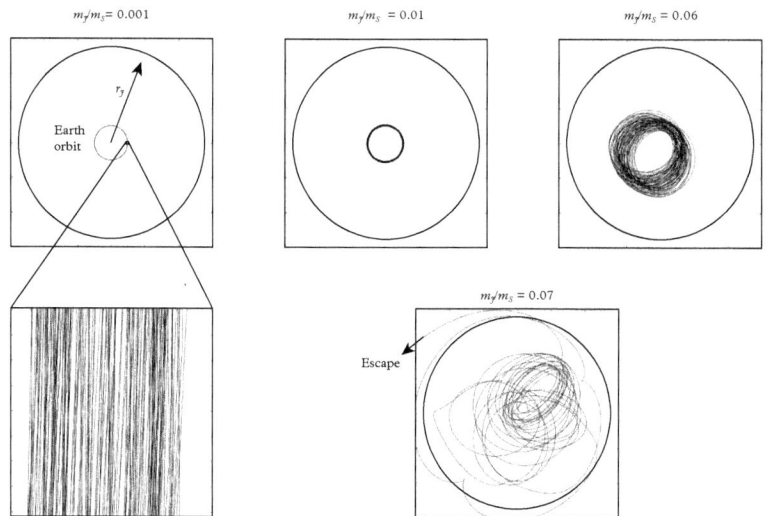

Figure 2.9 *Orbit of Earth as a function of the size of a Jupiter-like planet. The natural system has a Jupiter–Sun mass ratio of 0.001. As the size of Jupiter increases, the Earth orbit becomes unstable and can acquire escape velocity to escape from the Solar System.*

The discovery of chaos in the solar system was a watershed moment in the history of physics. The first hint at the existence of chaos was seen by the French mathematician Henri Poincaré in 1889 as he strove to prove whether the solar system was stable, participating in a mathematics competition sponsored by the king of Sweden. He won the competition by inventing a wide array of new mathematical tools that capture the qualitative behavior of dynamical systems (some of these are introduced in Chapter 3), but during the proof-reading of his winning manuscript, he discovered an error. As he frantically worked to correct his proof in time for printing, he uncovered an infinite nesting of patterns that showed that the restricted three-body problem was anything but stable, exhibiting complex behavior that was beyond his ability to follow. The published (and corrected) paper contained this first glimpse of chaos, which launched later mathematicians to explore deeper consequences.[9] The history of chaos theory has always been closely entwined with the physics of the solar system, but recent topics in the physics of chaos travel much farther, as will be introduced in Chapter 3 and pursued in many forms (evolutionary theory, nonlinear synchronization, dynamic networks, neural nets, and econophysics) through Chapters 4–8.

2.5   Phase space

Hamilton's canonical equations in Eq. (2.63) describe a flow, in the context of Chapter 1, in pairs of variables (q^a, p_a). The solutions to Hamilton's equations are trajectories in even-dimensional spaces with coordinate axes defined by (q^a, p_a). This space is called phase space. For systems with N degrees of freedom, the phase space is $2N$-dimensional. Phase space is more than a convenient tool for visualizing solutions to Hamilton's equations. This is because Hamilton's equations impose a specific symmetry (called symplectic) on the possible trajectories in phase space, and guarantees the conservation of phase space volume for Hamiltonian systems, known as Liouville's theorem.

For a dynamical system with one degree of freedom (Fig. 2.10), phase space is defined by two coordinate axes, one for the space variable q and the other for the momentum p. Trajectories in phase space are one-parameter curves (also known as parametric curves, with time as the parameter) as both position and momentum develop in time. A point on the trajectory at a specific time specifies the complete state of the system. Because energy is conserved, and is a constant of the motion for a conservative system, the energy isosurfaces have one dimension less than the dimension of the phase space. Energy isosurfaces can never cross, and so appear as contours in phase space.

As an example of trajectories in phase space, consider the one-dimensional pendulum. The Hamiltonian is

$$H = \frac{p^2}{2I} + mgL\left(1 - \cos q\right), \tag{2.113}$$

[9] For a history of Poincaré's discovery of chaos, see J. Barrow-Green, *Poincaré and the Three-Body Problem* (London Mathematical Society, 1997). Popular books on chaos in the solar system are Ivars Peterson, *Newton's Clockworks* (Macmillan, 1993) and F. Diacu and P. Holmes, *Celestial Encounters* (Princeton University Press, 1996).

2D Phase space

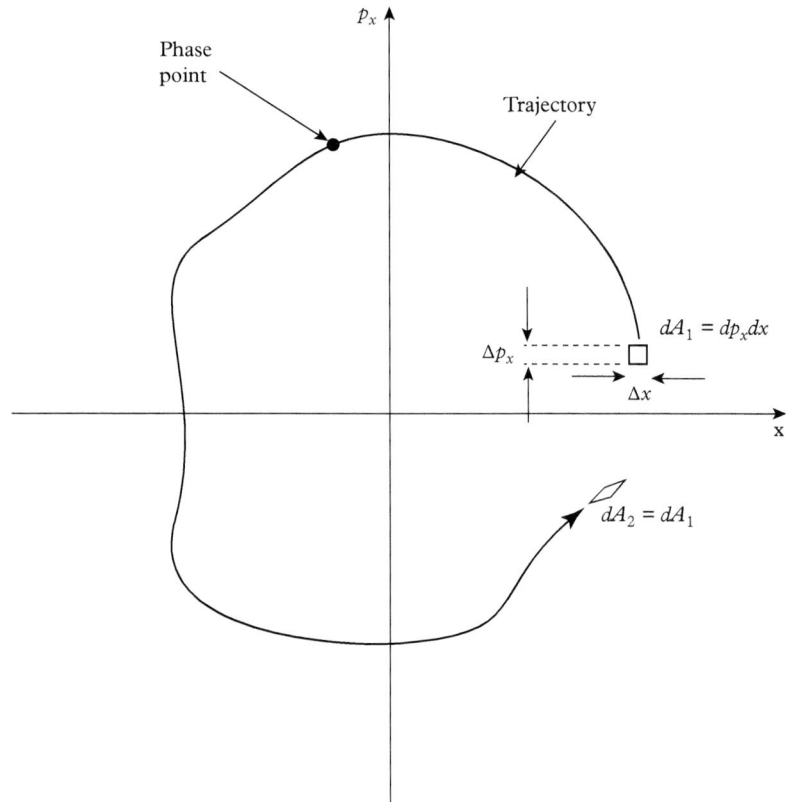

Figure 2.10 *Characteristic 2D phase space for a system with one degree of freedom. A phase point uniquely defines the state of the system that evolves on a state trajectory. An area of initial values evolves under time, and may distort, but does not change area for a conservative Hamiltonian system.*

where q is the angle and p is the angular momentum. The equations of motion are

$$\dot{q} = \frac{\partial H}{\partial p}$$
$$\dot{p} = -\frac{\partial H}{\partial q}$$

(2.114)

and the angular momentum is

$$p = \sqrt{2I\left(E - mgL\left(1 - \cos q\right)\right)},$$

(2.115)

where the trajectories in phase space are plotted in Fig. 2.11. Each curve corresponds to a different total energy. In this two-dimensional phase space, the trajectories lie on one-dimensional hypersurfaces.

Generalizing these examples to more degrees of freedom is straightforward. When there are N degrees of freedom, then the dimension of the phase space is $2N$. Within the $2N$-dimensional phase space, the isoenergy hypersurfaces have

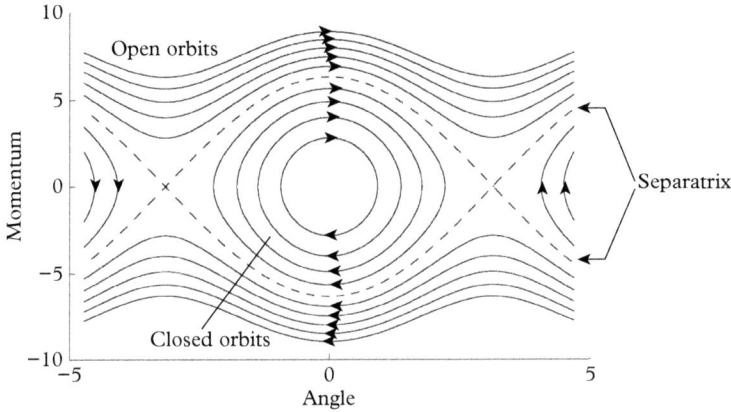

Figure 2.11 *Phase-space trajectories for a pendulum. The separatrix separates rotation of the pendulum (open orbits) from libration (closed orbits).*

dimension $2N - 1$, which are always odd-valued, because of energy conservation, which provides one constraint.

One of the important abstractions for many-particle systems is the idea of a single trajectory through a high-dimensional phase space. When the physical system consists of M particles, there are $N = 3M$ degrees of freedom and $2N = 6M$ dimensions to the phase space. The state of the system at a given instant in time is defined by a single point in the $6M$-dimensional phase space. As time progresses, this point moves through the phase space on a trajectory defined by the Hamiltonian equations.

2.5.1   Flows in phase space

In Chapter 1, a dynamical *flow* was defined generally as

$$\dot{q}^a = g^a \left(x^1, x^2, \ldots \right), \tag{2.116}$$

where the q^a are the dynamical variables and g^a can be any nonlinear vector function of the dynamical variables. The vector function g^a defines the local flow field, and the solution to the flow equation is $q^a(t)$, which defines a parametric curve (flow line) for each initial condition, with time as the parametric variable. Hamilton's equations clearly define a flow in a $2N$-dimensional phase space for N pairs of (q^a, p_a). However, there is a strong symmetry in Hamilton's equations that impose certain properties on the possible solutions. In other words, Hamilton's equations are not general flows, but are a restricted set of flows. This special symmetry can be used to construct a structure for Hamiltonian trajectories that is called symplectic geometry.

To begin constructing this structure, define a set of $2N$ coordinates η^a such that

$$\begin{aligned} \eta^a &= q^a \\ \eta^{a+N} &= p_a \end{aligned} \tag{2.117}$$

for a = 1:N. Therefore, the phase flow is described by

$$\left(\dot{\eta}^a, \dot{\eta}^{a+N}\right) = \left(\frac{\partial H}{\partial \eta^{a+N}}, -\frac{\partial H}{\partial \eta^a}\right), \tag{2.118}$$

where we have used Hamilton's equations for the time derivatives. We would like to define a flow on the single set of variables η^a. A compact notation to accomplish this is

$$\dot{\eta}^a = \omega^{ab}\frac{\partial H}{\partial \eta^b}, \tag{2.119}$$

where ω^{ab} is a skew-symmetric matrix

$$\omega^{ab} = -\omega^{ba}. \tag{2.120}$$

As an example, for three variables and their conjugate momenta, the skew-symmetric matrix is

$$\omega^{ab} = \begin{pmatrix} 0 & 0 & 0 & 1 & 0 & 0 \\ 0 & 0 & 0 & 0 & 1 & 0 \\ 0 & 0 & 0 & 0 & 0 & 1 \\ -1 & 0 & 0 & 0 & 0 & 0 \\ 0 & -1 & 0 & 0 & 0 & 0 \\ 0 & 0 & -1 & 0 & 0 & 0 \end{pmatrix}, \tag{2.121}$$

which is block-diagonal. The determinant of this matrix is

$$\left|\omega^{ab}\right| = 1. \tag{2.122}$$

Its product with itself is minus the identity

$$\omega^2 = -I \tag{2.123}$$

and it has following inverse properties

$$\left(\omega^{ab}\right)^{-1} = \left(\omega^{ab}\right)^T = \omega^{ba} = -\omega^{ab}. \tag{2.124}$$

The matrix ω^{ab} is of particular utility in evaluating areas in phase space. For instance an area in phase space is expressed as

$$\delta a = \omega^{ab} d\eta^a d\eta^b. \tag{2.125}$$

As an example, consider the parallelogram defined by the two vectors in a six-dimensional phase space

$$d\eta = \left(dq^1, dq^2, dq^3, dp_1, dp_2, dp_3 \right) \qquad (2.126)$$

and in matrix form Eq. (2.125) is

$$
\delta a = \begin{pmatrix} dq^1\,(1) \\ dq^2\,(1) \\ dq^3\,(1) \\ dp_1\,(1) \\ dp_2\,(1) \\ dp_3\,(1) \end{pmatrix}^T
\begin{pmatrix} 0 & 0 & 0 & 1 & 0 & 0 \\ 0 & 0 & 0 & 0 & 1 & 0 \\ 0 & 0 & 0 & 0 & 0 & 1 \\ -1 & 0 & 0 & 0 & 0 & 0 \\ 0 & -1 & 0 & 0 & 0 & 0 \\ 0 & 0 & -1 & 0 & 0 & 0 \end{pmatrix}
\begin{pmatrix} dq^1\,(2) \\ dq^2\,(2) \\ dq^3\,(2) \\ dp_1\,(2) \\ dp_2\,(2) \\ dp_3\,(2) \end{pmatrix}
$$
$$(2.127)$$
$$= \left[dq^1\,(1)\,dp_1\,(2) - dp_1\,(1)\,dq^1\,(2) \right]$$
$$+ \left[dq^2\,(1)\,dp_2\,(2) - dp_2\,(1)\,dq^2\,(2) \right]$$
$$+ \left[dq^3\,(1)\,dp_3\,(2) - dp_3\,(1)\,dq^3\,(2) \right]$$
$$= dq^a\,(1)\,dp_a\,(2) - dp_a\,(1)\,dq^a\,(2)$$

The relationship on the last line of Eq. (2.127) is more succinctly represented through a so-called "wedge" product, also known as an exterior Grassmann product, represented by

$$\delta a = dq^a \wedge dp_a. \qquad (2.128)$$

With this definition, there is no need to introduce the matrix ω^{ab}, and we could develop a consistent theory of the geometry of phase space without resorting to any rank-2 tensor, but we will not use the wedge-product notation further.

2.5.2   Liouville's theorem and conservation of phase space volume

Liouville's theorem is one of the important theorems derived for Hamiltonian mechanics. It states that volumes in phase space are invariant to the evolution of the system in time. The conservation of phase space volume follows from the geometry of phase space and Hamilton's equations. Indeed there are many invariants under Hamiltonian flows, including phase space areas and action integrals.

To gain an intuitive understanding of the time-evolution of a dynamical volume, consider a surface $S(t)$ enclosing a set of initial conditions in a volume of state space, as in Fig. 2.12. The set of initial conditions enclosed by the surface

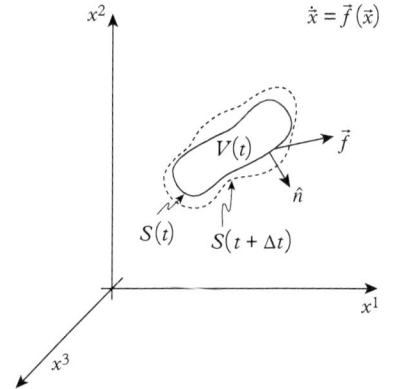

Figure 2.12 *Volume in state space evolves according to the flow equations.*

evolves in time governed by the flow equations $\dot{\vec{x}} = \vec{f}(\vec{x})$. At a later time the surface has changed to $S(t + dt)$ and encloses a new volume $V(t + dt)$. The change in volume per time is given by the net flux in or out of the surface

$$
\frac{dV}{dt} = \frac{V(t + dt) - V(t)}{dt} = \int_S \left(\vec{f} \cdot \hat{n} \right) dS
$$
$$
= \int_S \vec{f} \cdot d\vec{S}.
$$

$$(2.129)$$

By the divergence theorem, this becomes

$$
\frac{dV}{dt} = \int_V \vec{\nabla} \cdot \vec{f} \, dV \tag{2.130}
$$

integrated over the volume. Therefore, the time rate of change of a volume of states in state space is determined by the divergence of the flow equations integrated over the volume of states.

Example 2.4 Change of volume in state space

As an example, consider the flow equations

$$
\begin{aligned}
\dot{x} &= ax + by \\
\dot{y} &= by - ax \\
\dot{u} &= cu - dv \\
\dot{v} &= dv + cu.
\end{aligned} \tag{2.131}
$$

It has the divergence

$$
\begin{aligned}
\vec{\nabla} \cdot \vec{f} &= \frac{\partial f^x}{\partial x} + \frac{\partial f^y}{\partial y} + \frac{\partial f^u}{\partial u} + \frac{\partial f^v}{\partial v} \\
&= a + b + c + d
\end{aligned} \tag{2.132}
$$

and the rate of change in volume is

$$
\frac{dV}{V} = (a + b + c + d)dt \tag{2.133}
$$

with the solution

$$
V(t) = V(0) \, e^{(a+b+c+d)t}. \tag{2.134}
$$

Depending on the sign of the divergence, the volume grows exponentially or vanishes. However, if $\vec{\nabla} \cdot \vec{f} = 0$, then the volume does not change—it is conserved. Therefore, the vanishing of the divergence of the flow function is a strong test for volume conservation during dynamical evolution.

In the case of a Hamiltonian system, the flow is defined by

$$\dot{q}^a = \dot{\eta}^a = f^a\left(\eta^1, \eta^2, \ldots\right) = \frac{\partial H}{\partial \eta^{a+N}}$$

$$\dot{p}^a = \dot{\eta}^{a+N} = f^{a+N}\left(\eta^1, \eta^2, \ldots\right) = -\frac{\partial H}{\partial \eta^a}. \tag{2.135}$$

Therefore, for a small volume of initial conditions

$$\frac{1}{V}\frac{dV}{dt} = \vec{\nabla} \cdot \vec{f}$$

$$= \frac{\partial f^a}{\partial \eta^a}. \tag{2.136}$$

The expressions for f^a are given in Eq. (2.135) for the Hamiltonian H, that give

$$\frac{1}{V}\frac{dV}{dt} = \frac{\partial}{\partial \eta^a}\left(\frac{\partial H}{\partial \eta^{a+N}}\right) + \frac{\partial}{\partial \eta^{a+N}}\left(-\frac{\partial H}{\partial \eta^a}\right)$$

$$= \frac{\partial^2 H}{\partial \eta^a \partial \eta^{a+N}} - \frac{\partial^2 H}{\partial \eta^a \partial \eta^{a+N}} \tag{2.137}$$

$$= 0$$

and the divergence equals zero identically because of the relationship between the position and momentum partial derivatives of the Hamiltonian.[10]

The preceding derivation of Liouville's theorem uses a limiting approximation that is not required for the validity of the theorem. A more powerful proof relies on the special geometry of phase space imposed by the symplectic structure of Hamilton's equations with its associated mathematical tools and proofs. For instance, to show that area in phase space is invariant under time evolution described by Hamilton's equations, consider the time evolution of phase space coordinates from an initial set of coordinates to a new coordinate set

$$\eta^a \rightarrow X^a. \tag{2.138}$$

The new coordinates are related to the old by the coordinate transformation

$$dX^a = \mathcal{J}^a_b d\eta^b, \tag{2.139}$$

where $\mathcal{J}^a_b$ is the Jacobian of the coordinate transformation determined by the time evolution of the Hamiltonian dynamics. Starting from this equation we have

$$dX^a = \mathcal{J}^a_b d\eta^b$$

$$= \mathcal{J}^a_b \omega^{bc} \frac{\partial H}{\partial \eta^c} dt \tag{2.140}$$

[10] The conservation of phase space volume for a Hamiltonian system is called Liouville's theorem. Liouville was one of the leading French mathematicians in the first half of the 1800s, and he published a short paper in 1838 on a relationship among partial derivatives. Jacobi applied this theorem to Hamilton's equations in 1848, which was used by Boltzmann in 1871 to prove that phase space volume was conserved by the thermodynamics of a gas. It is remarkable that Liouville was completely unaware of the relevance of his 1838 theorem to dynamical systems, yet the conservation of phase space today carries his name. The history of phase space and of Liouville's theorem traces an arc through the central topics of 19th-century physics, and made some unexpected turns. A review of the history is given in Nolte (*Physics Today*, 2010).

using Eq. (2.119). Furthermore, by substituting in Eq. (2.139), this yields

$$
\begin{aligned}
dX^a &= \mathcal{J}_b^a \omega^{bc} \frac{\partial H}{\partial \eta^c} dt \\
&= \mathcal{J}_b^a \omega^{bc} \mathcal{J}_c^d \frac{\partial H}{\partial X^d} dt
\end{aligned}
\tag{2.141}
$$

and by taking the Hamiltonian (symplectic) transformation property of ω^{bc}

$$
\omega^{ad} = \mathcal{J}_b^a \omega^{bc} \mathcal{J}_c^d
\tag{2.142}
$$

gives

$$
dX^a = \omega^{ad} \frac{\partial H}{\partial X^d} dt,
\tag{2.143}
$$

which is just Hamilton's equations (Eq. (2.119)) in the new coordinates. The description of the evolution of an area in phase space is then

$$
\begin{aligned}
\delta a &= d\eta^a \omega^{ab} d\eta^b \\
&= -d\eta^a \omega_{ab} d\eta^b \\
&= -d\eta^a \mathcal{J}_a^c \omega_{cd} \mathcal{J}_b^d d\eta^b \\
&= -\left(d\eta^a \mathcal{J}_a^c\right) \omega_{cd} \left(\mathcal{J}_b^d d\eta^b\right) \\
&= -dX^c \omega_{cd} dX^d \\
&= dX^c \omega^{cd} dX^d \\
&= \delta A,
\end{aligned}
\tag{2.144}
$$

where δa is the original area, and δA is the evolved area. In other words, the new area after evolution through time is equal to the original area. Therefore, *areas are preserved under time evolution of the Hamiltonian system.*

   That volumes in phase space also remain invariant to time evolution can be seen through the relationship between volumes in original coordinates and transformed coordinates

$$
dQ^1 ... dQ^n dP_1 ... dP_n = |\mathcal{J}| \, dq^1 ... dq^n dp_1 ... dp_n,
\tag{2.145}
$$

where $|\mathcal{J}|$ is the Jacobian determinant. If $|\mathcal{J}| = \pm 1$, then the time evolution keeps the volumes in phase space invariant. Taking the determinant of the symplectic condition of Eq. (2.142) gives

$$
|\omega| = |\mathcal{J}|^2 |\omega| = \pm 1
\tag{2.146}
$$

and hence

$$
|\mathcal{J}| = \pm 1,
\tag{2.147}
$$

which is what is needed to show that time evolution keeps volumes of phase space invariant.

   The last two sections have demonstrated the utility of the symplectic representation of Hamilton's equations of motion. The introduction of the skew-symmetric matrix ω^{ab} facilitated several proofs on areas and volumes in phase space, namely that areas and volumes are invariant to time evolution of the Hamiltonian system. Similar arguments apply for canonical transformations and a dynamical quantity known as the Poisson bracket.

2.5.3   Poisson brackets

When asking how a dynamical quantity (some function of the p's and q's) varies along the trajectory in phase space, a new quantity emerges called the Poisson bracket. The Poisson bracket appears in several important contexts in classical mechanics, and has a direct analog in the commutator of quantum mechanics. The Poisson bracket is a useful tool with which to explore canonical transformations as well as integrals (constants) of motion.

If we have a dynamical quantity $G(q, p, t)$, then its time rate of change along a trajectory in phase space is

$$
\begin{aligned}
\frac{dG}{dt} &= \frac{\partial G}{\partial t} + \frac{\partial G}{\partial q^a}\dot{q}^a + \frac{\partial G}{\partial p_a}\dot{p}_a \\
&= \frac{\partial G}{\partial t} + \frac{\partial G}{\partial q^a}\frac{\partial H}{\partial p_a} - \frac{\partial G}{\partial p_a}\frac{\partial H}{\partial q^a} \\
&= \frac{\partial G}{\partial t} + \{G, H\},
\end{aligned}
\tag{2.148}
$$

where the canonical equations were used in the second step and the last expression in brackets is called the Poisson bracket. More generally, for two dynamical quantities F and G, the Poisson bracket is

$$
\textit{Poisson brackets:} \qquad \{G, F\} = \frac{\partial G}{\partial q^a}\frac{\partial F}{\partial p_a} - \frac{\partial G}{\partial p_a}\frac{\partial F}{\partial q^a},
\tag{2.149}
$$

where the implicit summation is over the degrees of freedom of the dynamical system. The definition of the Poisson bracket can be expressed more compactly using the symplectic matrix as

$$
\{G, F\} = \omega^{ab}\frac{\partial G}{\partial \eta^a}\frac{\partial F}{\partial \eta^b}
\tag{2.150}
$$

in which ω^{ab} creates antisymmetric combinations of the covariant derivatives of the two functions.

The Poisson bracket has several uses in classical mechanics. These are:

1. The time rate of change of a dynamical quantity is defined by its Poisson bracket with the Hamiltonian plus the partial derivative with time

$$
\frac{dG}{dt} = \frac{\partial G}{\partial t} + \{G, H\}.
\tag{2.151}
$$

2. The canonical equations can be written in terms of the Poisson bracket with the Hamiltonian as

$$\dot{q}^a = \{q^a, H\} \qquad \dot{p}_a = \{p_a, H\}. \tag{2.152}$$

3. The Poisson bracket is invariant under canonical transformation, and hence can be used to test if a transformation is canonical. For instance for the transformation $q, p \rightarrow Q(q, p), P(q, p)$ then

$$\{F(q, p), G(q, p)\}_{q, p} = \{F(Q, P), G(Q, P)\}_{Q, P}, \tag{2.153}$$

where the subscripts make the functional dependence explicit.

4. If G is an integral of the motion, then

$$\frac{\partial G}{\partial t} + \{G, H\} = 0 \tag{2.154}$$

and furthermore, if G does not depend explicitly on time then

$$\{G, H\} = 0. \tag{2.155}$$

5. If F and G are both integrals of the motion, then

$$\{F, G\} = W, \tag{2.156}$$

where W is another integral of the motion. This is known as Poisson's theorem.

6. If F_i and F_j are two dynamical quantities whose Poisson bracket vanishes,

$$\{F_i, F_j\} = 0, \tag{2.157}$$

then F and G are said to be in *involution*. This expression states that each quantity is invariant along the path of the other quantity. This is also a statement that the two dynamical functions are linearly independent. Clearly, each integral of the motion is in involution with the time-independent Hamiltonian function.

The Poisson bracket plays a central role in the correspondence between dynamics in classical systems and dynamics in quantum systems. In particular, Eq. (2.151) for classical systems has a direct correspondence to the quantum mechanical equation

$$\frac{d\hat{G}}{dt} = \frac{\partial \hat{G}}{\partial t} + \frac{1}{i\hbar}\left[\hat{G}, \hat{H}\right] \tag{2.158}$$

for a quantum operator $\hat{G}$ and Hamiltonian operator $\hat{H}$, where $\left[\hat{G}, \hat{H}\right]$ is the commutation relation. This correspondence suggests the plausible substitution

$$\{G, F\} \Rightarrow \frac{1}{i\hbar}\left[\hat{G}, \hat{F}\right] \tag{2.159}$$

when beginning with classical systems and going over to quantum mechanical dynamics. Conjugate variables in classical dynamics with non-vanishing Poisson brackets lead to non-vanishing commutator relations for the corresponding quantum operators. Perhaps the most famous of these is the canonical transformation for $q_i \Rightarrow Q_i$ and $p_j \Rightarrow P_j$

$$\{Q_i, P_j\} = \delta_{ij} \qquad \{Q_i, Q_j\} = 0 \qquad \{P_i, P_j\} = 0,$$

where the Poisson bracket is calculated relative to the original variables q_i and p_j. The corresponding quantum commutators are

$$[\hat{q}_i, \hat{p}_j] = i\hbar\delta_{ij} \qquad [\hat{q}_i, \hat{q}_j] = 0 \qquad [\hat{p}_i, \hat{p}_j] = 0,$$

which lead to Heisenberg's uncertainty principle.

2.6 Integrable systems and action–angle variables

When there are as many constants (integrals) of motion as there are degrees of freedom, then a system is said to be *integrable*. In this case, it is possible, in principle, to find a canonical transformation to action and angle variables for which the trajectories of the dynamical system are geodesics on the surface of a hypertorus.[11]

If a Hamiltonian system is integrable, then a canonical transformation exists that can transform the Hamiltonian into the set of equations

$$\begin{aligned} \dot{\mathcal{J}}_a &= -\frac{\partial H}{\partial \theta^a} = 0 \\ \dot{\theta}^a &= \frac{\partial H}{\partial \mathcal{J}_a} = \omega^a, \end{aligned}$$

(2.160)

where ω^a and $\mathcal{J}_a$ are both constants, and the Hamiltonian is not explicitly a function of the θ^a, meaning that the θ^a are ignorable coordinates the constants $\mathcal{J}_a$ are called the *action*, and the variables θ^a are called the *angle*. The angles are not physical angles, but rather describe the position of the trajectory point in phase space. The Hamiltonian takes on the very simple form

$$H = \omega^a \mathcal{J}_a$$

(2.161)

and the frequency is obtained as

$$\omega^a = \frac{\partial H}{\partial \mathcal{J}_a}.$$

(2.162)

[11] A hypertorus is the higher-dimensional generalization of a torus. Note that a regular torus has a 2D surface and is spanned by two periodic angular variables (see Fig. 2.14). An *n*-torus is spanned by *n* periodic angular variables. There can also be a 1-torus (see Fig. 2.13), which is just a circle.

The equations of motion for the system are simply

$$\theta^a(t) = \omega^a t + \omega_0^a. \tag{2.163}$$

The appropriate canonical transformation is achieved through the generating function $W(q, \theta)$ with the total differential

$$dW = p_a \, dq^a - \mathcal{J}_a d\theta^a. \tag{2.164}$$

Because the motion is periodic in θ, the generating function must also be a periodic function of θ. We choose a closed path k to have the period 2π, over which the integral of the total differential must vanish

$$\oint_k dW = 0 = \oint_k p_a \, dq^a - \mathcal{J}_k \oint_k d\theta^a \tag{2.165}$$

and hence

$$\mathcal{J}_k = \frac{1}{2\pi} \oint_k p_a \, dq^a, \tag{2.166}$$

where the closed path is around N mutually orthogonal directions in phase space. By Stoke's theorem, this means that the action

$$\mathcal{J}_k = \iint_{A_k} dp_a \, dq^a \tag{2.167}$$

is the area of the projection onto the kth plane in phase space.

Example 2.5 Action–angle coordinates for a harmonic oscillator

---

A simple 1D harmonic oscillator example helps make these ideas more concrete. The Hamiltonian is

$$H(q, p) = \frac{p^2}{2m} + \frac{1}{2}kx^2. \tag{2.168}$$

The equations of motion are

$$\dot{q} = \frac{\partial H}{\partial p} = \frac{p}{m}$$

$$\dot{p} = -\frac{\partial H}{\partial q} = -kq. \tag{2.169}$$

The action is

$$\mathcal{J} = \frac{H}{\omega} = \frac{p}{2m\omega} + \frac{kq^2}{2\omega}, \tag{2.170}$$

where $\omega = \sqrt{k/m}$ is the angular frequency of the oscillation. The trajectories in phase space are ellipses. Note that p and q do not have the same units. However, by normalizing them as

$$\begin{aligned} p' &= p/\sqrt{2m\omega} \\ q' &= q\sqrt{m\omega/2} \end{aligned} \tag{2.171}$$

then p' and q' have the same units, and the trajectories in phase space are circles of radius equal to $\sqrt{\mathcal{J}}$. The area enclosed by the trajectories in the new space is $A = \pi\mathcal{J}$, hence the area enclosed in the original phase space is $2\pi\mathcal{J}$.

The transformation between the (p, q) and the $(\mathcal{J}, \theta)$ are

$$\begin{aligned} p(\mathcal{J}, \theta) &= \sqrt{2m\omega\mathcal{J}} \sin\theta \\ q(\mathcal{J}, \theta) &= \sqrt{\frac{2\mathcal{J}}{m\omega}} \cos\theta \end{aligned} \tag{2.172}$$

with the inverse relations

$$\begin{aligned} \mathcal{J}(p, q) &= \frac{p^2}{2m\omega} + \frac{kq^2}{2\omega} \\ \theta(p, q) &= \tan^{-1}\left(\frac{p}{qm\omega}\right) \end{aligned} \tag{2.173}$$

Canonical transformation: $(q, p) \Rightarrow (\theta, \mathcal{J})$

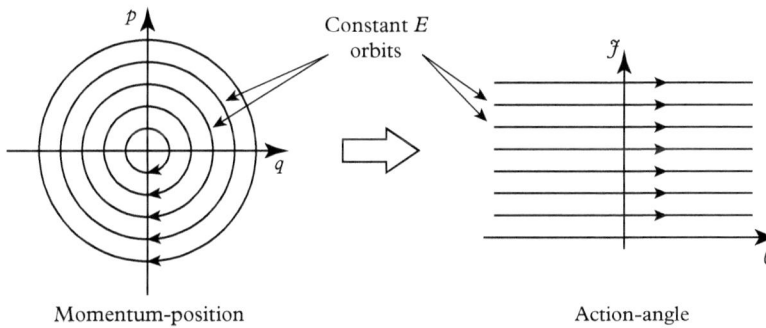

Figure 2.13 *Canonical transformation on a 1D harmonic oscillator taking (q, p) to the action–angle coordinates $(\theta, \mathcal{J})$. Each different orbit corresponds to a different total energy E.*

Example 2.6 Action–angle variables for the pendulum

A 1D pendulum provides a simple *nonlinear* example. The Hamiltonian is (see Eq. (2.113))

$$H = \frac{p_\phi^2}{2mL^2} + mgL\left(1 - \cos\phi\right),\tag{2.174}$$

where g is the acceleration due to gravity and p_ϕ is the angular momentum around the axis of motion. The angular momentum is

$$p_\phi = \pm\sqrt{2mL^2\left[E - mgL\left(1 - \cos\phi\right)\right]}\tag{2.175}$$

and the action is

$$\mathcal{J}(E) = \pm\frac{1}{2\pi}\oint\sqrt{2mL^2\left[E - mgL\left(1 - \cos\phi\right)\right]}\,d\phi.\tag{2.176}$$

This is an elliptic integral which is tabulated numerically.

The action can still be described by

$$\mathcal{J} = \frac{H}{\omega}$$
$$\theta = \omega t,\tag{2.177}$$

where $\omega = 2\pi/T(E)$ and $T(E)$ is the period of oscillation for a trajectory with energy E. But for the pendulum with arbitrary amplitude, $T(E)$ is no longer equal for all trajectories, and the motion is no longer harmonic. The period of oscillation $T(E)$ is a function of the energy of the pendulum. Note that the physical angle of the pendulum is $\phi(t)$ and is not a harmonic function, while $\theta(t)$ is the action angle and is a linear function of t.

The action $\mathcal{J}$ and angle θ present a new coordinate system to represent phase space. It has a very simple structure—even for the nonlinear pendulum—it is a torus. A natural coordinate system for the action–angle coordinates is polar. For a system with one degree of freedom, the system motion in phase space is a point on the $(\theta, \mathcal{J})$ phase plane describing a flow line that is a simple horizontal line, as shown on the right of Fig. 2.13.

For an integrable system with two degrees of freedom, the full phase space is four-dimensional (4D). However, since the system is integrable, it has two constants of the motion I_1 and I_2. Each action constant can be viewed as a fixed radius, with the angle variables θ_1 and θ_2 executing uniform circular motion. Two constants of the motion in 4D phase space define a 2D *hypersurface*. The structure of this 2D hypersurface is isomorphic to a torus. One angle variable defines the motion about the origin, while the other defines the motion about the center (core) of the torus. The full phase-space consists of infinitely many nested tori, like the layers of an onion, each for a specific choice of the constants of motion I_1 and I_2. None of these tori can intersect because of the uniqueness of the equations of motion. Once motion starts on a defined torus, it remains on that torus.

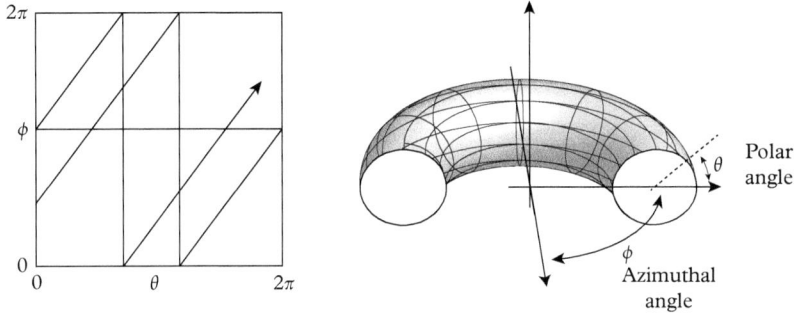

Figure 2.14 *Two equivalent (isomorphic) representations of a configuration space as a 2D torus. The configuration space has to angular variables.*

For integrable systems with many degrees of freedom, the action–angle formalism produces an angle coordinate for each degree of freedom, and each angle coordinate forms the surface of a hypertorus.

As an example, consider two independent linear harmonic oscillators that represent a system with two degrees of freedom. The Hamiltonian is

$$H = \frac{1}{2m_1}\left((p_1)^2 + m_1^2\omega_1^2\left(q^1\right)^2\right) + \frac{1}{2m_2}\left((p_2)^2 + m_2^2\omega_2^2\left(q^2\right)^2\right). \qquad (2.178)$$

This is a completely separable problem in which there are two frequencies of oscillation ω_1 and ω_2 and two angle variables θ_1 and θ_2 that increase linearly modulo 2π. This Hamiltonian also represents a single 2D anisotropic harmonic oscillator. The two angles define a configuration space that is equivalent to the surface of a 2D torus (see Fig. 2.14).

The ratio of the two frequencies plays an important role in the interpretation of the qualitative nature of the dynamics. If they have a ratio equal to a rational fraction, then the motion repeats itself after some period of time, and not all possible points in the configuration space are accessible for a given initial condition. On the other hand, if the ratio of frequencies is equal to an irrational number, then the motion never repeats exactly, and the orbit can come arbitrarily close to any point in the configuration space for a sufficiently long time. If the two oscillators are coupled by a non-integral term in the Hamiltonian, then qualitatively different behavior occurs for commensurate vs. incommensurate (rational vs. irrational) ratios of the frequencies. We will see in Chapter 3 that new fixed points will appear in the phase plane, and chaos can emerge. In Chapter 4, we will see the possibility of synchronization among "nearby" frequencies.

2.7 Summary

In this chapter, *Hamilton's principle of stationary action* was shown to be a powerful and fundamental aspect of physical law—dynamical systems follow trajectories in *phase space* for which the *action* along the trajectory is an extremum. The action is defined as the time average of the difference between kinetic and potential energies, which is also the time average of the *Lagrangian* ($L = T - U$). Once the Lagrangian is defined, the *Euler equation* of variational calculus leads immediately to the *Euler–Lagrange equations* of dynamics. Hamiltonian dynamics are derived from the Lagrange equations through the *Legendre transform* that expresses the equations of dynamics as first-order ordinary differential equations of the *Hamiltonian* ($H = T + U$), which is a function of the generalized coordinates and their *conjugate momenta*. Consequences of the Lagrangian and Hamiltonian equations of dynamics are the *conservation of energy* and *conservation of momentum* with applications to collisions and orbital dynamics. The Hamiltonian equations of motion are *integrable* when there are as many integrals of the motion as degrees of freedom. In this case *action–angle coordinates* can be defined that reduce the dynamical motions to phase-space trajectories on a hyperdimensional torus.

Euler equation

In the calculus of variations, the integral of a function f integrated between two endpoints is minimized when the following condition is satisfied

$$\left(\frac{\partial f}{\partial y} \right) - \frac{d}{dx} \left(\frac{\partial f}{\partial y'} \right) = 0. \tag{2.9}$$

This equation determines the condition for stationarity of an integral.

Euler–Lagrange equations

The equations of motion are defined by the Lagrangian function $L = T - U$ of generalized coordinates q^a

$$\frac{\partial L}{\partial q^a} - \frac{d}{dt} \left(\frac{\partial L}{\partial \dot{q}^a} \right) = 0. \tag{2.20}$$

Conjugate momenta

The dynamic momenta p_a that are conjugate to generalized coordinates q^a are

$$p_a = \frac{\partial H}{\partial \dot{q}^a}. \tag{2.58}$$

Hamilton's canonical equations

Hamilton's equations of motion for coordinate-momentum pairs are defined by the Hamiltonian function $H = T + U$ as

$$\dot{q}^a = \frac{\partial H}{\partial p_a}$$

$$\dot{p}_a = -\frac{\partial H}{\partial q^a}. \tag{2.63}$$

Effective potential in central-force motion

Central-force motion in a potential $V(r)$ can be reduced to motion in an effective potential

$$\mathcal{V}(r) = V(r) + \frac{\ell^2}{2\mu r^2} \tag{2.91}$$

expressed in terms of the reduced mass μ and the conserved angular momentum ℓ.

Hamilton's equations in symplectic notation

The symplectic symmetry of the Hamiltonian between conjugate coordinate-momentum pairs allows a simpler expression of Hamilton's equations.

$$\dot{\eta}^a = \omega^{ab} \frac{\partial H}{\partial \eta^b}. \tag{2.119}$$

Action–Angle variables

Integrable systems can be reduced to action–angle variables with simple equations of motion on a hyperdimensional torus in phase space

$$\dot{\mathcal{J}}_a = -\frac{\partial H}{\partial \theta^a} = 0$$
$$\dot{\theta}^a = \frac{\partial H}{\partial \mathcal{J}_a} = \omega^a. \tag{2.160}$$

Action

Action is an integral around a trajectory in phase space

$$\mathcal{J}_k = \frac{1}{2\pi} \oint_k p_a \, dq^a. \tag{2.166}$$

2.8   Bibliography

V. I. Arnold, *Mathematical Methods of Classical Mechanics*, 2nd ed (Springer-Verlag, New York, 1989).

T. Frankel, *The Geometry of Physics: An Introduction* (Cambridge, 2003).
   Good graduate-level text.

Goldstein, *Classical Mechanics* (Addison-Wesley, 2001).
   One of the most popular main-stream texts for undergrad/grad students.

Lanczos, *The Variational Principles of Mechanics* (Dover, 1949).

D. S. Lemons, *Perfect Form: Variational Principles, Methods, and Applications in Elementary Physics* (Princeton, 1997).
   A simple and very clear description of the variational principle.

D. D. Nolte, The tangled tale of phase space, *Physics Today*, vol. 63, pp. 33–38, Apr 2010.
   An account of the history of phase space and Liouville's Theorem.

R. Talman, *Geometric Dynamics* (Wiley, 2000).
   One of the most accessible texts on the geometric approach to mechanics.

See Chapters 18 and 19 in Robert H. Wasserman, *Tensors and Manifolds: With Applications to Mechanics and Relativity* (Oxford University Press, 1992).

2.9 Homework exercises

Analytic problems

1. **Variational calculus:** Find $I(\varepsilon)$ for $y = x$ and $\eta = \sin x$ for the function $f = \left(\dfrac{dy}{dx}\right)^2$ where $x_1 = 0$ and $x_2 = 2\pi$.

2. **Variational calculus:** Find $I(\varepsilon)$ for $y = x$ and $\eta = \sin x$ for the function $f = \sqrt{1 + \left(\dfrac{dy}{dx}\right)^2}$ where $x_1 = 0$ and $x_2 = 2\pi$.

3. **Geodesic:** Prove that a geodesic on the plane is a straight line.

4. **Geodesic:** Prove that a geodesic curve on a right cylinder is a helix.

5. **Parabolic wire:** Derive the Hamiltonian and Hamilton's equations of motion for a bead sliding on a frictionless parabolic wire under gravity. Draw the isosurfaces in phase space.

6. **Spherical surface:** Derive the Hamiltonian and Hamilton's equations of motion for a bead constrained to slide on a frictionless spherical surface under gravity. Render a set of isosurfaces of the motion in phase space.

7. **Motion on cylinder:** Apply Hamilton's equations to a particle constrained to move on the surface of a cylinder, subject to the force $\vec{F} = -k\vec{r}$ where the position vector is relative to a point on the central axis of the cylinder.

8. **Disk rolling down an inclined plane:** Use the Euler–Lagrange equations to derive the acceleration of a solid disk rolling without slipping down an inclined plane.

9. **Particle on a cone:** Derive the equations of motion for a particle constrained to move on a cone. Find the equation for $\ddot{r}$ but do not solve. Find constants of the motion.

10. **Double pendulum:** A double pendulum consists of two ideal pendulums (mass at the end of a massless rod) connected in series (the top of the second pendulum is attached to the mass of the first). The lengths of the pendulums

are L_1 and L_2, and the masses are M_1 and M_2. The problem has two generalized coordinates (the two angular deflections). Derive the Euler–Lagrange equations of motion and solve them for small deflections.

11. **Constrained harmonic oscillators:** Two ideal one-dimensional harmonic oscillators with masses M_1 and M_2 and spring constants k_1 and k_2 are subject to the constraint $q_1^2 + q_1^2 = q^2$ where q is a constant. Derive the equations of motion for the system.

12. **Elastic collision:** For the elastic collision in Fig. 2.3, derive an expression for the relationship between ϕ and ψ.

13. **Central force law:** Consider a central force law $f = kr$. What is the character of the solutions (what type of functions) occur for $k < 0$? For $k > 0$?

14. **Virial theorem:** Calculate $\langle E \rangle$ and $\langle U \rangle$ for a circular orbit in an inverse-square force law. What is the relationship between these two quantities?

15. **Three-body problem:** Derive the equations of motion and the period for three equal masses that are positioned at the vertex points of an equilateral triangle.

16. **Raising and lowering operators:** Derive the Poisson bracket $\{\hat{a}, \hat{a}^*\}$ for the raising and lowering operators in Eq. (2.77)

17. **Poisson brackets:** Verify the six properties of the Poisson brackets in Eqs. (2.151) to (2.157).

Computational project[12]

18. **Restricted three-body problem:** Do a restricted three-body simulation with a small mass placed at the Lagrange points of the Sun–Jupiter system. The radius of the Lagrange points (for small mass ratio) is

$$r = R_{SJ} \sqrt[3]{\frac{m}{3M_\odot}}.$$

[12] Matlab programming examples are found at www.works.bepress.com/ddnolte.

Part 2

Nonlinear Dynamics

Proportional response to a stimulus, also known as linear response theory, is a common and powerful principle of physics. It describes the harmonic oscillator and supports the principle of superposition that is used across physics from free-body diagrams to the interference of electromagnetic waves. However, even something as simple as a pendulum deviates from linear behavior, and no harmonic oscillator remains linear under all conditions. Indeed, little in the world is purely linear, and nonlinearity opens up wide new ranges of behavior that span from world-wide weather patterns to the synchronization of the flashes of fireflies in summer.

This section introduces the mathematical tools that are used to analyze and understand the behavior of nonlinear systems. The introduction to *nonlinear dynamics* and *chaos* uses a state space approach, identifying *fixed points* and *nullclines* that constrain the behavior of *flow lines* in their neighborhoods. The emergence of *deterministic chaos*, as the dimensionality of the dynamics increases, is a fundamental result that has far-reaching importance to science and society. The *synchronization* of coupled nonlinear oscillators is a central paradigm in this book, illustrating how collective behavior can emerge from the interactions of many individual nonlinear elements. This topic leads naturally to the emergence of complex behavior on connected *networks* of individual elements.

Nonlinear Dynamics and Chaos

3

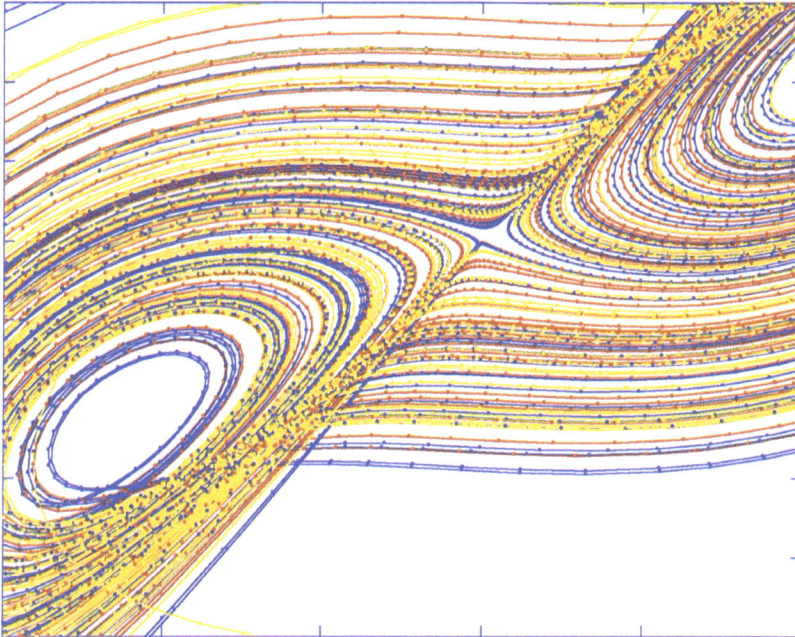

A fragment of the Lorenz attractor

Linear physics is the bedrock upon which all introductory physics and engineering courses are built. The power of linearity comes from proportionality and additivity. With proportionality, all systems respond in a manner proportional to the strength of the stimulus. With additivity, known as the principle of linear superposition, the behavior of complicated systems can be decomposed into their fundamental elements. In linear systems, the whole *is* quite simply the sum of the parts. But the world is not linear. While many systems behave approximately linearly for small displacements, all systems are bounded by natural constraints (the length of a spring, the size of the system) and cannot maintain linearity to all levels of excitation. Some systems are fundamentally nonlinear, even for their smallest excursions. And when these systems have several degrees of freedom, most of the intuition and mathematical tools we use for linear systems no longer apply, and an entirely new toolbox is needed to understand these types of systems.

This chapter introduces a number of important nonlinear problems and develops the tools needed to understand their behavior. There are many well-known nonlinear problems in two variables (such as the pendulum and the van der Pol oscillator), but a fundamental change in behavior takes place in the transition from two variables to three variables. The key feature of dynamics in three or more variables is the possibility of non-regular (chaotic) motion. Although this motion is unpredictable in detail, it has well-defined statistical and geometric properties. Analytic results can be extracted, such as the exponential separation of trajectories from arbitrarily close initial conditions, called sensitivity to initial conditions (SIC), or the geometric structure of repeated visits of the system trajectories to certain parts of the dynamical space, known as strange attractors.

Nonlinear systems can have numerous signatures, but not all signatures may occur in the same system. Chaos is certainly one dramatic signature, as well as sudden jumps in system behavior, known as bifurcations. But there are other more subtle signatures that can signify nonlinear dependence, such as the coupling of amplitude to frequency. Anyone who has spent time in boats is familiar with this specific type of nonlinear signature for water waves: big waves have long periods and small waves have short periods. Amplitude–frequency coupling is an essential signature of nonlinearity, and in some complex systems it is the origin of the nonlinear property of synchronization, which is the subject of Chapter 5. A short list of nonlinear signatures is given in Table 3.1. These all make nonlinear physics a rich field of study with many beautiful phenomena.

Table 3.1 *Nonlinear system signatures*

- Breakdown of linear superposition
- Exponential sensitivity to initial conditions (SIC)
- Sudden discontinuous changes (thresholds, jumps, bifurcations, intermittency)
- Amplitude-frequency coupling, frequency entrainment or phase locking (synchronization)
- Hysteresis
- Order within chaos (islands of stability)
- Self-similarity (fractals and scaling)
- Emergence (large-scale structure/order arising from local interactions among many parts)

3.1 One-variable dynamical systems

One-dimensional (1D) dynamical systems may not seem to be of great complexity, but they do demonstrate some of the types of behavior that are easily generalized to two dimensions and higher. For instance, the identification of the steady states (fixed points) of a 1D system illustrates the importance of linearization as well as the classification of types of fixed points.

A 1D dynamical system has the simple flow equation

$$\dot{x} = f(x). \tag{3.1}$$

When the function is zero, then

$$\dot{x} = f(x^*) = 0 \tag{3.2}$$

and the position x^* defines a fixed point of the flow. This is because the time rate of change of the variable at that point is zero, and so the system point cannot move away from it. However, one may still ask about the stability of a fixed point.

To answer this question, the system is linearized around the fixed point through the expansion

$$f(x) = f(x^*) + \left(x - x^*\right)f'\left(x^*\right) + \cdots, \tag{3.3}$$

where $f(x^*) = 0$ and therefore

$$f(x) = \left(x - x^*\right)f'(x^*) \tag{3.4}$$

to lowest order. With a change in variable,

$$u = x - x^*, \tag{3.5}$$

the differential equation becomes

$$\dot{u} = f'u, \tag{3.6}$$

which has the solution

$$u(t) = u(0)e^{\lambda t}, \tag{3.7}$$

where

$$\lambda = \left.\frac{df(x)}{dx}\right|_{x^*} \tag{3.8}$$

is called the Lyapunov exponent. Clearly, one can make the classification

$$\begin{aligned} \lambda < 0 \quad &\text{Stable Node (Attractor)} \\ \lambda > 0 \quad &\text{Unstable Node (Repellor),} \end{aligned} \tag{3.9}$$

where for a stable node, small excursions experience a restoring force with negative feedback like a spring, while for an unstable node a small excursion is amplified with positive feedback (see Fig. 3.1).

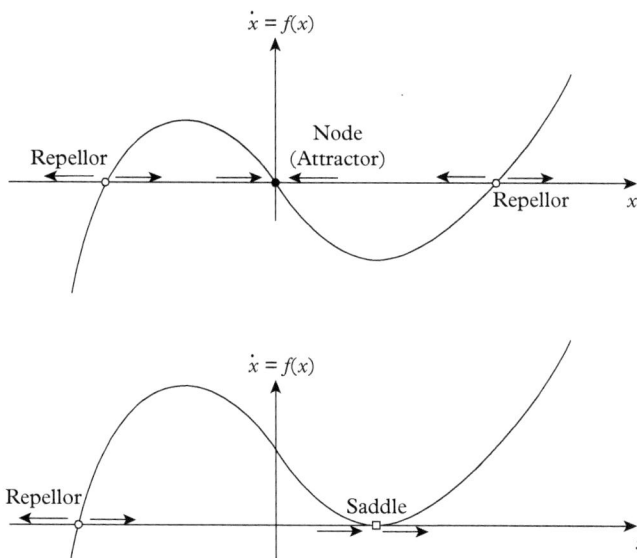

Figure 3.1 *Fixed points in a 1D system. A stable node, or attractor, attracts nearby trajectories, while repellors repel them. A marginal case, called a saddle, occurs when an attractor and a repellor merge.*

3.2 Two-variable dynamical systems

The behavior of two-variable dynamical systems opens the door to much more interesting behavior than one dimension. The equations are still simple, the visualizations are easy, and two-dimensions are important because of the stark contrast they set against the behavior of dynamical systems in three dimensions.

3.2.1 Two-dimensional fixed points

Consider a nonlinear system with a two-variable flow

$$\dot{x} = f(x, y)$$
$$\dot{y} = g(x, y). \tag{3.10}$$

The fixed point at (x^*, y^*) is defined by the equations

$$f(x^*, y^*) = 0$$
$$g(x^*, y^*) = 0 \tag{3.11}$$

at which all the time derivatives equal zero. The fixed points are the equilibrium points of the dynamical system. A system that starts with an initial condition equal to the fixed point cannot move away from it, because there is no time rate of change in the variables at that special point. On the other hand, the equilibrium condition at the fixed points can be stable, unstable, or mixed (saddle points). To find out what their stability is, these equations are expanded around the fixed point by the change of variables

$$u = x - x^*$$
$$v = y - y^*. \tag{3.12}$$

The expanded equations are

$$\dot{u} = u\frac{\partial f}{\partial x} + v\frac{\partial f}{\partial y} + O\left(u^2, v^2, uv\right)$$
$$\dot{v} = u\frac{\partial g}{\partial x} + v\frac{\partial g}{\partial y} + O\left(u^2, v^2, uv\right). \tag{3.13}$$

Neglecting the higher order terms, the linearized equations are

$$\begin{pmatrix} \dot{u} \\ \dot{v} \end{pmatrix} = \mathcal{J} \begin{pmatrix} u \\ v \end{pmatrix}, \tag{3.14}$$

where the Jacobian matrix is

$$\mathcal{J} = \begin{pmatrix} \dfrac{\partial f}{\partial x} & \dfrac{\partial f}{\partial y} \\[2mm] \dfrac{\partial g}{\partial x} & \dfrac{\partial g}{\partial y} \end{pmatrix}. \tag{3.15}$$

The small nonlinear terms are not important in cases of robust fixed points. However, in the case of marginal fixed points, these terms do lead to relevant effects.

The characteristic equation for the Jacobian is

$$\det\left(\mathcal{J} - \lambda I\right) = 0. \tag{3.16}$$

This has the eigenvalues (Lyapunov exponents) in two dimensions of

$$\lambda_{1,2} = \frac{1}{2}\tau \pm \frac{1}{2}\sqrt{\tau^2 - 4\Delta}, \tag{3.17}$$

where

$$\tau = trace(\mathcal{J})$$
$$\Delta = \det(\mathcal{J}) \tag{3.18}$$

and eigenvectors are defined by the eigenvalue equation

$$\mathcal{J}\mathbf{v} = \lambda\mathbf{v}. \tag{3.19}$$

The general solutions near the fixed point are therefore

$$\mathbf{x}(t) = c_1 e^{\lambda_1 t}\mathbf{v}_1 + c_2 e^{\lambda_2 t}\mathbf{v}_2, \tag{3.20}$$

where the solution is exponential in the characteristic values of the Jacobian. The exponents can be real or complex, positive or negative, leading to a range of behavior for flows near fixed points.

3.2.2 Phase portraits

The behavior of any two-dimensional (2D) dynamical system is easy (and fun) to visualize using the technique of "phase portraits." The general 2D system of Eq. (3.10) specifies velocities with two components $(\dot{x}, \dot{y})$, and every point on the (x, y) plane has a velocity vector attached to it. In addition, every point serves as a possible condition, or state, of the system, and a trajectory parallel to the tangent vector[1] emanates from every point. A phase portrait is defined as the (x, y) state-space superposed with a velocity field plus the trajectories (also called flow lines). There are special curves and points on the state-space plane. For instance, the equation $\dot{x} = f(x, y) = 0$ defines a curve on which every point has no x-component to the velocity. This is called the *x-nullcline*. Similarly, there is a *y-nullcline* for the points that have no y-component to the velocity. These x- and y-nullclines intersect at the "fixed points" of the system, where the velocity vector vanishes completely. A phase portrait also has one or more curves of a type known as a *separatrix*. A separatrix is a curve that is not crossed by any flow line. Separatrixes conveniently divide up the state space into regions that are dynamically separated. If the system has an initial value in one part of the state space, it can never cross over into another region that is separated by a separatrix.

An example of a phase portrait with vector fields, flow lines, nullclines, and separatrixes is shown in Fig. 3.2 for the system of equations

$$\dot{x} = x + e^{-y}$$
$$\dot{y} = -y. \tag{3.21}$$

The nullclines are defined by the two equations when the time derivatives are equal to zero, as

$$x = -e^{-y} \qquad x\text{–nullcline}$$
$$y = 0 \qquad y\text{–nullcline.} \tag{3.22}$$

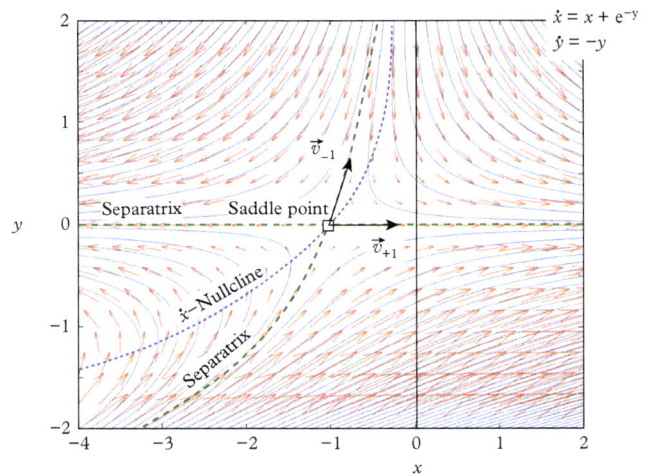

Figure 3.2 *Phase portrait showing vector fields, flow lines, nullclines, and separatrixes. The intersection of the nullclines defines a fixed point, which is the steady-state solution of the system. Redrawn from Strogatz (1994).*

A fixed point occurs where the nullclines intersect at $(x^*, y^*) = (-1, 0)$. The separatrixes are the flow lines emanating from the fixed point in directions determined by the eigenvectors of the Jacobian matrix. The Jacobian matrix at the fixed point is

$$\mathcal{J} = \begin{pmatrix} 1 & -1 \\ 0 & -1 \end{pmatrix}. \tag{3.23}$$

This matrix eigenvalues $\lambda = (1, -1)$ corresponding to a *saddle point* where one set of solutions converges on the fixed point along one line and emerges from the fixed point along the other. The eigenvectors of the matrix are

$$\vec{v}_{+1} = \begin{pmatrix} 1 \\ 0 \end{pmatrix}$$
$$\vec{v}_{-1} = \frac{1}{\sqrt{5}} \begin{pmatrix} 1 \\ 2 \end{pmatrix}, \tag{3.24}$$

which point along the lines of fastest divergence and convergence, respectively. These lines are the separatrixes that divide up the flow into distinct regions. The separated regions are called basins, in which the flow lines are confined. In this example one of the separatrixes coincides with the $\dot{y}$ nullcline. The separatrixes for this saddle point are also known as the *unstable and stable manifolds*, for the diverging and converging trajectories, respectively.

An important consequence of the uniqueness of every velocity vector attached to every point on the phase portrait is an important theorem called the non-crossing theorem. This theorem states that no trajectory in the state space can cross itself in finite time. Because every point in space has a unique velocity vector, if a trajectory did cross itself at a point, that would mean that there were be two distinct velocity vectors at that point, one for each trajectory, which is not allowed. In 2D phase portraits, the non-crossing theorem guarantees that there are well-defined basins separated by separatrixes. However, we will see that in three-dimensional (3D) phase portraits, that while the non-crossing theorem still holds, basins dissolve because trajectories can pass over and under each other, like strands of spaghetti, opening the possibility of chaos. For the same reasons, chaotic behavior is precluded from 2D dynamical systems.

3.2.3   Types of 2D fixed points

Because of the numerous possible types of roots of the characteristic equation, there are numerous possible types of fixed points in two dimensions. In general, these can be classified among one of six types of behaviors. Two are stable, two are unstable, one is a saddle point and a special case is purely oscillatory without decay (called a center). These are shown, with the relevant signs of the Lyapunov exponent, in Fig. 3.3.

2D Fixed point classification

Unstable node

Stable node

$Re(\lambda) > 0$
$Im(\lambda) = 0$

$Re(\lambda) < 0$
$Im(\lambda) = 0$

Unstable spiral

Stable spiral

$Re(\lambda) > 0$
$Im(\lambda) \neq 0$

$Re(\lambda) < 0$
$Im(\lambda) \neq 0$

Center

Saddle point

$Re(\lambda) = 0$
$Im(\lambda) \neq 0$

$Re(\lambda) = \pm C_{1,2}$

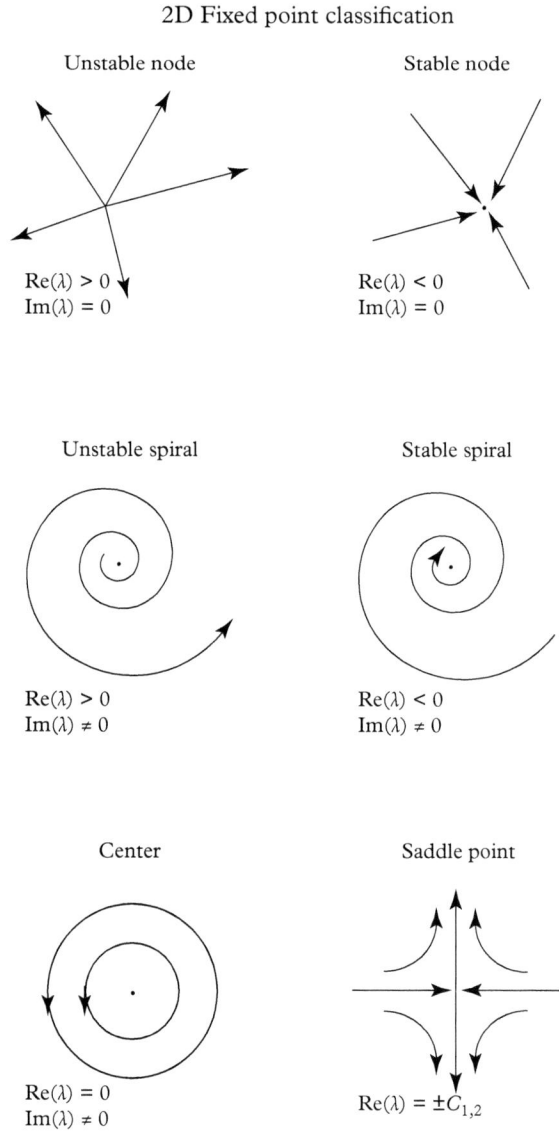

Figure 3.3 *Fixed point classification in 2D state-space flows according to the Lyapunov exponents.*

An equivalent means of classifying these 2D fixed points is in a (Δ, τ) diagram shown in Fig. 3.4. The broad areas in Fig. 3.4 are the major types of fixed points. The fixed points that occur on the parabolic curve are borderline cases that can be sensitive to the nonlinear terms that were neglected in the linearization.

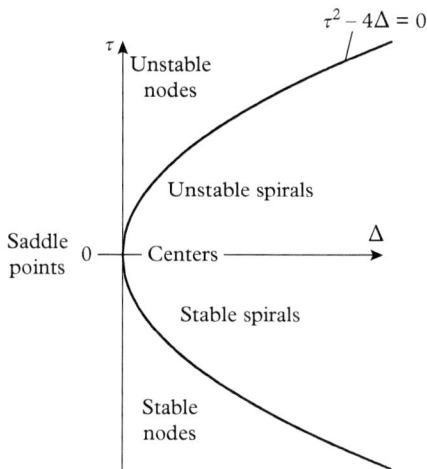

Figure 3.4 *Trace-determinant space of the characteristic values of the Jacobian of the linearized 2D system.*

Example 3.1

For the system of Eq. (3.21)

$$\dot{x} = x + e^{-y}$$
$$\dot{y} = -y. \tag{3.25}$$

There is a fixed point at $(x^*, y^*) = (-1, 0)$. The Jacobian matrix is

$$\mathcal{J} = \begin{pmatrix} 1 & -e^{-y} \\ 0 & -1 \end{pmatrix}\Bigg|_{(-1,0)} = \begin{pmatrix} 1 & -1 \\ 0 & -1 \end{pmatrix} \quad \begin{array}{l} \tau = 0 \\ \Delta = -1. \end{array} \tag{3.26}$$

From Fig. 3.4, the fixed point is a saddle point.

Example 3.2 Marginal case of a center and spirals

This is an important example because it shows how a simple parameterization of the Cartesian equations into polar coordinates immediately identifies a constant of the motion. The flow is

$$\dot{x} = -y + ax\left(x^2 + y^2\right)$$
$$\dot{y} = x + ay\left(x^2 + y^2\right). \tag{3.27}$$

continued

Example 3.2 *continued*

There is a fixed point at the origin $(x^*, y^*) = 0$, and the Jacobian matrix is

$$\mathcal{J} = \begin{pmatrix} 3ax^2 + ay^2 & -1 + 2axy \\ 1 + 2ayx & ax^2 + 3ay^2 \end{pmatrix}\Bigg|_{(0,0)} = \begin{pmatrix} 0 & -1 \\ 1 & 0 \end{pmatrix} \tag{3.28}$$

for which $\tau = 0$, $\Delta = 1$. The eigenvalues are $\lambda = \pm i$ and the origin is a center for $a = 0$. But for small finite values of a, the origin is a fixed point for a spiral. When $a < 0$, the spiral is stable. When $a > 0$, the spiral is unstable.

Example 3.3 Rabbits vs. sheep: competition of species

The following flow describes the problem of rabbits and sheep competing for the same grazing land (adapted from Strogatz (1994)). In this simple model the rabbit concentration is given by x, and the sheep concentration is given by y. The rabbits breed faster than the sheep. As the rabbit population increases, it consumes grass needed for the sheep. Sheep, however, consume more grass per individual.

$$\begin{aligned} \dot{x} &= x(3 - x - 2y) \\ \dot{y} &= y(2 - x - y). \end{aligned} \tag{3.29}$$

These equations are examples of Lotka–Volterra equations that are explored in more detail in Chapter 7 on Evolutionary Dynamics. The nullclines are

$$\begin{aligned} x(3 - x - 2y) &= 0 \\ y(2 - x - y) &= 0 \end{aligned} \tag{3.30}$$

The stable fixed points are at $(0, 2)$, $(3, 0)$. The unstable fixed points are at $(0, 0)$, $(1, 1)$. At each of the fixed points, the Jacobian matrix is

$$(0,0) \quad \mathcal{J} = \begin{pmatrix} 3 & 0 \\ 0 & 2 \end{pmatrix} \quad \tau = 5 \quad \Delta = 6 \quad \text{unstable node}$$

$$(0,2) \quad \mathcal{J} = \begin{pmatrix} -1 & 0 \\ -2 & -2 \end{pmatrix} \quad \tau = -3 \quad \Delta = 2 \quad \text{stable node}$$

$$(3,0) \quad \mathcal{J} = \begin{pmatrix} -3 & -6 \\ 0 & -1 \end{pmatrix} \quad \tau = -4 \quad \Delta = 3 \quad \text{stable node}$$

$$(1,1) \quad \mathcal{J} = \begin{pmatrix} -1 & -2 \\ -1 & -1 \end{pmatrix} \quad \tau = -2 \quad \Delta = -1 \quad \text{saddle point}$$

The vector flow of the rabbit vs. sheep model is shown in Fig. 3.5. There are two stable fixed points to which the dynamics are attracted. There is an unstable fixed point at the origin, and a saddle point in the middle. The two stable fixed points are solutions either entirely of rabbits or entirely of sheep. This model shows a dynamic model in which rabbits and sheep cannot coexist.

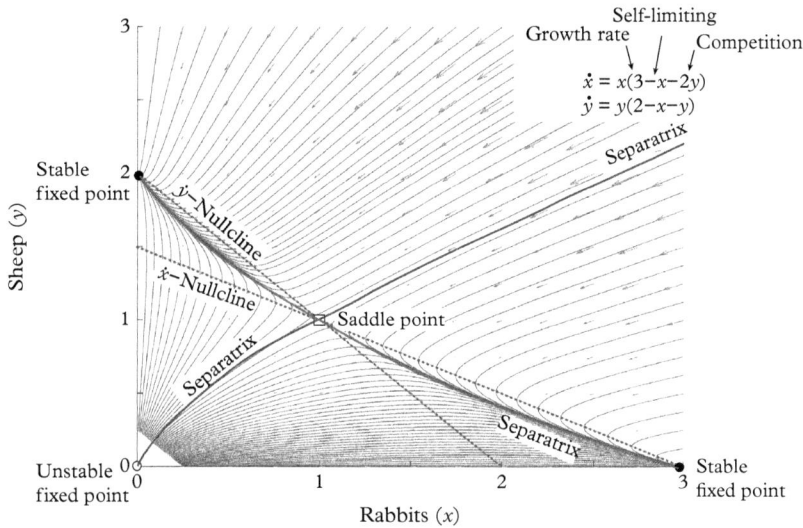

Figure 3.5 *Phase portrait for the rabbit vs. sheep example. The rabbits are on the x-axis, and the sheep are on the y-axis. Redrawn from Strogatz (1994).*

Example 3.3 is a case of what is known as "winner-take-all" dynamics in which one species drives the other to extinction. Either one can come out as the winner depending on the initial conditions. Coexistence is not possible for these parameters, because there is no symbiotic relationship—it is a purely competitive relationship. On the other hand, by choosing other values of the parameters that support symbiosis, co-existence can be possible. The outcomes of competition among species competing to survive, or among corporations competing for customers, are topics in the study of evolutionary dynamics and econophysics, which are explored in detail in Chapters 7 and 8.

3.2.4 Separatrixes: stable and unstable manifolds

The separatrixes in Fig. 3.5 divide up the phase plane into distinct regions. No flow lines cross a separatrix. They are the boundaries between qualitatively different stream line behavior. The separatrixes cross at the saddle fixed point and are the lines of fastest approach and fastest divergence from the fixed point. Because of the singular property of separatrixes, they are also known as the stable and unstable manifolds of the fixed point. The behavior of stable and unstable manifolds in nonlinear dynamics is closely linked to the emergence of chaos, and played a key role in the historical discovery of chaos in Hamiltonian systems by Henri Poincaré in 1889 when he was exploring the three-body problem. In this rabbit–sheep example, one of the separatrixes lies close to the y- and x-nullclines, but the other separatrix is very different from either nullcline.

The separatrixes are obtained through the eigenvectors of the Jacobian matrix at the saddle fixed point. The Jacobian matrix eigenvalues and eigenvectors at the saddle point in the rabbit vs. sheep model are

$$\mathcal{J} = \begin{pmatrix} -1 & -2 \\ -1 & -1 \end{pmatrix} \quad \lambda = \left\{ \begin{array}{l} \sqrt{2}-1 \\ -\left(\sqrt{2}+1\right) \end{array} \right. \quad v_1 = \frac{1}{\sqrt{3}} \begin{pmatrix} \sqrt{2} \\ -1 \end{pmatrix} \quad v_2 = \frac{1}{\sqrt{3}} \begin{pmatrix} \sqrt{2} \\ 1 \end{pmatrix}.$$

These eigenvectors define the directions of fastest divergence and convergence at the fixed point, respectively. The unstable manifold is found by numerically solving the equations of motion for initial conditions

$$\begin{pmatrix} x_1 \\ y_1 \end{pmatrix} = \pm \varepsilon v_1, \tag{3.31}$$

where ε is a small value. Because the speed of solution goes to zero at the fixed point, the equations can alternatively be solved in terms of the path length element ds. For this example the dynamical equations become

$$\begin{aligned} \frac{dx}{ds} &= \frac{x(3-x-2y)}{\sqrt{x^2(3-x-2y)^2 + y^2(2-x-y)^2}} \\ \frac{dy}{ds} &= \frac{y(2-x-y)}{\sqrt{x^2(3-x-2y)^2 + y^2(2-x-y)^2}} \end{aligned} \tag{3.32}$$

and the unstable manifold is the trajectory arising from the initial conditions at $\pm \varepsilon v_1$.

To find the stable manifold requires an additional step. This is because any initial condition that is a small step along the direction of the eigenvector v_2 will only reconverge on the fixed point without generating a curve. Therefore, the dynamical equations need to be time-reversed (path-reversed):

$$\begin{aligned} \frac{dx}{ds} &= -\frac{x(3-x-2y)}{\sqrt{x^2(3-x-2y)^2 + y^2(2-x-y)^2}} \\ \frac{dy}{ds} &= -\frac{y(2-x-y)}{\sqrt{x^2(3-x-2y)^2 + y^2(2-x-y)^2}}, \end{aligned} \tag{3.33}$$

where the minus signs reverse the stable and unstable manifolds and the stable manifold is found as the trajectory traced away from the initial conditions. The stable manifold is now found by taking the initial conditions

$$\begin{pmatrix} x_2 \\ y_2 \end{pmatrix} = \pm \varepsilon v_2 \tag{3.34}$$

and the trajectory traces out the path-reversed stable manifold.

The stable and unstable manifolds are shown in Fig. 3.5. The unstable manifold emerges from the saddle fixed point and performs a trajectory to the two stable fixed points. The stable manifold arises from the unstable fixed point at $(0, 0)$, or arises at infinity, and converges on the saddle. These stable and unstable manifolds are the separatrixes for this rabbit-sheep system, dividing up the phase plane into four regions that confine the flow lines. A closely related concept is that of a "basin of attraction." In this rabbit–sheep system there are two stable fixed points and two basins of attraction. These fixed points attract all of the flow lines in the two basins bounded by the stable manifold. Basins of attraction play an important role in recurrent neural networks and applications of associative memory, to be discussed in Chapter 6.

3.2.5 Limit cycles

Fixed points are not the only possible steady states of a dynamical system. It is also possible for a system to oscillate (or orbit) repetitively. This type of steady-state solution is called a limit cycle. In two dimensions there are three types of limit cycles shown in Fig. 3.6. A stable limit cycle attracts trajectories, while an unstable limit cycle repels trajectories. The saddle limit cycle is rare; it attracts from one side and repels on the other.

A classic example of a limit cycle is the van der Pol oscillator, which is a nonlinear oscillator developed originally to describe the dynamics of space-charge effects in vacuum tubes. It displays self-sustained oscillations as the system gain is balanced by nonlinear dissipation.

3.2.5.1 Van der Pol oscillator

The van der Pol oscillator equation is

$$\ddot{x} = 2\mu\dot{x}\left(1 - \beta x^2\right) - \omega_0^2 x, \tag{3.35}$$

where μ is positive and provides the system gain to support self-sustained oscillations. The natural frequency in the absence of the nonlinear term is $\omega_0{}^2$. The nonlinear term in x^2 provides self-limiting behavior to balance the gain and keeps the oscillations finite.

Stable limit cycle   Saddle limit cycle   Unstable limit cycle

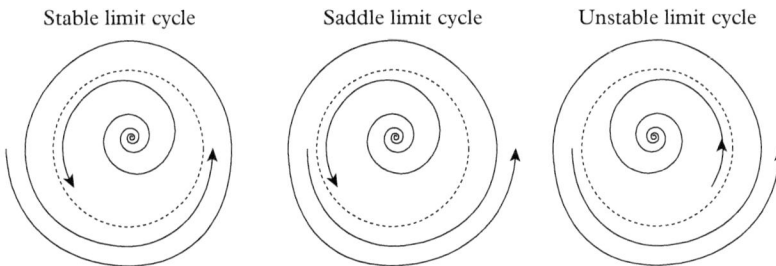

Figure 3.6 *There are three types of 2D limit cycles: stable, saddle, and unstable. The saddle is rare.*

A dimensionless flow representation of the oscillator is

$$\dot{x} = y$$
$$\dot{y} = -x + y\left(1 - x^2\right). \tag{3.36}$$

There is a fixed point at $(x^*, y^*) = 0$ which has the Jacobian matrix

$$\mathcal{J} = \begin{pmatrix} 0 & 1 \\ -1 & 1 \end{pmatrix} \qquad \begin{array}{c} \tau = 1 \\ \Delta = 1 \end{array} \tag{3.37}$$

with the characteristic equation and eigenvalues

$$-\lambda(1 - \lambda) + 1 = 0$$
$$\lambda^2 - \lambda + 1 = 0 \tag{3.38}$$
$$\lambda = \frac{1 \pm i\sqrt{3}}{2}.$$

Therefore, the origin is a fixed point for an unstable spiral. However, the spiral does not grow without bound, but instead approaches the limit cycle, that is stable.

A phase portrait for the van der Pol oscillator is shown in Fig. 3.7 with $\beta = 1$, $\mu = 0.5$, and $\omega_0^2 = 1$. The stable limit cycle is the attractor of all initial conditions as all stream lines converge on the cycle. Several limit cycles for different values of the parameters are shown in Fig. 3.8. Smaller β allows the gain to produce larger oscillations. Larger μ produces strongly nonlinear oscillations. Two examples of

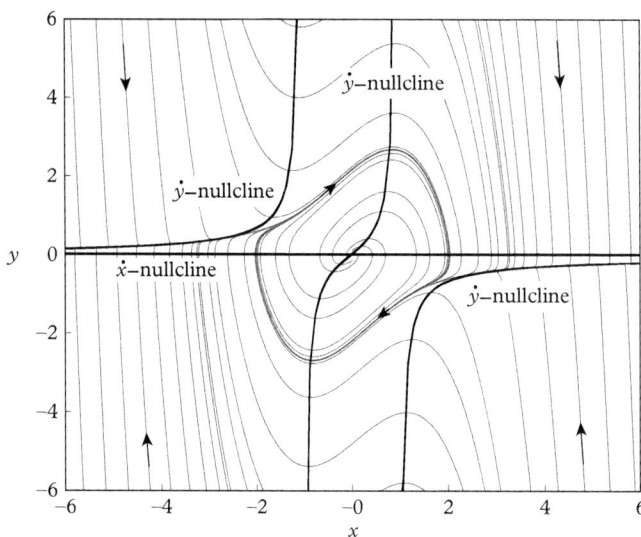

Figure 3.7 *Stream lines and velocity vectors for the van der Pol oscillator with* $\beta = 1$, $\mu = 0.5$, *and* $\omega_0^2 = 1$.

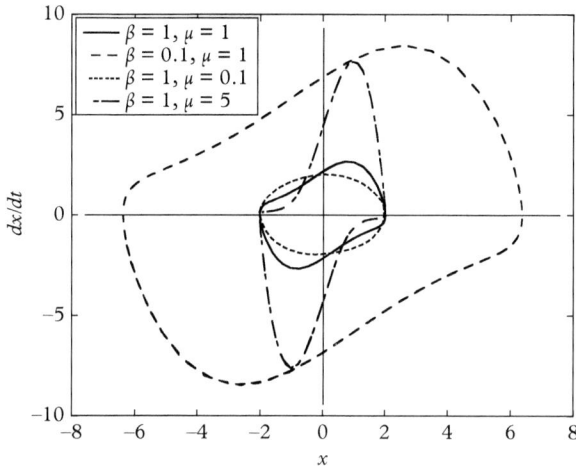

Figure 3.8 *Van der Pol limit cycles in state space for different parameters. Smaller β allows the gain to produce larger oscillations. Larger μ produces strongly nonlinear oscillations.*

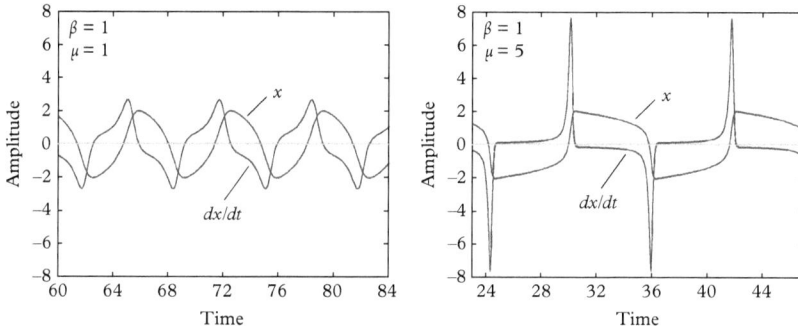

Figure 3.9 *Time series of* x *and* dx/dt *for two sets of parameters for the van der Pol oscillator.*

the time series of a van der Pol oscillator show *x* and *dx/dt* as functions of time in Fig. 3.9. The oscillations shown in Fig. 3.9(b) are highly nonlinear, driven by a large value of μ. The frequencies are noticeably different between Fig. 3.9(a) and (b).

3.2.5.2 *Poincaré sections (first-return map)*

A limit cycle is not a fixed point, but it is a steady state of the system. Like fixed points, it comes in different types of stability. Poincaré recognized these similarities and devised a way to convert a limit cycle in 2D into an equivalent fixed point in 1D. This method introduces a discrete map called the first-return map. The Poincaré section is shown schematically in Fig. 3.10. In this case the Poincaré section is a line that intersects the limit cycle. As the trajectory approaches the limit cycle, a series of intersection points with the Poincaré section is generated, as shown on the left. The sequence of points converges on the limit cycle for large numbers of iterations.

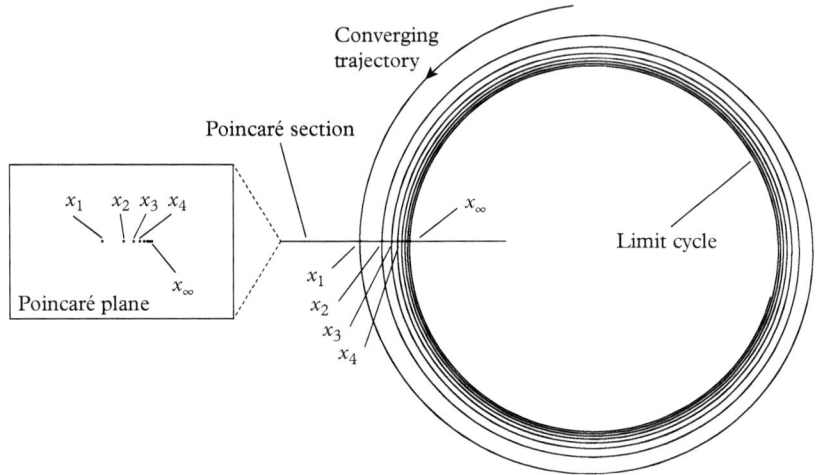

Figure 3.10 *First-return map on a Poincaré section. The limit cycle in 2D is converted to a fixed point of a discrete map in 1D.*

The first-return map for a 2D limit cycle is a sequence of points with the recurrence relation

$$x_{n+1} = F(x_n). \tag{3.39}$$

This map has a fixed point x^* when

$$x^* = F\left(x^*\right). \tag{3.40}$$

Expanding about the fixed point, we have

$$x_n - x^* = \left.\frac{dF}{dx}\right|_{x^*} (x_{n-1} - x^*)$$

$$\Delta x_{n+1} = \left.\frac{dF}{dx}\right|_{x^*} \Delta x_n \tag{3.41}$$

that iterates to

$$\Delta x_{n+1} = \left.\frac{dF}{dx}\right|_{x^*} \Delta x_n = \left.\frac{dF}{dx}\right|_{x^*} \left(\left.\frac{dF}{dx}\right|_{x^*} \Delta x_{n-1} \right) = \dots = \left(\left.\frac{dF}{dx}\right|_{x^*} \right)^n \Delta x_0, \tag{3.42}$$

which can be rewritten in terms of a multiplier

$$\Delta x_{n+1} = M^n \Delta x_0, \tag{3.43}$$

where the Floquet multiplier is defined as

$$M = \left.\frac{dF}{dx}\right|_{x^*} \quad \text{Floquet multiplier} \tag{3.44}$$

A limit cycle of the full dynamics is a fixed point of the map which can be classified according to

Multiplier	Type of cycle
$M > 1$	Repelling cycle
$M < 1$	Attracting cycle
$M = 1$	Saddle cycle

Example 3.4 First-return map (Poincaré section)

Consider the flow

$$\dot{r} = r\left(1 - r^2\right)$$
$$\dot{\theta} = 1. \tag{3.45}$$

This flow has a constant angular velocity, and hence it is easy to construct a first-return map, because the first return time is $t = 2\pi$. Rewrite the flow expression into

$$\frac{dr}{dt} = r\left(1 - r^2\right)$$
$$\frac{dr}{r\left(1 - r^2\right)} = dt \tag{3.46}$$

and integrate to give

$$\int_{r_0}^{r_1} \frac{dr}{r\left(1 - r^2\right)} = \int_0^{2\pi} dt = 2\pi. \tag{3.47}$$

This is solved for r_1 in terms of r_0

$$r_1 = \frac{1}{\sqrt{1 + e^{-4\pi}\left(\dfrac{1}{r_0^2} - 1\right)}} = F(r_0), \tag{3.48}$$

which gives the first-return map

$$F(r) = \left(1 + e^{-4\pi}\left(\frac{1}{r^2} - 1\right)\right)^{-1/2}. \tag{3.49}$$

This has a fixed point at $r^* = 1$. Linearizing around the fixed point

$$r = 1 + \eta$$
$$\dot{r} = \dot{\eta} = (1 + \eta)\left(1 - (1 + \eta)^2\right) \approx -2\eta \tag{3.50}$$

continued

Example 3.4 *continued*

has an exponential solution

$$\eta\left(t\right) = \eta_0 e^{-2t} = \eta_0 e^{-4\pi} \tag{3.51}$$

giving the Floquet multiplier

$$M = e^{-4\pi}, \tag{3.52}$$

which is less than one, and hence the limit cycle is stable.

3.2.5.3 *Andronov–Hopf bifurcations*

Limit cycles can occur when there is a balance between competing terms in the flow equations. In the case of the van der Pol oscillator, it is the balance between the gain parameter and the nonlinear self-limiting term in the equation. Therefore, one can think of the system as having a *control parameter* that can be used to tune one of the competing terms and hence change the balance. In this situation, there is often a threshold value for the control parameter at which the qualitative behavior of the system changes abruptly. This is called a *bifurcation.*

Consider a flow in (r, θ) where there is a control parameter c that varies continuously from negative values through zero to positive values

$$\begin{aligned} \dot{r} &= r\left(c - r^2\right) \\ \dot{\theta} &= 1. \end{aligned} \tag{3.53}$$

The fixed points of the flow are

$$\begin{aligned} r^* &= 0 \\ r^* &= \sqrt{c}, \end{aligned} \tag{3.54}$$

where the first is a node at the origin and the second is the radius of a limit cycle. The Lyapunov exponent of the node at $r^* = 0$ is $\lambda = c$, and hence the node is stable if $c < 0$ and unstable if $c > 0$. Furthermore, the limit cycle exists only when the argument of the square-root is positive, which also occurs when the control parameter $c > 0$, and hence the unstable node is bounded by the limit cycle, just as in the case of the van der Pol oscillator in Eq. (3.35). Therefore, as the control parameter c increases from negative values and crosses zero, the stable node at the origin converts to an unstable spiral that is bounded by a limit cycle. The radius of the limit cycle increases with increasing magnitude c. Therefore, there is a qualitative and sudden change in behavior of the system at the threshold value $c = 0$. This constitutes a bifurcation for this flow.

The system dynamics are plotted as a function of time in Fig. 3.11 as the control parameter slowly increases as a function of time from negative values through the bifurcation threshold to positive values, and then slowly decreases again

$\dot{r} = r(c - r^2)$
$\dot{\theta} = 1$

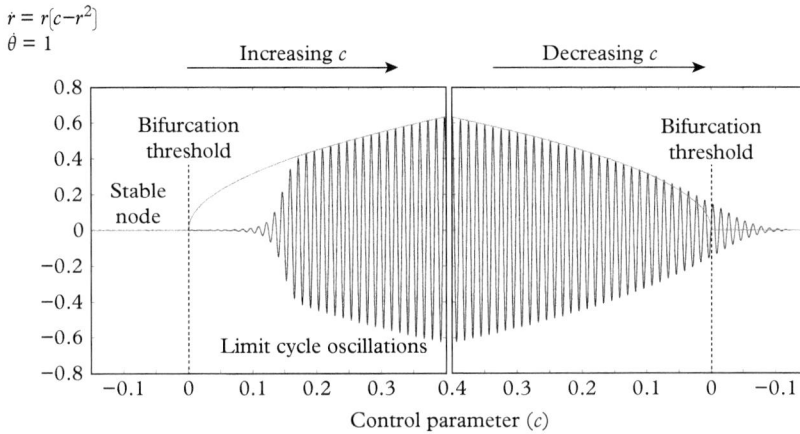

Figure 3.11 *The system dynamics as the control parameter, c, increases slowly from negative (stable node) through the Hopf bifurcation at c = 0 to the stable limit cycle. The parameter then decreases slowly through the bifurcation threshold again. The hysteresis between the onset (extinction) of the oscillations is caused by finite-time effects.*

through the bifurcation threshold. The horizontal axis is the control parameter, parameterized by the time. The amplitude of the limit-cycle oscillations varies as $\sqrt{c}$, shown as the continuous curve in the figure. In the numerical simulation, the system shows a hysteresis with a delayed onset of the oscillations with increasing c, and a delayed extinction with decreasing c. This hysteresis is due to critical slowing down near the threshold and to the finite rate of the sweeping control parameter. The Lyapunov exponent, which is the relaxation rate, approaches zero at the threshold, and there is not enough time for the system to relax as the control parameter is changed through successive values. Once the system is far enough from the threshold, the oscillation amplitude follows the square-root prediction.

An interesting thing happens to the system dynamics when the sign of the r^2 term in Eq. (3.53) is flipped to $(c + r^2)$. The Lyapunov exponent is the same, and the origin is still a stable node for $c < 0$ and unstable for $c > 0$. But in this case, the limit cycle only exists for $c < 0$, and furthermore the stability of the limit cycle changes to an unstable limit cycle. If the control parameter c starts at negative values, the central node is stable, but as $c > 0$, the system catastrophically diverges. In the example with $(c - r^2)$, the system always finds a stable solution regardless of the sign of the control parameter, converging on a stable limit cycle in one case or a stable node in the other. But in the example with $(c + r^2)$, the system only leads to a stable solution in the special case when $c < 0$, and the initial conditions are inside the limit cycle radius. For all other situations, the system diverges. Although each example shows a critical bifurcation threshold, known as an *Andronov–Hopf bifurcation* or sometimes just *Hopf bifurcation*, the system behavior at the bifurcation is qualitatively different. In the first globally stable case, the bifurcation is called *supercritical*, while in the second the bifurcation is called *subcritical*.

There are many types of bifurcations that can occur in nonlinear systems that go beyond the simple Hopf bifurcation examples here. Many of these are classified depending on whether the transitions are continuous or discontinuous, whether or not they lead to system hysteresis, and on their global stability properties.[2]

3.2.5.4 Homoclinic orbits

A special type of cycle on the phase plane is a homoclinic orbit. A homoclinic orbit occurs when the unstable manifold of a saddle point turns back and becomes the stable manifold of the same saddle point. An example of a homoclinic orbit is seen in an anisotropic oscillator with a spring constant that is a function of how much it is stretched. Consider the oscillator equation

$$m\ddot{x} + k(1 + x)x = 0 \tag{3.55}$$

in which the spring constant is a function of the displacement $k_{eff}(x) = k(1 + x)$ that becomes increasingly stiffer when the mass displaces to the right, but less stiff when the mass displaces to the left. The 2D flow is

$$\dot{x} = v$$
$$\dot{v} = -\frac{k}{m}(1 + x)x. \tag{3.56}$$

For small amplitudes, this behaves like an ordinary harmonic oscillator with a center at $(x, \dot{x}) = (0, 0)$. For larger amplitudes, there is a saddle fixed point at $(x, \dot{x}) = (-1, 0)$ where the spring constant vanishes. The state-space is shown in Fig. 3.12. The trajectory that originates from the saddle point turns back on itself

[2] Bifurcations tend to have descriptive names, such as pitchfork, flip, or saddle-node bifurcations, among others.

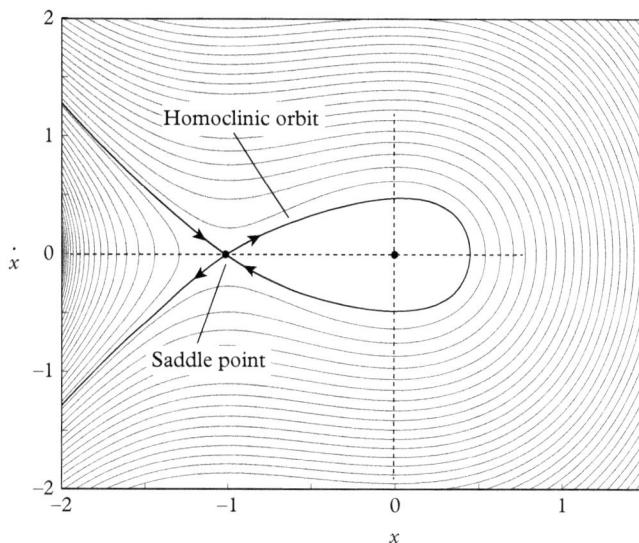

Figure 3.12 *Example of a homoclinic orbit for an anisotropic oscillator.*

and reenters the saddle point. The period of this orbit is infinite, because the phase point leaves and reenters the saddle point with zero velocity. Compare this state-space plot to the state space of the pendulum in Fig. 2.11.

Example 3.5 Saddle-node bifurcation

A model that shows the evolution of a homoclinic orbit from a saddle point, through a process known as a saddle-node bifurcation, is

$$\dot{x} = ay + bx\left(c - y^2\right)$$
$$\dot{y} = -x + \alpha, \tag{3.57}$$

where there are two fixed points at

$$(x_1, x_2) = \left(\alpha, \frac{1}{2b\alpha}\left(a + \sqrt{a^2 + 4b^2\alpha^2 c}\right)\right) \tag{3.58}$$

$$(x_1, x_2) = \left(\alpha, \frac{1}{2b\alpha}\left(a - \sqrt{a^2 + 4b^2\alpha^2 c}\right)\right).$$

The Jacobian matrix is

$$\mathcal{J} = \begin{pmatrix} b\left(c - y^2\right) & a - 2bxy \\ -1 & 0 \end{pmatrix}. \tag{3.59}$$

As a specific example consider

$$\dot{x} = 0.7y + 0.7x\left(0.5 - y^2\right)$$
$$\dot{y} = -x + \alpha, \tag{3.60}$$

where α is the parameter that controls the evolution of the homoclinic orbit. The stable and unstable manifolds are shown in Fig. 3.13 for three values of α. For $\alpha = 0.6$, one unstable manifold converges on a limit cycle around the unstable fixed point. For $\alpha = 0.75$, one of the unstable manifolds turns back to become the stable manifold of the same saddle point. This creates a homoclinic orbit. The limit cycle is destroyed when α passes the critical value 0.75, representing a bifurcation. For $\alpha = 0.9$, one branch of the unstable manifold converges on the other branch, and the unstable fixed point gives rise to one of the stable manifolds.

In this example, the homoclinic orbit occurs for a special value of α, which makes it a special case for this nonlinear system. However, in integrable Hamiltonian systems, homoclinic orbits are common. As non-integrable perturbations are turned on, the homoclinic orbits are perturbed and, in higher dimensional systems, give rise to Hamiltonian chaos. This situation occurs for the three-body problem which was where Poincaré first saw that homoclinic orbits could display chaotic behavior.

continued

Example 3.5 *continued*

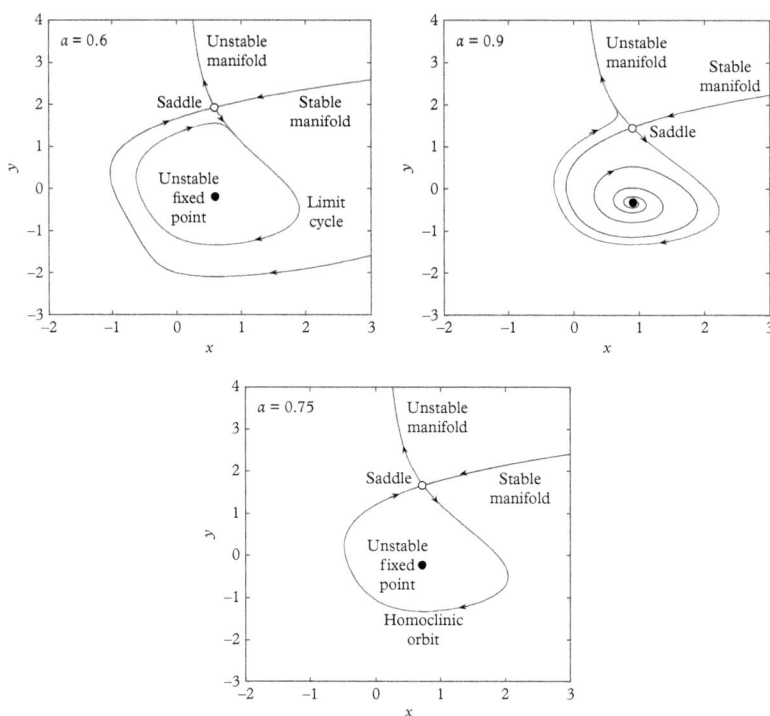

Figure 3.13 *Evolution of a homoclinic orbit from a saddle point as the control parameter α in Eq. (3.60) is changed. For α = 0.6, the unstable manifold converges on a limit cycle around an unstable fixed point. For α = 0.75, the unstable manifold folds back as the stable manifold in a homoclinic orbit. For α = 0.9, the unstable manifold converges with the other unstable branch, while the stable manifold arises from the unstable fixed point. Redrawn from Medio (1992).*

3.3 Discrete iterative maps

Continuous-time flows are computationally expensive, while discrete maps are usually much easier to implement. Furthermore, it is often possible to find simple iterative maps that capture the essential behavior of a flow, especially the properties of the Poincaré map, which are themselves discrete maps. An iterative map is a rule that takes a current value and projects it to the next value, which becomes the new current value that projects to the next value, and so on iteratively. The map is usually defined through a function

$$x_{n+1} = f(x_n) \tag{3.61}$$

that takes the value x_n and produces the next value x_{n+1}. The iterative character is easy to see through the chain

$$x_{n+1} = f\left(f\left(x_{n-1}\right)\right) = f\left(f\left(f\left(x_{n-2}\right)\right)\right) = \ldots = f^{(n)}\left(x_1\right), \tag{3.62}$$

which begins with an initial value x_1. This type of mapping can lead to stable answers, for which $x^* = x_{n+1} = x_n$, which are the fixed points of the mapping. Or it can lead to seemingly random values of x_n, depending on the properties of the mapping function $f(x)$. The seemingly random behavior is called "chaos," but it is often not truly random, having specific structure that relates back to the properties of $f(x)$. Indeed, by studying the chaotic progression of numbers, some information about $f(x)$ can be extracted—that is, obtaining information out of chaos.

3.3.1   One-dimensional logistic map

A simple model that shows the transition to chaos through a bifurcation cascade is the logistic map. This is a discrete map in one variable that is iterated from an initial condition. The iterative equation is

$$x_{n+1} = rx_n(1 - x_n) \tag{3.63}$$

for $0 < r < 4$. When $r < 3$, the equation has a single fixed point. However, at $r = 3.0$, the steady-state solution bifurcates into two stable branches, and these bifurcate again at $r = 3.45$ into four, evolving into fully developed chaos at $r = 3.57$. The iterative mapping is illustrated in Fig. 3.14. For $r = 2.7$ there is a single fixed point that attracts the iterations. For $r = 3.2$ the mapping shows a period-two cycle, and for $r = 3.5$ it shows a period-four cycle. For $r = 3.82$, there is no repetition, and the system is in chaos. However, for a slightly larger value of r at $r = 3.85$, there is a period-three cycle. The appearance of period-three cycles in the midst of chaos is a common characteristic of these quadratic maps.

The bifurcation diagram is shown in Fig. 3.15 as the "gain" parameter r is increased. Note that the bifurcation cascade is repeated at numerous different scales, shown under several magnifications. This self-similarity of the bifurcation graph is a feature of fractals, which are commonly encountered in chaotic dynamics.

3.3.2   Feigenbaum number and universality

The logistic map is only one of a virtually limitless number of iterated maps. Each different mapping function produces a different graphical solution with different bifurcation cascades. But if every nonlinear system is different, then what type of over-arching principles can be applied to chaos? The answer is that there are common themes among wide classes of nonlinear systems, including quantitative

Fixed point   Period two   Period four

$r = 2.7$ $r = 3.2$ $r = 3.5$

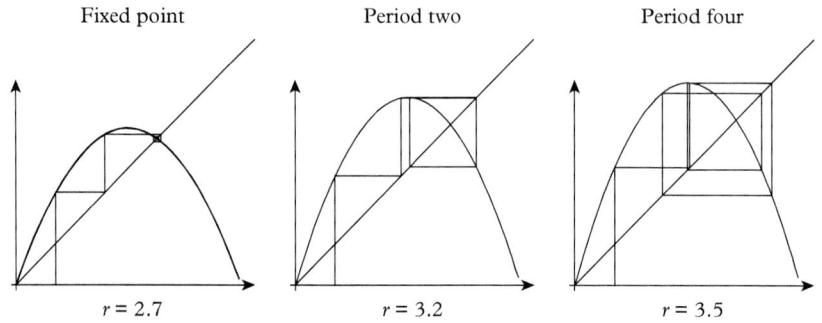

Figure 3.14 *Graphical solution to the logistic map for increasing gain value r. The initial condition is iterated. In the case of r = 2.7, the iterations converge on a single fixed point. At higher values of the "gain" parameter r there are period-two and period-four cycles. Full chaos occurs for r = 3.82 with no repeating pattern. However, at r = 3.85 a period-three cycle appears.*

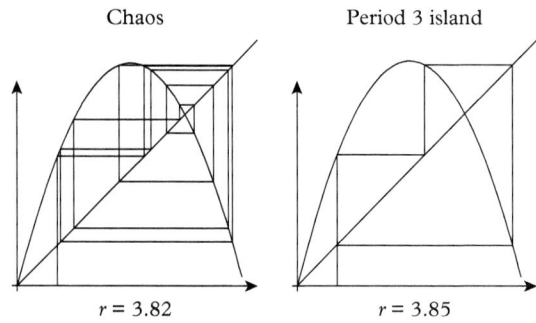

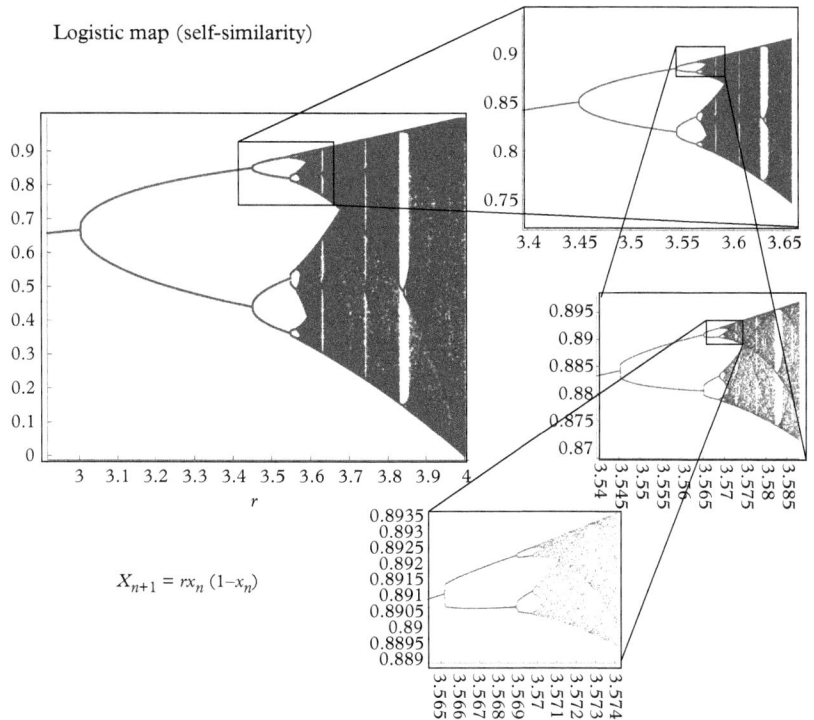

Chaos   Period 3 island

$r = 3.82$ $r = 3.85$

Logistic map (self-similarity)

$$X_{n+1} = rx_n (1-x_n)$$

Figure 3.15 *Bifurcation diagram of the logistic map. The sub-areas are self-similar, with the same features appearing as for the larger scale.*

Feigenbaum ratio

$$\delta = \lim_{n\to\infty} \frac{S_{n+2} - S_{n+1}}{S_{n-1} - S_n} = 4.6692$$

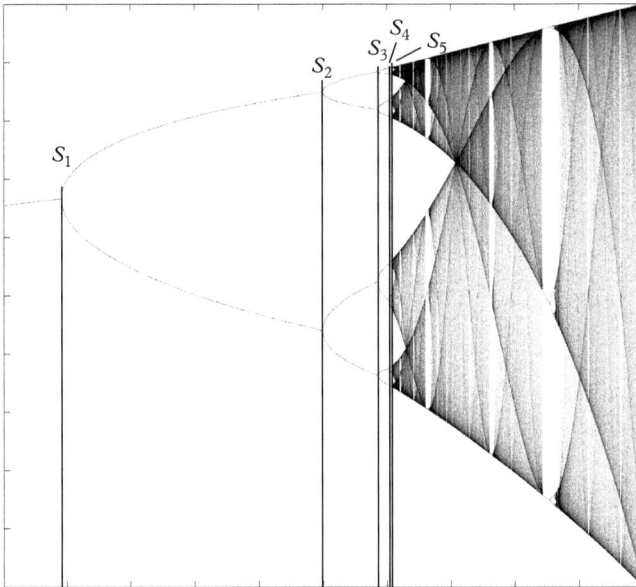

Figure 3.16 *Bifurcation cascade showing the sequence of thresholds, S_1, S_2, S_3, S_4, etc., that generate the Feigenbaum ratio.*

similarities that in some cases become identities. For instance, the logistic map has an inverted quadratic dependence represented by the parabola in Fig. 3.16. One of the discoveries in the field of chaos was the universal behavior of the bifurcation cascade for *every* iterated map that has a lowest-order quadratic dependence. The bifurcation thresholds of r have a fixed ratio in the limit as the bifurcations approach chaos:

$$\delta = \lim_{n\to\infty} \frac{S_{n+2} - S_{n+1}}{S_{n+1} - S_n} = 4.66920160910299067185320382.$$

This Feigenbaum ratio is the same for all quadratic maps (which tend to be the most common). Therefore, in the face of the apparent randomness of chaotic systems, universals do emerge.

3.4 Three-dimensional state space and chaos

The character of nonlinear dynamical systems changes dramatically when going from 2D to 3D state space. Dynamic systems are subject to the no-crossing theorem, which states that no trajectory can cross itself (in finite time). However, in a

3D state space there is plenty of "room" for trajectories to move above and below and around other trajectories without intersecting. Hence, the addition of a third dimension opens the possibility for chaotic behavior.

Because the transition to chaos occurs in the transition from 2D to 3D, the most "interesting" chaotic systems tend to have low dimensionality of three or four dimensions. These systems can totter on the edge of chaos, and are tipped into or out of chaos depending on the parameters defining the dynamics. In systems of high dimension, on the other hand, chaos is more easily displayed, as for instance in the motions of a gas molecule in a gas at thermodynamic equilibrium. Therefore, most of the examples of chaos in this chapter are dynamical systems with three or four state-space dimensions.

3.4.1 Stability and fixed-point classification

Flows in 3D take the form

$$
\begin{aligned}
\dot{x}_1 &= f_1(x_1, x_2, x_3) \\
\dot{x}_2 &= f_2(x_1, x_2, x_3) \\
\dot{x}_3 &= f_3(x_1, x_2, x_3).
\end{aligned}
\tag{3.64}
$$

The Jacobian matrix is

$$
\mathfrak{J} =
\begin{pmatrix}
\dfrac{\partial f_1}{\partial x_1} & \dfrac{\partial f_1}{\partial x_2} & \dfrac{\partial f_1}{\partial x_3} \\[2ex]
\dfrac{\partial f_2}{\partial x_1} & \dfrac{\partial f_2}{\partial x_2} & \dfrac{\partial f_2}{\partial x_3} \\[2ex]
\dfrac{\partial f_3}{\partial x_1} & \dfrac{\partial f_3}{\partial x_2} & \dfrac{\partial f_3}{\partial x_3}
\end{pmatrix}
\tag{3.65}
$$

with characteristic values given by

$$
\det(\mathfrak{J} - \lambda I) = 0,
\tag{3.66}
$$

which yields a cubic equation

$$
\lambda^3 + p\lambda^2 + q\lambda + r = 0,
\tag{3.67}
$$

which is solved for three eigenvalues and eigenvectors.

3.4.1.1 3D fixed point classifications

Fixed points in 3D are classified according to the signs of the real and imaginary parts of the characteristic values of the Jacobian matrix. Table 3.2 shows the possible combinations of the characteristic values and the associated fixed-point classification. The behavior of trajectories near the fixed point is illustrated in Fig. 3.17. Attracting or repelling nodes have three eigenvalues and eigenvectors.

Table 3.2 *3D fixed point classification.*

Characteristic values	Fixed point
$(-, -, -)$	Attracting node
$(-, -r + ia, -r - ia)$	Stable spiral
$(+, +, +)$	Repellor
$(+, +r + ia, +r - ia)$	Unstable spiral
$(-, -, +)$	Saddle point index 1
$(-, +, +)$	Saddle point index 2
$(-r + ia, -r - ia, +)$	Spiral saddle index 1
$(-, +r + ia, +r - ia)$	Spiral saddle index 2

Stable or unstable spirals spiral around the fixed point as 3D trajectories. However, saddle points in 3D have more possibilities, depending on the number of positive or negative eigenvalues and are denoted by indices. For the spiral saddles, the trajectory can have stable or unstable spirals within two of the dimensions, with unstable or stable behavior, respectively, in the dimension perpendicular to the spiral surface.

3.4.1.2 Limit cycles and Poincaré sections

For a 3D flow, the Poincaré section is a plane and the first-return map is two-dimensional. The map is given by

$$x_1^{n+1} = F_1\left(x_1^n, x_2^n\right) \quad \text{with fixed points at} \quad x_1^* = F_1\left(x_1^*, x_2^*\right)$$
$$x_2^{n+1} = F_2\left(x_1^n, x_2^n\right) \qquad\qquad\qquad x_2^* = F_2\left(x_1^*, x_2^*\right)$$

Fixed points in the Poincaré plane are limit cycles of 3D trajectories. The behavior of limit cycles in 3D flows is much more varied. Trajectories can now move above and below each other like spaghetti, providing discrete mappings that are no longer monotonic (as they must be for 2D flows) as shown in Fig. 3.18. The stability analysis is similar to the analysis for 2D limit cycles. The Floquet matrix is

$$\mathcal{J}_M = \begin{pmatrix} \dfrac{\partial F_1}{\partial x} & \dfrac{\partial F_1}{\partial y} \\ \dfrac{\partial F_2}{\partial x} & \dfrac{\partial F_2}{\partial y} \end{pmatrix}\Bigg|_{(x^*, y^*)} \tag{3.68}$$

and the characteristic equation

$$|\mathcal{J}_M - MI| = 0 \tag{3.69}$$

3D Fixed point classification

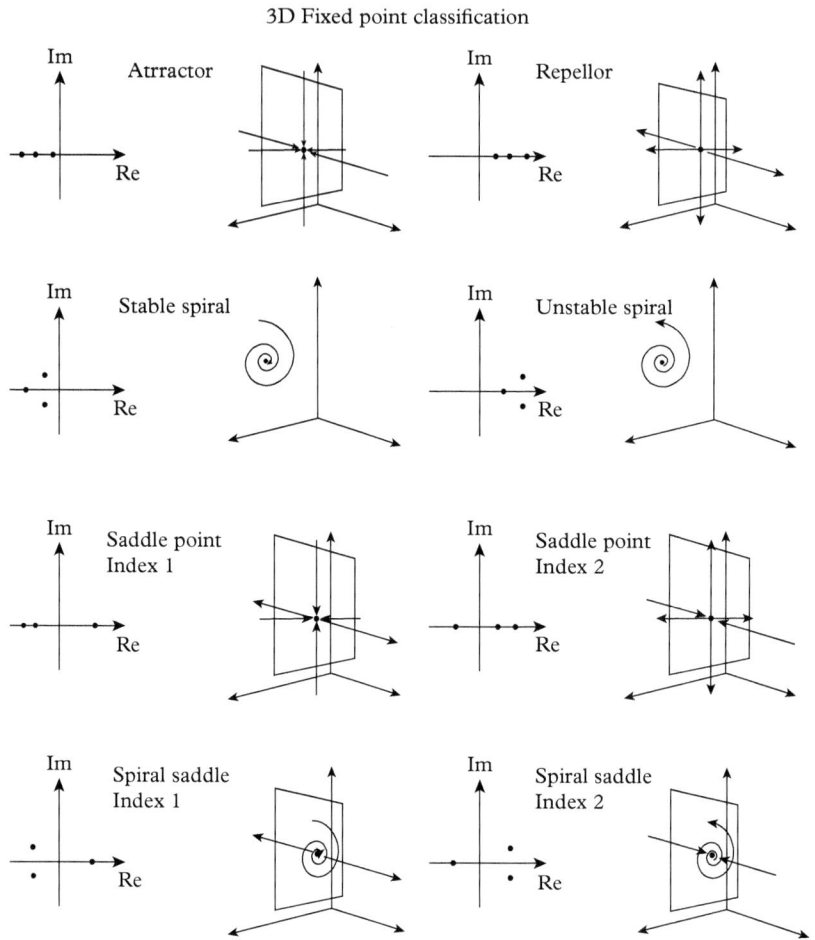

Figure 3.17 *Classification of 3D fixed points. The solutions to the determinant on the complex plane are shown on the left with a picture of the trajectories near the fixed point on the right.*

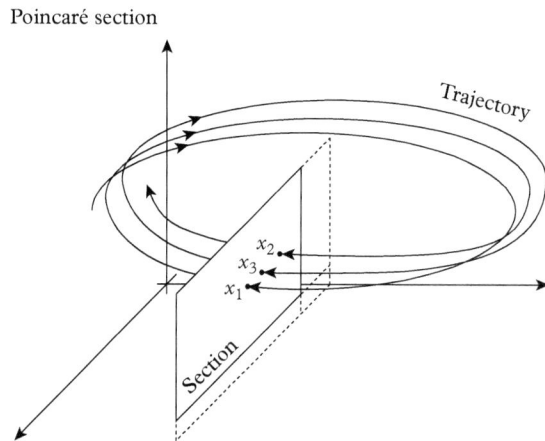

Figure 3.18 *Poincaré section for a 3D limit cycle. The discrete map no longer needs to be monotonic, because successive trajectories can pass above or below each other.*

Limit cycle classification

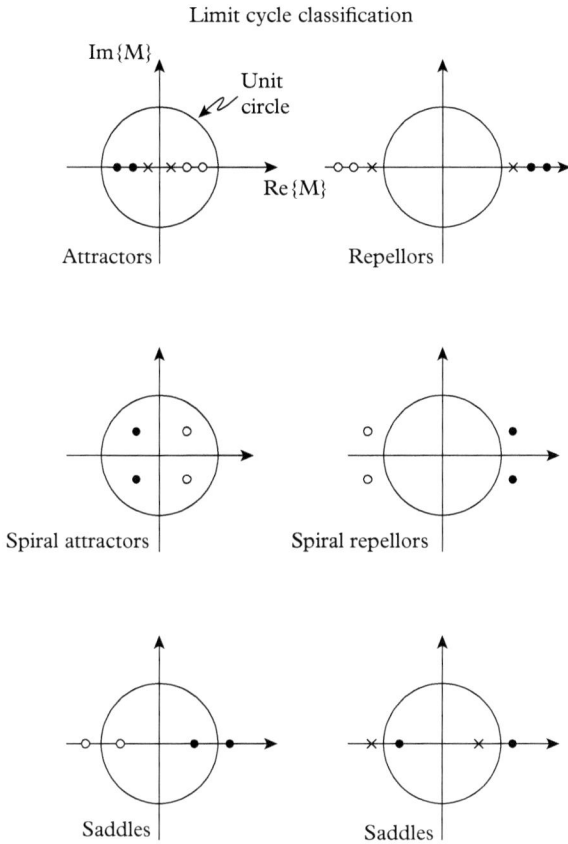

Figure 3.19 *Floquet multipliers for fixed points defining limit cycles in 3D. Pairs of solutions are shown in each case. Multipliers outside the unit circle are repellors, while multipliers inside the circle are attractors. Negative values lead to alternating signs in the approach to the limit cycle.*

gives the Floquet multipliers

$$M_{1,2} = \frac{1}{2}\tau \pm \frac{1}{2}\sqrt{\tau^2 - 4\Delta}. \qquad (3.70)$$

The classification of the limit cycles depends on the magnitude of the multipliers relative to the unit circle on the complex plane, as shown in Fig. 3.19. Limit cycles can have attracting or repelling maps, and spirals or saddles. Multipliers outside the unit circle lead to unstable limit cycles, while inside the unit circle lead to stable limit cycles. The signs of the eigenvalues determine whether the asymptotic approach to the fixed point is monotonic (if the M is positive) or alternates (if the M is negative). Nodes and saddles have real-valued eigenvectors pointing along stable and unstable manifolds. Complex multipliers describe spirals (stable or unstable).

3.4.2   Non-autonomous (driven) flows

Non-autonomous flows have explicit time dependence

$$\dot{\mathbf{x}} = \mathbf{F}(\mathbf{x}, t). \qquad (3.71)$$

The time variable introduces a state space of one dimension higher than defined by $\mathbf{x}$. The new dimension is time, although time is not on an equal footing with $\mathbf{x}$ because time cannot be controlled. However, the higher dimensionality does allow different dynamics. For instance, in a 2D state space, chaos is not allowed because of the non-crossing theorem. However, in a driven 2D system, the extra dimension of time lifts this restriction and chaos is thus possible.

There are many ways to introduce a new variable related to time. For instance the new variable may be introduced as

$$\begin{aligned} z &= t \\ \dot{z} &= 1. \end{aligned} \qquad (3.72)$$

On the other hand, for $\theta = \omega t$, a natural variable to describe the dynamics is

$$\begin{aligned} z &= \omega t \\ \dot{z} &= \omega \end{aligned} \qquad (3.73)$$

and the angle can be plotted as modulus 2π. Both of these substitutions convert the non-autonomous flows into autonomous flows.

In the case of a harmonic forcing function, the new variable can be

$$\begin{aligned} z &= \sin \omega t \\ \dot{z} &= \omega \cos \omega t. \end{aligned} \qquad (3.74)$$

This representation has the benefit that trajectories are bounded along the new dimension, while in the first cases the trajectories are not.

In the following examples, 2D systems are driven by a time-dependent term that makes the flow non-autonomous and introduces the possibility for these systems to display chaotic behavior.

Example 3.6 The driven damped pendulum

The driven damped pendulum is described by the equations

$$\begin{aligned} \dot{x} &= y \\ \dot{y} &= a \sin z - by - \sin x \\ \dot{z} &= \omega \end{aligned} \qquad (3.75)$$

for the drive amplitude a with $b = \gamma/mL^2$. This system displays no sustained self-oscillation (because of the damping), but can be driven resonantly at the natural frequency $\omega_0^2 = g/L = 1$, for which amplitudes become large and the nonlinear angular term dominates. Note that z is the surrogate for time, increasing the dimensionality of the dynamics from 2D to 3D and allowing the transition to chaos.

The results for a driven damped pendulum for $b = 0.05$, $a = 0.7$, and $\omega = 0.7$ are shown in Fig. 3.20. The time dependence of the amplitude is plotted in Fig. 3.20(a), and phase space trajectories are plotted in Fig. 3.20(b). A Poincaré section is shown in Fig. 3.20(c) "strobed" at the drive period. Note the character of the section is not completely random. The Poincaré section remains relatively sparse, with noticeable gaps (known as *lacuna*). This Poincaré section represents part of the attractor of the nonlinear dynamics, which has a wispy or filamentary nature that is described by fractal geometry. Clearly, the additional time variable has allowed the system to exhibit chaotic behavior.

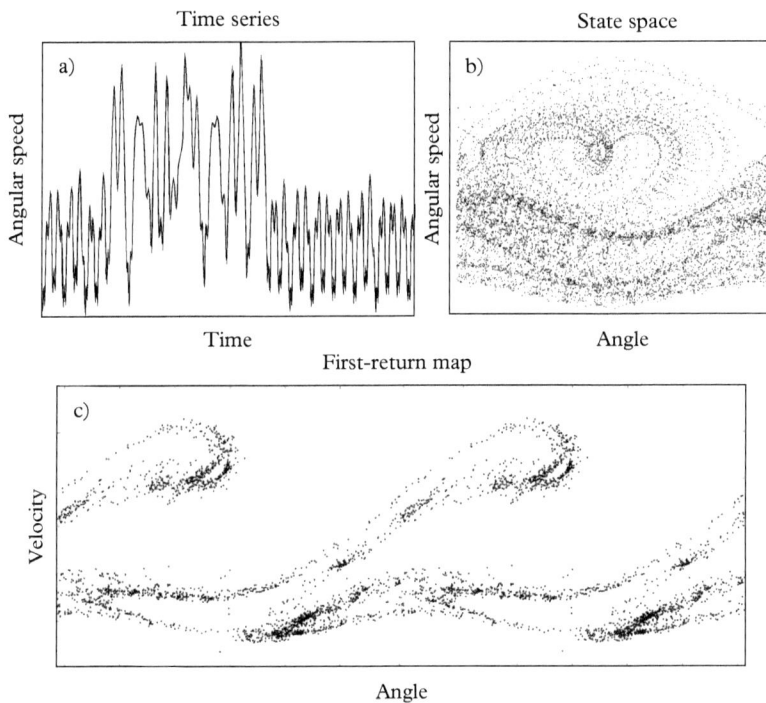

Figure 3.20 *Dynamics of a driven-damped oscillator for b = 0.1, ω = 0.7, and a = 0.7: (a) angle variable vs. time; (b) state space; (c) Poincaré section sampled at times that are integer multiples of the drive period.*

Example 3.7 The driven damped Duffing model

The double-well potential

$$V(x) = \frac{1}{4}x^4 - \frac{1}{2}x^2 \tag{3.76}$$

is frequently encountered in dynamical systems, exhibiting two degenerate minima. In the absence of damping, or a driving term, it has the dynamical equation

$$\ddot{x} - x + x^3 = 0. \tag{3.77}$$

When damped and driven harmonically, this produces the flow

$$\begin{aligned} \dot{x} &= y \\ \dot{y} &= a \sin z - by + x - x^3 \\ \dot{z} &= 1, \end{aligned} \tag{3.78}$$

which can show chaotic motion for certain ranges of the parameters. (This model is explored in the exercises.)

3.4.3 3D autonomous dynamical models

The previous examples of 3D chaotic systems acquired their third dimension through a time-dependent drive on an otherwise non-autonomous flow. In this section, autonomous flows in 3D can show self-sustained chaotic behavior. These flows are similar to the autonomous van der Pol oscillator that self-oscillates without any harmonic drive, where the system trajectories are limit cycles that are 1D submanifolds of the 2D dynamical space. In the case of 3D autonomous chaotic systems, the system trajectories are likewise limited to submanifolds of the full dynamical space. However, these attracting submanifolds are not limit cycles, but instead are submanifolds with complex geometry and possibly fractional dimensions (fractals).

Example 3.8 The Lorenz model

Perhaps the most famous nonlinear dynamical system of three variables is the Lorenz model proposed by Edward Lorenz in 1963 for convective systems

$$\begin{aligned} \dot{x} &= p\,(y-x) \\ \dot{y} &= rx - xz - y \\ \dot{z} &= xy - bz, \end{aligned} \tag{3.79}$$

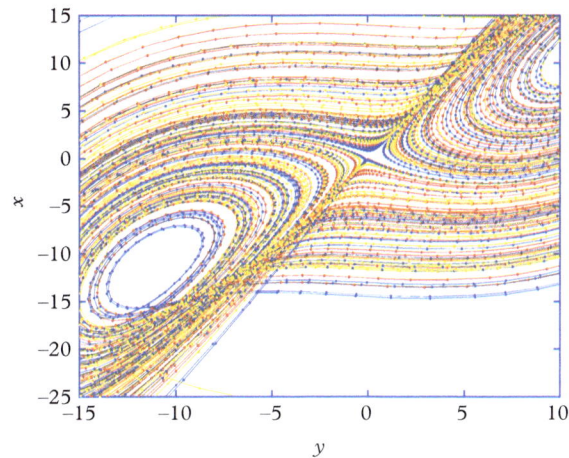

Figure 3.21 *The x–y projection of numerous trajectories on the attracting submanifold for the Lorenz system.*

where all the variables are dimensionless. They can be roughly identified in terms of hydrodynamic flow, where x is the stream function, y is the change in temperature in a convection cell, and z is the deviation from linear temperature variation in the cell. This model contains the barest essentials of buoyancy and Navier Stokes and is subject to instability for high Rayleigh number. A numerical solution to the equations is shown as an x–y projection in Fig. 3.21 for $p = 10$, $r = 100$, and $b = 8/3$. This figure shows part of the famous "Lorenz Butterfly," centered on one of the fixed points for the system at $(0, 0)$ which is a saddle point.

The time-evolution of the system trajectory is chaotic, never repeating in exactly the same way, but also never deviating away from the butterfly. The butterfly is known as a *strange attractor*. It has zero volume, but is embedded within the 3D state space. All initial conditions converge onto the butterfly, but never in the same way, and two neighboring points on the butterfly diverge from each other exponentially. A strange attractor is obviously not a fixed point, but it is nevertheless a steady property of the dynamical system. The existence of a strange attractor is a hallmark of chaos.

Example 3.9 Rössler system

A modification of the Lorenz system, but with simpler nonlinearity, was proposed in 1976 by O. E. Rössler as a system that also showed strong chaotic behavior. The flow is

$$\dot{x} = -y - z$$
$$\dot{y} = x + ay \tag{3.80}$$
$$\dot{z} = b + z(x - c).$$

continued

Example 3.9 *continued*

Note that the only nonlinearity is in the zx term of the last equation. The phase properties of this system are controlled by the parameter a. For instance, when $a = 0.15$, even though the system is chaotic, the phase of the system is well-defined as

$$\phi = \arctan\left(\frac{y}{x}\right) \tag{3.81}$$

and evolves regularly in time, while at $a = 0.3$ the system has no regular phase, as shown in Fig. 3.22. This model is useful when studying synchronization of chaotic systems because synchronization properties can be contrasted between the phase-coherent and the phase-incoherent regimes.

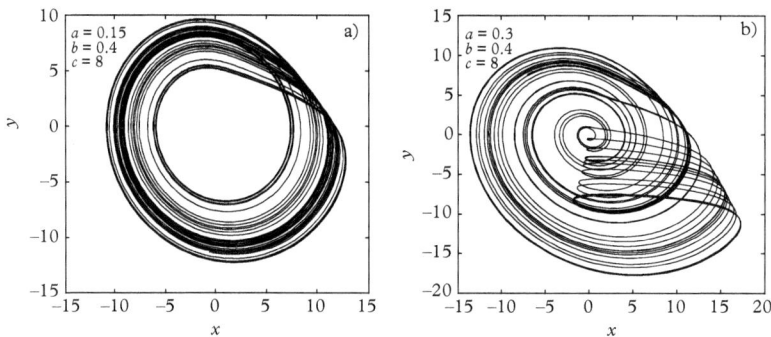

Figure 3.22 *Projections of the Rössler attractor on the x–y plane for (a) a = 0.15 and (b) a = 0.3. The phase is well-defined in the first case, but not the second.*

3.4.4 Evolving volumes in state space

The dynamical flow equations can be used to show how phase-space volume changes as a dynamical system evolves. For a flow

$$\dot{\vec{x}} = \vec{f}(\vec{x}) \tag{3.82}$$

the volume change is

$$V(t + dt) = V(t) + dt \int_S \vec{f} \cdot \hat{n}\, dS \tag{3.83}$$

evaluated on the surface S enclosing the original conditions, from which

$$\frac{dV}{dt} = \frac{V(t+dt) - V(t)}{dt} = \int_S \vec{f} \cdot \hat{n}\, dS = \int_S \vec{f} \cdot d\vec{S} = \int_V \vec{\nabla} \cdot \vec{f}\, dV \tag{3.84}$$

that uses the divergence theorem. Therefore

$$\dot{V} = \int_V \vec{\nabla} \cdot \vec{f} dV. \tag{3.85}$$

By finding the divergence of the flow, it is possible to find if the phase-space volume is increasing or decreasing.

Example 3.10 The Lorenz system

For the Lorenz system

$$\begin{aligned} \dot{x} &= \sigma(y-x) & f_1 &= \sigma(y-x) \\ \dot{y} &= rx - y - xz & f_2 &= rx - y - xz \\ \dot{z} &= xy - bz & f_3 &= xy - bz \end{aligned} \tag{3.86}$$

and

$$\begin{aligned} \vec{\nabla} \cdot \vec{f} &= \frac{\partial f_1}{\partial x} + \frac{\partial f_1}{\partial y} + \frac{\partial f_1}{\partial z} \\ &= -\sigma - 1 - b < 0 \end{aligned} \tag{3.87}$$

for $\sigma, b > 0$. The volume rate of change is then

$$\dot{V} = -(\sigma + 1 + b)\, V \tag{3.88}$$

and the volume changes according to

$$V(t) = V(0)\, e^{-(\sigma + 1 + b)t}. \tag{3.89}$$

This example illustrates that the volume of phase space shrinks exponentially to zero as $t \to \infty$. Note that this implies that the volume of the attractor is zero, or more specifically, that the dimension of the attractor is smaller than the dimension of the embedding space. This contraction in the volume is specifically caused by dissipation.

3.5 Fractals and strange attractors

In 2D dissipative systems, there can be stable nodes and limit cycles that attract all nearby trajectories and to which all nearby trajectories relax exponentially. When the dimension of the state space expands to 3D, there can still be stable nodes and limit cycles, but in addition there now are attracting sets imbedded in the state space. These sets tend to be diffuse collections of points with varying density, including some regions where no points occur at all. These sets are embedded in the dynamical space, but do not fill it, with a measured volume that is vanishingly small. When trajectories relax to these stable but ethereal sets, they are called strange attractors. Strange attractors are submanifolds of reduced, but not necessarily of integer, dimensions. A manifold of fractional dimension is called a fractal. Many chaotic attractors are fractal.

Fractals are self-similar structures that look the same at all scales (e.g., see Fig. 3.15). If one zooms in on a small region of a fractal, it looks just the same as it did on a larger scale. One example of a fractal, known as the Koch curve, is shown in Fig. 3.23. The Koch curve begins in generation 1 with $N_0 = 4$ elements. These are shrunk by a factor of $b = 1/3$ to become the four elements of the next generation, and so on. The number of elements varies with the observation scale according to the equation

$$N(b) = N_0 b^{-D}, \tag{3.90}$$

where D is called the fractal dimension. In the example of the Koch curve, the fractal dimension is

$$D = -\frac{\ln(N_0)}{\ln(1/b)} = \frac{\ln(4)}{\ln(3)} = 1.26, \tag{3.91}$$

Koch curve

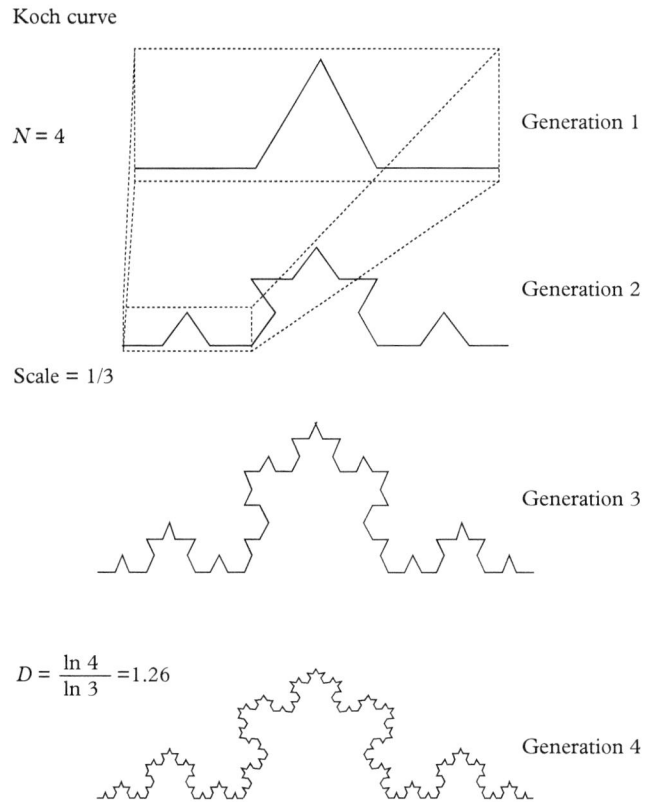

$N = 4$

Generation 1

Scale = 1/3

Generation 2

Generation 3

Figure 3.23 *Generation of a Koch curve. The fractal dimension is D = ln(4)/ln(3) = 1.26. At each stage, four elements are reduced in size by a factor of 3. The "length" of the curve approaches infinity as the features get smaller and smaller. But the scaling of the length with size is determined uniquely by the fractal dimension.*

$$D = \frac{\ln 4}{\ln 3} = 1.26$$

Generation 4

Box counting

$$D = \frac{\ln 4}{\ln 3} = 1.26$$

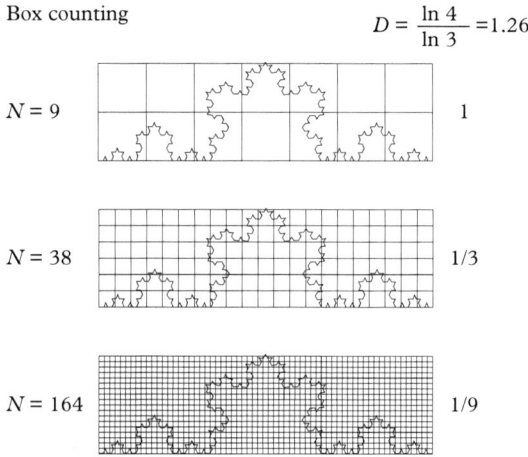

$N = 9$ 1

$N = 38$ 1/3

$N = 164$ 1/9

Figure 3.24 *Calculation of the fractal dimension using box counting. At each generation, the size of the grid is reduced by a factor of 3. The number of boxes that contain some part of the fractal curve increases as $N \propto b^{-D}$, where b is the scale.*

which is a number less than 2. The fractal is embedded in 2D, but has a fractional dimension that is greater than $D = 1$ for a line, but less than $D = 2$ for a plane.

From the construction of the Koch curve, it is clear that the number of elements of scale b varies as b^{-D}. However, when confronted by a fractal of unknown structure, one of the simplest methods to find the fractal dimension is through box counting. This method is shown in Fig. 3.24. The fractal set is covered by a set of boxes of size b, and the number of boxes that contain at least one point of the fractal set are counted. As the boxes are reduced in size, the number of covering boxes increases as b^{-D}. To be numerically accurate, this method must be iterated over several orders of magnitude. The number of boxes covering a fractal has this characteristic power law dependence, as shown in Fig. 3.25, and the fractal dimension is obtained as the slope.

The analogy of the scaling of a fractal with fractional dimension is sometimes more easily visualized from the way that the mass enclosed in a surface increases as the surface expands. For instance in systems with Euclidean dimension E, the mass enclosed by a "sphere" of radius r increases as r^E. The mass of a line increases as r, of a plane as r^2 and of a volume as r^3. In these cases the mass dimension is equal to the Euclidean dimension. This idea of a mass dimension is generalized easily to fractal objects, such as the aggregate shown in Fig. 3.26. This aggregate is known as a diffusion-limited aggregate (DLA), and sometimes occurs in colloidal systems. As the radius of the circle increases, the mass within the circle increases as the power $r^{1.7}$. Therefore, this DLA has a mass exponent that behaves as if it were a fractional-dimensional object with fractal dimension $D = 1.7$.

Another type of fractal is known as a *dust*. Fractal dusts are common for chaotic discrete maps that consist of sets of points. An example of a fractal dust is shown in Fig. 3.27. The iterative fractal construction begins with $N = 5$ sections

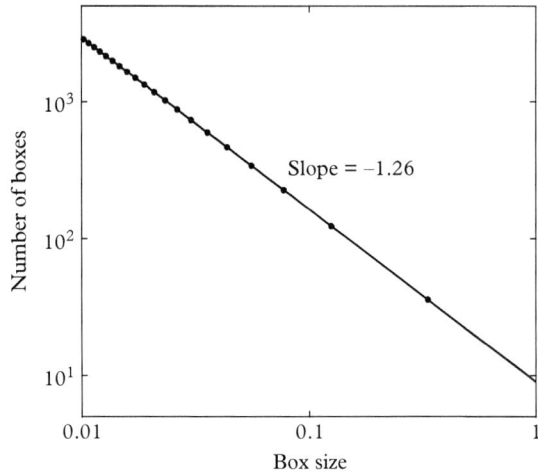

Figure 3.25 *Number of covering boxes as a function of box size. The negative slope is the fractal dimension D = 1.26.*

Diffusion-limited aggregate (DLA)

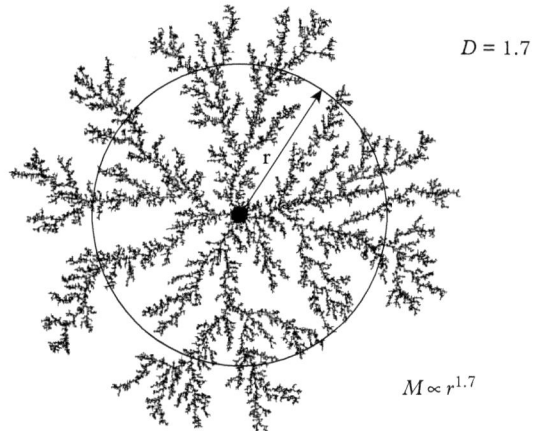

$D = 1.7$

$M \propto r^{1.7}$

Figure 3.26 *A diffusion-limited aggregate (DLA) is a mass fractal with fractal dimension D = 1.7.*

of a square, each with side length equal to 3 units. At the next iteration, each occupied sub square is converted again to $N = 5$ smaller sub squares, with a scale ratio of 3. When this process goes to infinity, what remains is a fractal dust. It is a set of points embedded in 2D. The fractal dimension in this example is equal to

$$D = \frac{\ln 5}{\ln 3} = 1.465. \qquad (3.92)$$

The fractal dimensions of several common chaotic attractors are: Lorenz attractor $D = 2.06$, Rössler attractor $D = 2.01$, Logistic map $D = 0.538$, and Hénon map $D = 1.26$. Each fractal structure has an Euclidean embedding dimension $E \geq D$.

Fractal dust $\qquad D = \dfrac{\ln 5}{\ln 3} = 1.465$

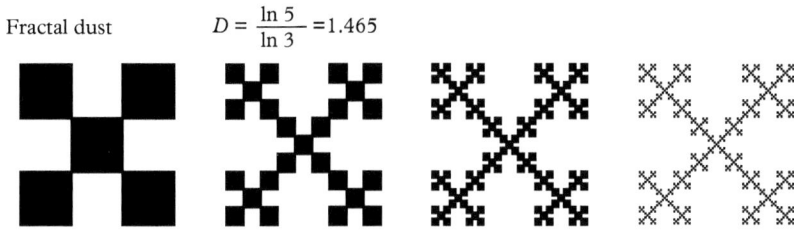

Figure 3.27 *Example of a fractal "dust." The Euclidean dimension is E = 2, but the fractal dimension is 1.465. In the limit of infinite generations, the area of points goes to zero, but the mass dimension approaches the fractal dimension. Strange attractors of a Poincaré section have the character of a dust.*

3.6 Hamiltonian chaos

Hamiltonian chaos occurs in non-integrable conservative systems with two or more degrees of freedom. Hamiltonian chaos differs from dissipative chaos chiefly because phase-space volume is conserved as the system evolves, and attractors cannot exist. There are no "transients" in Hamiltonian systems.

For a Hamiltonian system the state space is the phase space. The space has $2N$ dimensions defined by the set

$$\{q_1, p_1, q_2, p_2, \ldots, q_N, p_N\} \,. \tag{3.93}$$

Hamilton's equations are

$$\frac{dq_a}{dt} = \frac{\partial H(q, p)}{\partial p_a}$$
$$\frac{dp_a}{dt} = -\frac{\partial H(q, p)}{\partial q_a} \,. \tag{3.94}$$

Because there is no dissipation, the volume in phase space occupied by a set of points cannot change. For instance, consider a 1D system

$$\frac{1}{V}\frac{dV}{dt} = \vec{\nabla} \cdot \vec{f} = \frac{\partial f_1}{\partial x_1} + \frac{\partial f_2}{\partial x_2}$$
$$= \frac{\partial}{\partial q_1}\left(\frac{\partial H}{\partial p_1}\right) + \frac{\partial}{\partial p_1}\left(-\frac{\partial H}{\partial q_1}\right) \tag{3.95}$$
$$= \frac{\partial^2 H}{\partial q_1 \partial p_1} - \frac{\partial^2 H}{\partial p_1 \partial q_1} = 0,$$

which is easily generalized to $2N$ dimensions. Therefore, phase space volume is conserved for a Hamiltonian system as required by Liouville's theorem. The impossibility of attractors raises interesting questions about the nature of chaos in conservative, i.e., Hamiltonian, systems in contrast to dissipative systems.

3.6.1   Area-preserving maps

Just as Hamiltonian flows preserve volume in phase space, discrete maps of these systems conserve areas on the Poincaré plane. These 2D iterative maps are called area-preserving maps. The ratio of infinitesimal areas between original and transformed coordinates is equal to the determinant of the Jacobian matrix of the transformation. If the Jacobian is equal to ± 1, then areas are preserved under the transformation.

For example, consider the Hénon map

$$f(x, y) = 1 - Cx^2 + y$$
$$g(x, y) = Bx \tag{3.96}$$

with the Jacobian matrix

$$\mathfrak{J} = \begin{pmatrix} -2Cx & 1 \\ B & 0 \end{pmatrix} \tag{3.97}$$

that has $\det(\mathfrak{J}) = -B$. The Hénon map is area-preserving when $B = \pm 1$. Because the map is non-dissipative, there is no strange attractor. Instead, every initial condition leads to a distinct set of points, called orbits. Orbits accessible to one initial condition are not accessible to other initial conditions. Therefore, to map out the behavior of the area-preserving map, it is necessary accumulate many orbits for many initial conditions.

A map with beautifully complex structure is the Lozi map

$$x_{n+1} = 1 + y_n - C|x_n|$$
$$y_{n+1} = Bx_n, \tag{3.98}$$

which is area-preserving when $B = -1$. A portion of the phase plane is shown in Fig. 3.28 for $B = -1$ and $C = 0.5$. The map used 200 randomly selected initial conditions, and the orbits for each initial condition were iterated 500 times. In the figure, there are orbits within orbits. Some orbits are regular, and others are chaotic.

3.6.2   The standard map

The relationships between continuous dynamical systems and their discrete maps are sometimes difficult to identify. However, a familiar discrete map, called the standard map, arises from a randomly kicked dumbbell rotator. The system has angular momentum $\mathfrak{J}$ and a phase angle ϕ. The strength of the angular momentum kick is given by the perturbation parameter ε, and the torque of the kick is a function of the phase angle ϕ. The system is characterized by the discrete map

$$\mathfrak{J}_{n+1} = \mathfrak{J}_n + \varepsilon \sin \phi_n$$
$$\phi_{n+1} = (\phi_n + \mathfrak{J}_{n+1}) \bmod 2\pi \tag{3.99}$$

Figure 3.28 *A portion of the iterated Lozi map for B = −1 and C = 0.5. Each set of points with the same hue is from a separate orbit.*

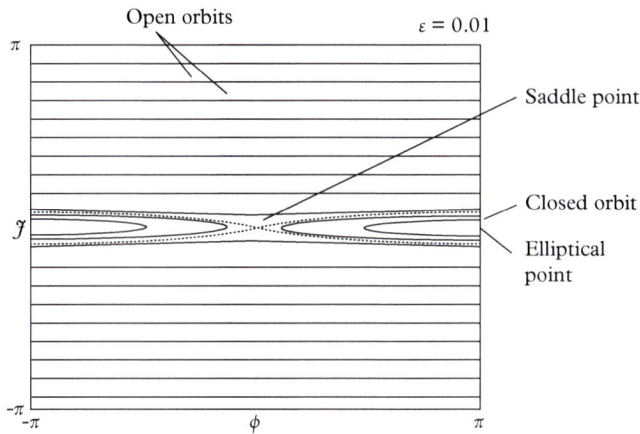

Figure 3.29 *The (φ, 𝒥) space of the standard map for at ε = 0.01. Most orbits are open (free rotation of the rotator), but a hyperbolic saddle point has appeared even at this low perturbation that separates closed orbits (oscillations).*

in which the rotator is "strobed," or observed, at regular periods of 2π. When $\varepsilon = 0$, the orbits on the $(\phi, \mathcal{J})$ plane are simply horizontal lines—the rotator spins with regular motion at a speed determined by $\mathcal{J}$. These horizontal lines represent open (freely rotating) orbits (compare with Fig. 2.13). As the perturbation ε increases slightly from zero, as shown in Fig. 3.29, the horizontal lines are perturbed, and closed orbits open up near $\mathcal{J} = 0$. There is a hyperbolic fixed point (a saddle point) in the center $(0, 0)$, and a single elliptical fixed point (a center) at the edges $(-\pi, 0)$ and $(\pi, 0)$. As the perturbation increases further, more closed orbits appear around $\mathcal{J} = 0$ and orbits begin to emerge at other values of $\mathcal{J}$.

There is an interesting pattern to the increasing emergence of closed orbits (and the eventual emergence of chaos at large perturbation). This can be seen

Winding
number

Hyperbolic point

Open orbits

Closed orbits

$\mathcal{J}$

θ

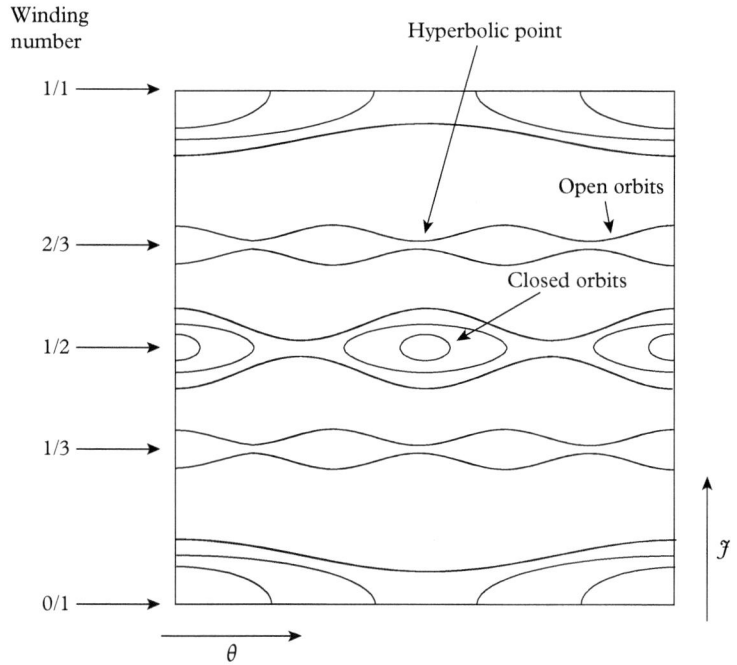

Figure 3.30 *Open and closed orbits in the standard map. This example is at low perturbation where the fundamental and 1:2 orbits have separated into closed orbits. The 2:3 and 1:3 orbits are still open. The open orbits convert to closed orbits when the hyperbolic points touch.*

most easily by rescaling the $(\phi, \mathcal{J})$ axes to plot winding number versus angle, as shown in Fig. 3.30. The standard map represents the Poincaré section of perturbed motion on a torus in (θ, ϕ). The winding number is the ratio of the periods of θ and ϕ. When these periods have the ratio of small numbers, like 1:2 or 2:3, these are called "resonances." Orbits like these are highly sensitive to the perturbation, because a change in the period of one variable easily phase-locks with a change in the other variable. This is a type of synchronization, a topic that will be explored extensively in Chapter 5. In Fig. 3.30, the 0:1, 1:1, and 1:2 winding numbers are shown with closed orbits. The orbits around winding numbers 1:3 and 2:3 are still open, but are trending towards creating new hyperbolic fixed points with increasing ε, when new closed orbits will appear. As the perturbation increases further, open orbits around winding number ratios of successively larger integer fractions convert to chains of closed orbits that look like islands, where the number of islands in a chain is equal to the denominator of the fraction. This break-up of the open orbits for successive resonances is the basis of a famous theorem in nonlinear physics, known as the KAM theorem (named after Kolmogorov, Arnold and Moser). Based on this theorem, the stability of an orbit against break-up is determined by its distance from small integer resonances. The most stable orbit—the one farthest from small-integer ratios and hence the one that is "most" irrational—is the orbit with a winding number equal to the golden mean $\phi = 1.618$.

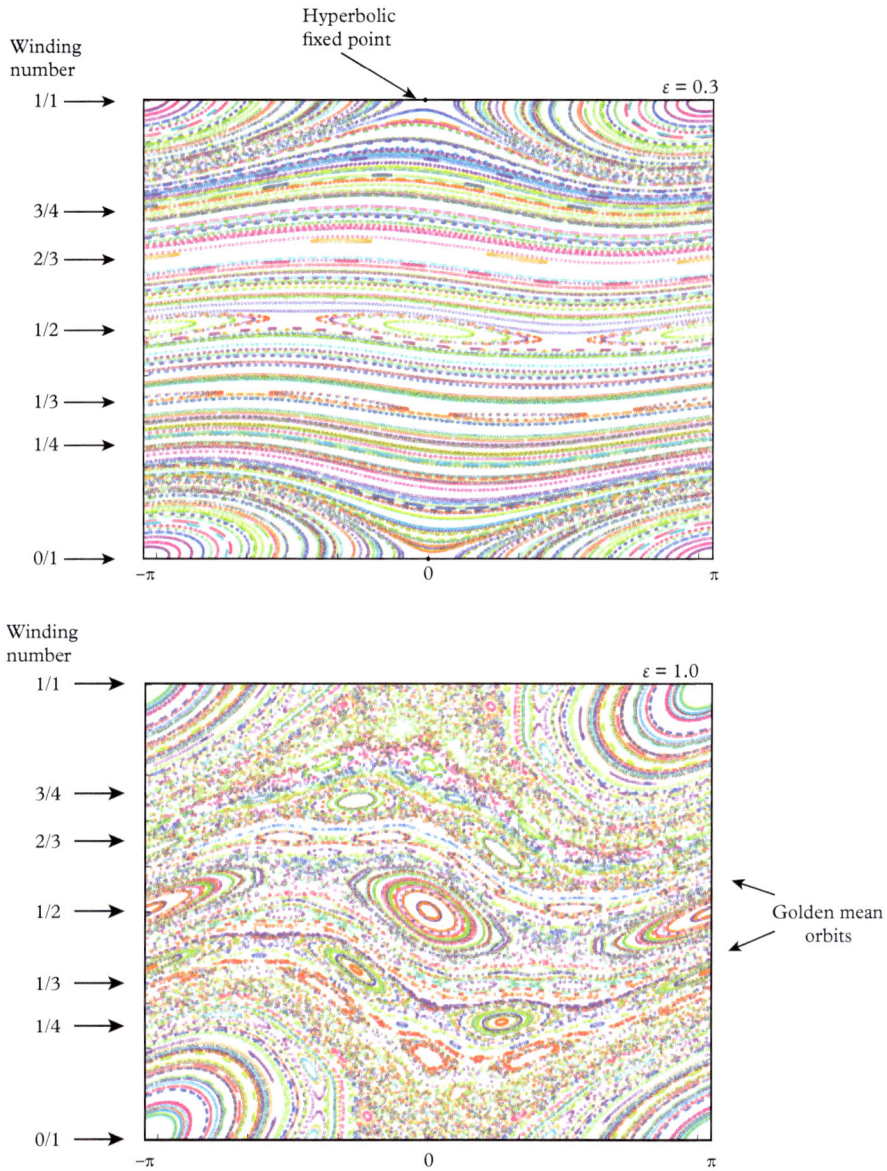

Figure 3.31 *Simulations of the standard map for (a) ε = 0.3 and (b) ε = 1.0. For weak perturbation, the low resonances have separated into closed orbits with many open orbits remaining. For strong perturbation, there is fully developed chaos that evolved from the fundamental hyperbolic fixed point, with many resonances.*

The cascade of break-ups of open orbits into chains of islands with increasing perturbation strength eventually leads to chaos. Chaos emerges first at the hyperbolic saddle point of the lowest resonance 0:1 and 1:1. From computer simulations, the threshold is $\varepsilon_c = 0.97$. Examples of simulations are shown in Fig. 3.31 for $\varepsilon = 0.3$ and $\varepsilon = 1.0$. For $\varepsilon = 0.3$, there are open and closed orbits that have opened at winding numbers 0:1, 1:4, 1:3, 1:2, 2:3, 3:4, and 1:1. The primary hyperbolic fixed point in this plot is at $(0, 0)$ and $(0, 1)$. For $\varepsilon = 1.0$, many more open orbits have broken up, and chaos has emerged at the primary hyperbolic point, as well as at secondary and higher hyperbolic points. With even further increase in the perturbation, eventually only the primary elliptic point will remain as an island in a sea of chaos. Finally, even this disappears.

3.6.3   Integrable vs. non-integrable Hamiltonian systems

In an integrable system the number of constants of motion is equal to the number of degrees of freedom, and no chaos is possible. However, if a non-integrable perturbation is added to an integrable system with state-space dimension equal to or higher than 3D, then chaos *is* possible. Constants of motion occur when the Hamiltonian is cyclic in a coordinate q^a:

$$\dot{p}_a = -\frac{\partial H}{\partial q^a} = 0, \tag{3.100}$$

for which the Hamiltonian is not explicitly a function of the q^a. However, the p_a often are not the constants of motion, and appropriate canonical transformations are needed to define the action variables $\mathcal{J}_a$ that are the constants of motion defined by

$$\dot{\theta}_a = \frac{\partial H(\theta, \mathcal{J})}{\partial \mathcal{J}_a}$$
$$\dot{\mathcal{J}}_a = -\frac{\partial H(\theta, \mathcal{J})}{\partial \theta_a}. \tag{3.101}$$

Integrable systems live on invariant tori in phase space. For example, a two-degrees of freedom system has a four-dimensional (4D) phase space. There are two constants of motion that are cyclic. The constants are $\omega_1 = \frac{\partial H}{\partial \mathcal{J}_1}$ and $\omega_2 = \frac{\partial H}{\partial \mathcal{J}_2}$. The trajectories reside on a series of nested 2D tori with a separate torus for each pair (ω_1, ω_2). The intersection of this torus with the Poincaré plane defines an ellipse. There are different ellipses for different constants of motion, creating a set of infinitely nested ellipses.

When a "small" non-integrable Hamiltonian is added as a perturbation to an integrable Hamiltonian, the trajectories start to move off the invariant tori. The tori dissolve in succession as the non-integrable part increases. When the tori dissolve, every different initial condition produces a unique trajectory that may have chaotic first-returns or regular first-returns. The complete set of trajectories, however, may separate into distinct probability regions in the Poincaré section, with some regions having vanishingly low probabilities of being occupied and others having high. In general, the high-probability regions are the remnants of the invariant tori, and the low-probability regions are islands enclosed by the remnant tori.

Non-integrable terms can be added to an integrable Hamiltonian through a perturbation as

$$H_{Tot} = H_{int} + \varepsilon H_{nonint} \qquad (3.102)$$

and the motion can be studied as the perturbation strength increases. Alternatively, terms that are higher-order in the variables can be added to the integrable Hamiltonian. For small energies, these terms are small, but with increasing total energy the state point increasingly probes the non-integrable higher-order terms of the potential and causes trajectories to move off the invariant tori. This is the situation for the Hénon–Heiles Hamiltonian.

3.6.4   The Hénon–Heiles Hamiltonian

A non-integrable Hamiltonian system that displays Hamiltonian chaos is the Hénon–Heiles Hamiltonian model of star motion in the plane of a galaxy. The Hamiltonian is

$$H = \frac{1}{2}p_x^2 + \frac{1}{2}x^2 + \frac{1}{2}p_y^2 + \frac{1}{2}y^2 + \left[x^2y - \frac{1}{3}y^3\right] \qquad (3.103)$$

with a potential function

$$V(x, y) = \frac{1}{2}x^2 + \frac{1}{2}y^2 + x^2y - \frac{1}{3}y^3. \qquad (3.104)$$

This is a harmonic potential with additional cubic terms that break azimuthal symmetry. For small total energies, the harmonic terms dominate because the small-amplitude motions keep the cubic contributions small. With increasing energy, the potential becomes successively less harmonic and less integrable, generating more complex motion.

The equations for the 4D flow arise from Hamilton's equations

$$
\begin{aligned}
\dot{x} &= \frac{\partial H}{\partial p_x} = p_x \\
\dot{y} &= \frac{\partial H}{\partial p_y} = p_y \\
\dot{p}_x &= -\frac{\partial H}{\partial x} = -x - 2xy \\
\dot{p}_y &= -\frac{\partial H}{\partial y} = -y - x^2 + y^2.
\end{aligned}
\qquad (3.105)
$$

The four dimensions are reduced by one because energy conservation provides a single constraint. The Poincaré map visualizes the system in two dimensions. The Poincaré map is established on the y–p_y plane. The constant energy is then

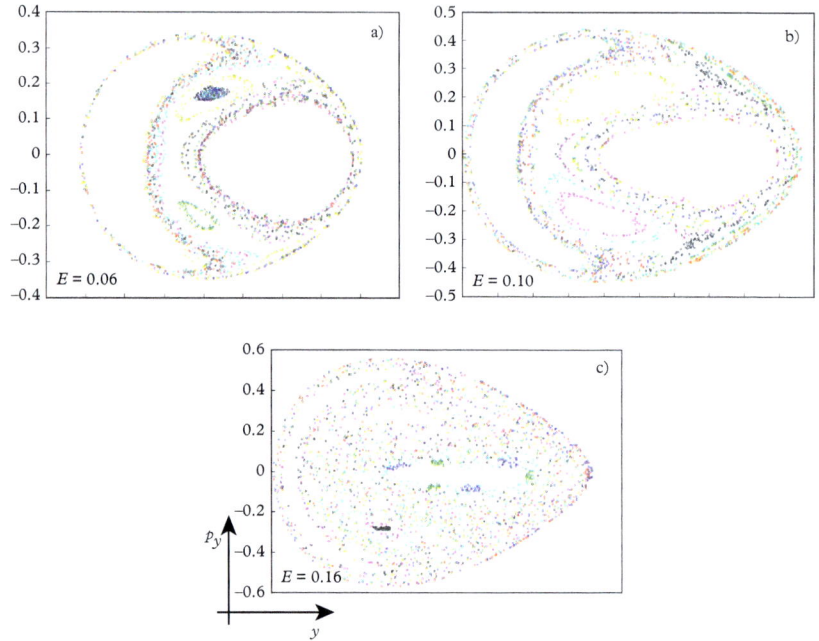

Figure 3.32 *(a) Poincaré section of the Hénon–Heiles model for E = 0.06. (b) Poincaré section of the Hénon–Heiles model for E = 0.10. (c) Poincaré section of the Hénon–Heiles model for E = 0.16.*

$$E = \frac{1}{2}p_x^2 + \frac{1}{2}p_y^2 + \frac{1}{2}y^2 - \frac{1}{3}y^3. \tag{3.106}$$

Examples of the 2D Poincaré sections for the Hénon–Heiles Hamiltonian are shown in Fig. 3.32 for constant energies $E = 0.06, 0.10,$ and 0.16. At low energy, there are several recognizable invariant nested tori that survive from the integrable Hamiltonian in the low-energy limit. However, as the energy increases, and the non-integrability increases, these disintegrate into regions of diffuse sets of points. At the highest energy, there are few remnants of the invariant tori.

Hamiltonian chaos is distinct from the dissipative chaos that occurs in systems like the Lorenz model. There is no dissipation in Hamiltonian chaos and hence there are no attractors, either regular or strange, as each initial condition leads to a distinct orbit. Instead, the structure of the Poincaré section of an area-preserving map reflects the density of states within phase space. Some regions of phase space are visited regularly, while others can be visited only with low probability. Hamiltonian chaos lies at the border between macroscopic physics and microscopic physics. Many quantum systems have no intrinsic dissipation and are described by Hamiltonians. Trajectories in quantum systems are not distinct as in classical systems, and are characterized by constructive and destructive interference of standing waves. Nonetheless, potentials that exhibit classical chaos in the macroscopic regime tend to show wild interference patterns, known as quantum scars, in the quantum regime. In this way, quantum chaos carries over continuously to classical chaos, and the classical and quantum systems share many aspects in common.

3.7 Summary and glossary

Nonlinear dynamics and *chaos* represent a large field for both students and researchers and many good books are devoted entirely to the topic. In this chapter, the barest essentials were presented, based on the geometric structure of *phase portraits* and *stability analysis* that classifies the character of *fixed points* based on the properties of *Lyapunov exponents*. *Limit cycles* are an important type of periodic orbit that occur in many nonlinear systems, and their stability is analyzed according to *Floquet multipliers*. Chaos is not possible in 2D state space because of the non-crossing theorem. However, if a time dependence is added to an *autonomous* 2D state space system to make it *non-autonomous*, then chaos can emerge, as in the case of the *driven damped pendulum*. Autonomous systems with 3D state spaces can exhibit chaos, with famous examples in the *Lorenz* and *Rössler* models. Dissipative chaos often has *strange attractors* with *fractal dimensions*. Non-dissipative Hamiltonian systems are also capable of chaos, but of a different character than dissipative systems. *Liouville's theorem* guarantees that the volume of phase space remains constant, but the phase-space volume can be twisted and folded, as in *area-preserving maps* like the *standard map*. *Integrable* Hamiltonians with invariant tori in phase-space (of dimension greater than two) show no chaos, but when non-integrable terms are added to the potential function, chaos emerges. Stronger perturbations cause break-up of more of the invariant tori according to the *KAM theory*.

Nonlinear dynamics has a language all its own. Here is a glossary to help decipher texts on the subject.

Attractor. A lower-dimensional hypersurface in state-space to which the trajectories of a dissipative dynamical system relax.

Autonomous. A flow that has no explicit time dependence.

Basin. A sub-region of state-space containing a set of trajectories that remain within that sub-region.

Bifurcation. A sudden change of a system behavior caused by an infinitesimal change of a control parameter.

Degrees of freedom. The number of conditions needed to define a trajectory. Also, for a general flow, it is the number of independent variables.

Fixed point. A position in state-space of a flow that satisfies $\dot{\vec{x}} = \vec{f}(\vec{x}) = 0$.

Floquet multiplier. Characteristic values of the Jacobian matrix of the first return map.

Flow. A multi-dimensional dynamical system that has the equation $\dot{\vec{x}} = \vec{f}(\vec{x})$.

Fractal. A geometric set of points that is self-similar at all scales. Fractals scale as if they had fractional dimension (hence the term *fractal*).

Heteroclinic. Two orbits that cross in state space (in infinite time).

Homoclinic. An orbit that crosses itself in state space (in infinite time).

Integrable. A Hamiltonian system with as many constants of motion as the number of independent generalized coordinates.

Invariant tori. N-dimensional sub-surfaces in $2N$-dimensional phase space on which trajectories of integrable Hamiltonian systems reside.

Limit cycle. A periodic attractor in state space.

Lyapunov exponent. Characteristic values of the Jacobian matrix that define the stability of fixed points and the rate of separation of nearby trajectories.

Manifold. A lower-dimensional subspace of state space. The stable manifold of a saddle point is the trajectory of the eigenvector of the Jacobian matrix that has a negative Lyapunov exponent. The unstable manifold of a saddle point is the trajectory of the eigenvector of the Jacobian matrix that has a positive Lyapunov exponent.

Node. A fixed point with either all negative (stable attractor) or all positive (unstable repellor) Lyapunov exponents.

Non-autonomous. A flow with explicit time dependence.

Non-integrable. A Hamiltonian system with more degrees of freedom than constants of motion. Non-integrable systems can exhibit Hamiltonian chaos.

Nullcline. A manifold in state space along which one of the time derivatives equals zero.

Phase space. A special case of state space for Hamiltonian or Lagrangian systems consisting of $2N$ dimensions for N independent generalized coordinates.

Poincaré section. A set of points on a lower-dimensional surface within state space that intersects dynamical trajectories.

Repellor. An unstable fixed point with all positive Lyapunov exponents.

Saddle point. A fixed point with negative and positive Lyapunov exponents.

Spirals. Attractor or repellor with complex Lyapunov exponents.

Stars. An attractor or a repellor with equal Lyapunov exponents.

State space. A space whose dimension equals the number of explicit variables of a dynamical flow and within which the trajectories of the dynamical system reside.

3.8 Bibliography

D. Acheson, *From Calculus to Chaos* (Oxford University Press, 1997).

K. T. Alligood, T. D. Sauer and J. A. Yorke, *Chaos: An Introduction to Dynamical Systems* (Springer, 1997).

D. Arrowsmith and C. M. Place, *Dynamical Systems: Differential Equations, Maps and Chaotic Behavior* (Chapman Hall/CRC, 1992).

J. Barrow-Green, Poincaré and the three-body problem, In *History of Mathematics*, Vol. 11 (American Mathematical Society, 1997).

Brian Davies, *Exploring Chaos* (Perseus Books, 1999).

Robert C. Hilborn, *Chaos and Nonlinear Dynamics* (Oxford University Press, 2004).

Philip Holmes, Poincaré, celestial mechanics, dynamical systems theory and chaos, *Physics Reports*, vol. 193, pp. 137–163, 1990.

A. E. Jackson, *Perspectives of Nonlinear Dynamics* (Cambridge University Press, 1989).

E. N. Lorenz, *The Essence of Chaos* (University of Washington Press, 1993).

A. Medio, *Chaotic Dynamics: Theory and Applications to Economics* (Cambridge University Press, New York, 1992).

E. Ott, *Chaos in Dynamical Systems* (Cambridge University Press, 1993).

C. Robinson, *Dynamical Systems: Stability, Symbolic Dynamics and Chaos* (CRC Press, 1995).

Hans-Jürgen Stöckmann, *Quantum Chaos: An Introduction* (Cambridge University Press, 1999).

Steven H. Strogatz, *Nonlinear Dynamics and Chaos* (Westview, 1994).

3.9 Homework exercises

Analytic problems

1. **Second-order dynamics to coupled first-order:** Convert the van der Pol oscillator equation

$$\ddot{x} = 2\mu\dot{x}\left(1 - \beta x^2\right) - \omega_0^2 x$$

   into two coupled first-order flows.

2. **Fixed-point classification:** Find the nullclines and fixed points of the following 2D dynamics. Analyze the fixed points and classify them.

   (a) $\dot{x} = x(x-y)$; $\dot{y} = y(2x-y)$ (b) $\dot{x} = x - x^3$; $\dot{y} = -y$ (c) $\dot{x} = x^2 - y$; $\dot{y} = x - y$ (d) $\dot{x} = x(2-x-y)$; $\dot{y} = x - y$

3. **Phase portraits:** Sketch the nullclines and phase portraits of each of the following (Do not use a computer . . . sketch by hand). Identify and classify all fixed points.

   (a) $\dot{x} = x - y ; \dot{y} = 1 - e^x$ (b) $\dot{x} = x - x^3 ; \dot{y} = -y$
   (c) $\dot{x} = x(x-y) ; \dot{y} = y(2x-y)$ (d) $\dot{x} = y ; \dot{y} = x(1+y) - 1$
   (e) $\dot{x} = x(2-x-y) ; \dot{y} = x - y$ (f) $\dot{x} = x^2 - y ; \dot{y} = x - y$
   (g) $\dot{x} = xy - 1 ; \dot{y} = x - y^3$ (h) $\dot{x} = \sin y ; \dot{y} = x - x^3$

4. **Gravitational equilibrium:** A particle of unit mass moves along a line joining two stationary masses, m_1 and m_2, which are separated by a distance a. Find the particle equilibrium position and stability based on a linearized analysis of

$$\ddot{x} = \frac{Gm_2}{(x-a)^2} - \frac{Gm_1}{x^2}.$$

Sketch the phase portrait by hand (do not use a computer).

5. **Linearization:** Classify the fixed point at the origin for the system

$$\dot{x} = -x^3 - y$$
$$\dot{y} = x.$$

Linearization fails to classify it correctly, so prove otherwise that it is a spiral.

6. **Centers as marginal cases:** Convert the equations of Example 3.2 into polar coordinates and show functionally that the solutions are spirals when $a \neq 0$.

7. **Spirals:** Consider the two systems

$$\dot{r} = r(1-r^2) \qquad \dot{r} = r(1-r)$$
$$\dot{\theta} = 1 \qquad\qquad \dot{\theta} = 1$$

   (a) Find the fixed points of the flows. Evaluate their stability. What are the dimensionalities of these systems?
   (b) Construct the Poincaré sections, find the limit-cycle fixed points, and evaluate their stabilities.

8. **Symbiotic relationship:** Modify the parameters in the expressions in parentheses of Eq. (3.29) of Example 3.3 to establish a stable node.

9. **Floquet multiplier:** Consider the vector field given in polar coordinates by

$$\dot{r} = r - r^2$$
$$\dot{\theta} = 1.$$

Find the Floquet multiplier for the periodic orbit and classify its stability. Use the positive x-axis as the section coordinate.

10. **3D fixed point:** For the Lorenz model

$$\dot{x} = p(y-x) \qquad p = 10$$
$$\dot{y} = rx - xz - y \qquad b = 8/3$$
$$\dot{z} = xy - bz \qquad r = 28$$

Analyze and classify the fixed point at $(0, 0, 0)$. What is the largest Lyapunov exponent?

11. **3D fixed point:** For the Rössler model

$$\dot{x} = -y - z \qquad\quad a$$
$$\dot{y} = x + ay \qquad\quad b = 0.4$$
$$\dot{z} = b + z(x-c) \qquad c = 8$$

Linearize and classify the fixed point at $(0, 0, 0)$ as a function of a.

12. **Fractal dust:** In the 2D fractal dust, what is the fractal dimension if
    (a) $N = 3$ and $b = 1/3$
    (b) $N = 8$ and $b = 1/3$
    (c) $N = 9$ and $b = 1/3$
13. **Hénon map:**
    (a) Find all the fixed points for the Hénon map and show they exist only for $C > C_0$ for a certain C_0.
    (b) A fixed point of a map is linearly stable if and only if all eigenvalues of the Jacobian satisfy $|\lambda| < 1$. Determine the stability of the fixed points of the Hénon map, as a function of B and C.
14. **Canonical transformation:** (a) For a simple harmonic oscillator of mass m and spring constant k, find the canonical transformation from (q, p) to $(\theta, \mathcal{J})$. Show that the phase-space trajectories are circles. (b) Show that the radius of the circular trajectory is $\sqrt{\mathcal{J}}$. (c) Show that in the original phase space the area enclosed by the ellipse is equal to $2\pi\mathcal{J}$.
15. **Hénon–Heiles:** For the Hénon–Heiles Hamiltonian, find the explicit expression for the 4D flow

$$\left(\dot{x} \quad \dot{p}_x \quad \dot{y} \quad \dot{p}_y \right) = \vec{f}\left(x, p_x, y, p_y \right).$$

Show explicitly that these equations lead to no volume contraction in phase space.

Computational projects[3]

16. **Poincaré return map:** Explore the Poincaré section for the van der Pol oscillator. Numerically find the Floquet multiplier.
17. **Andronov–Hopf bifurcation:** Explore the state space of the subcritical Hopf bifurcation.

$$\dot{r} = r\left(c + r^2 \right) \qquad \dot{\theta} = 1.$$

18. **Logistic map:** Explore the logistic map bifurcation plot. Find and zoom in on period 3 and period 5 cycles.
19. **Bifurcation maps:** Create a bifurcation map of
    (a) $x_{n+1} = r \cos x_n$
    (b) $x_{n+1} = r x_n - x_n^3$
20. **Amplitude–frequency coupling:** For the van der Pol oscillator

$$\ddot{x} = 2\mu\dot{x}\left(1 - \beta x^2 \right) - \omega_0^2 x \qquad \begin{matrix} \omega_0^2 = 1 \\ \beta = 1 \end{matrix}$$

find the period of oscillation as a function of the gain parameter μ.
21. **Driven damped pendulum:** Numerically plot the Poincaré section for the driven damped pendulum for $c = 0.05$, $F = 0.7$, and $\omega = 0.7$.

[3] Matlab programming examples are found at www.works.bepress.com/ddnolte.

22. **Duffing model:** For the damped-driven Duffing model, explore the parameters that cause the system to exhibit chaos.

23. **Driven van der Pol:** For the van der Pol equation

$$\dot{x} = y$$
$$\dot{y} = b \sin z - \mu \left(x^2 - a \right) - x$$
$$\dot{z} = 1.$$

Explore the case for $a = 1$, $\mu = 3$, and vary b from 2.5 to 3.13. Notice the sub-harmonic cascade. Is there synchronization?

24. **Lorenz butterfly:** Explore the "butterfly" by changing the value r. Can you destroy the butterfly?

25. **Chaotic phase oscillator:** Track the phase (modulo 2 pi) as a function of time for the Rössler attractor for $a = 0.15$ and again for $a = 0.3$.

26. **Evolving volumes:** For the Rössler system, evaluate the change in volume of a small initial volume of initial conditions. You will need to evaluate (average) the divergence around the trajectory because it is not a constant (unlike for the Lorenz model).

27. **Hénon map:** Show (numerically) that one fixed point is always unstable, while the other is stable for C slightly larger than C_0. What happens when $\lambda = -1$ at $C_1 = \frac{3}{4}(1 - B)^2$. Create a bifurcation diagram for the Hénon map as a function of C for $B = 0.3$.

28. **Lozi map:** Numerically explore the "Lozi map" and test the limit when $B \rightarrow -1$

$$x_{n+1} = 1 + y_n - C \, |x_n|$$
$$y_{n+1} = Bx_n$$

What do you notice in the limit?

29. **Hamiltonian chaos:** Using the methods of Poincaré sections and invariant KAM tori, explore the dynamics of trajectories of a particle subject to the following potential

$$V(x, y) = \frac{1}{2}\left(y - 2x^2\right)^2 + \frac{1}{2}x^2.$$

Explore the similarities and differences with the Hénon–Heiles model, first analytically, and then numerically.

30. **Chua's circuit:** Explore the attractor of Chua's diode circuit

$$\dot{x} = \alpha \, (y - h(x))$$
$$\dot{y} = x - y + z$$
$$\dot{z} = -\beta y,$$

where the single-variable nonlinear function $h(x)$ is

$$h(x) = m_1 x + \frac{1}{2} \, (m_0 - m_1) \, [|x + 1| - |x - 1|]$$

Look for the "double scroll" attractor.

Coupled Oscillators and Synchronization

Saturn's rings from Cassini Orbiter
Photograph from <http://photojournal.jpl.nasa.gov/catalog/PIA07873>. Image credit:
NASA/JPL.

Synchronization of the motions of separate, but interacting, systems is a ubiquitous phenomenon that ranges from the phase-locking of two pendulum clocks on a wall (discovered by Christian Huygens in 1665), to the pace-maker cells in the heart that keep it pumping and keep you alive. Other examples include the rings of Saturn, atoms in lasers, circadian rhythms, neural cells, electronic oscillators, and biochemical cycles. Synchronization cannot occur among linear coupled oscillators for which the normal modes of oscillation are linear combinations of the modes of the separate oscillators. In nonlinear oscillators, on the other hand, frequencies are "pulled" or shifted. Two different isolated frequencies can even merge, be "entrained," to a single "compromise" frequency in which both systems are frequency-locked to each other.

The previous chapter described the dynamics of individual nonlinear systems in state space. The dimensions of such systems can be large, with many degrees of freedom. The dynamic variables are all strongly coupled, and it is generally not possible to isolate a subset of some variables from the others. On the other hand, in many complex systems it is possible to identify numerous semi-autonomous subsystems contained within larger systems. For instance there are animal species within a broader ecosystem, or compartmentalized biochemical reactions within cells, or individual neurons in a neural network. The subsystems are often non-linear, and may be autonomous oscillators or even chaotic. The key element here is the identification of discrete units, the quasi-autonomous systems, within the broader interacting networks.

This concept of distinguishable systems that are coupled within wide interaction networks lies at the boundary between two worlds. On the one hand, there is the fully holistic viewpoint in which complex systems of many variables and high dimension cannot be separated into subunits without destroying the essential behavior that is to be understood. On the other hand, there is the reductionist viewpoint in which each component of the system is isolated and studied individually. Reality often lies between, in which individual systems and their properties are understood in isolation, but, when coupled, lead to emergent properties of the interacting network that are often surprising.

4.1 Coupled linear oscillators

Linear oscillators that have linear coupling cannot be synchronized. Even if all the oscillators are identical, this is a condition of neutral stability in the sense that a perturbation to the system will neither grow nor decay. The steady-state solutions for N coupled oscillators consist of a spectrum of N distinct oscillation frequencies called eigenfrequencies. Because of the coupling, the eigenfrequencies are not the same as the original uncoupled frequencies. New frequencies of the collective oscillation modes arise, but there is no overall synchronization of the phases of motion. Synchronization requires nonlinearity, as we shall see in the next section. Here we start with a study of the general behavior and properties of coupled linear oscillators.

4.1.1 Two coupled linear oscillators

A simple example of two coupled linear oscillators is composed of two masses attached by springs to two walls, with a third spring attaching the masses, as shown in Fig. 4.1.

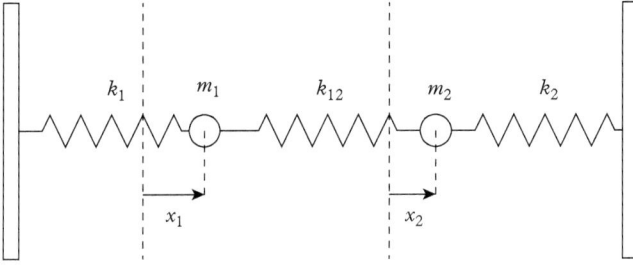

Figure 4.1 *Two equal masses on equal springs coupled by a third spring.*

If the masses $m_1 = m_2 = M$ are equal, and the spring constants attaching them to the walls $k_1 = k_2 = k$ are equal, the equations of motion for the two masses are

$$M\ddot{x}_1 + (k + k_{12})\, x_1 - k_{12}x_2 = 0$$
$$M\ddot{x}_2 + (k + k_{12})\, x_2 - k_{12}x_1 = 0. \tag{4.1}$$

We seek solutions for the normal modes of oscillation when both masses move with the same frequency ω. The assumed solutions are

$$x_1\,(t) = B_1 e^{i\omega t}$$
$$x_2\,(t) = B_2 e^{i\omega t}, \tag{4.2}$$

which leads to the coupled equations

$$\left(-M\omega^2 + k + k_{12}\right) B_1 - k_{12}B_2 = 0$$
$$-k_{12}B_1 + \left(-M\omega^2 + k + k_{12}\right) B_2 = 0. \tag{4.3}$$

For a nontrivial solution to exist, the determinant must vanish

$$\begin{vmatrix} \left(-M\omega^2 + k + k_{12}\right) & -k_{12} \\ -k_{12} & \left(-M\omega^2 + k + k_{12}\right) \end{vmatrix} = 0, \tag{4.4}$$

which is the characteristic (or secular) equation. Solving for the eigenfrequencies yields

$$\omega_1 = \sqrt{\frac{k + 2k_{12}}{M}} \qquad \omega_2 = \sqrt{\frac{k}{M}} \tag{4.5}$$

and the normal modes of oscillation are

$$\eta_1 = x_1 - x_2 \qquad \eta_2 = x_1 + x_2 \tag{4.6}$$

that define asymmetric and symmetric modes, respectively. In the symmetric mode, the two masses move together and there is no relative displacement, which is why they oscillate at the same frequency as the uncoupled masses. For the asymmetric mode, the frequency of oscillation is larger than the uncoupled frequency because of the additional stiffness of the coupling spring.

For two *unequal* masses with equal spring constants, the equations are more involved. The coupled equations are

$$\left(-\omega^2 + \omega_1^2 + \frac{k_{12}}{m_1}\right) B_1 - \frac{k_{12}}{m_1} B_2 = 0$$

$$-\frac{k_{12}}{m_2} B_1 + \left(-\omega^2 + \omega_2^2 + \frac{k_{12}}{m_2}\right) B_2 = 0$$

(4.7)

for the isolated frequencies

$$\omega_1^2 = \frac{k}{m_1} \qquad \omega_2^2 = \frac{k}{m_2}.$$

(4.8)

The normal mode eigenfrequencies are solutions to

$$\omega^2 = \frac{1}{2}\left(\omega_2^2 + \frac{k_{12}}{m_2} + \omega_1^2 + \frac{k_{12}}{m_1}\right) \pm \frac{1}{2}\sqrt{\left(\omega_2^2 + \frac{k_{12}}{m_2} - \omega_1^2 - \frac{k_{12}}{m_1}\right)^2 + 4\frac{k_{12}}{m_1}\frac{k_{12}}{m_2}}. \quad (4.9)$$

The eigenfrequencies are shown in Fig. 4.2 as a function of the coupling k_{12} for the case when $\omega_1 = 10$ and $\omega_2 = 11$. There are two normal modes—the symmetric and the asymmetric modes, as before. At zero coupling, the eigenfrequencies are

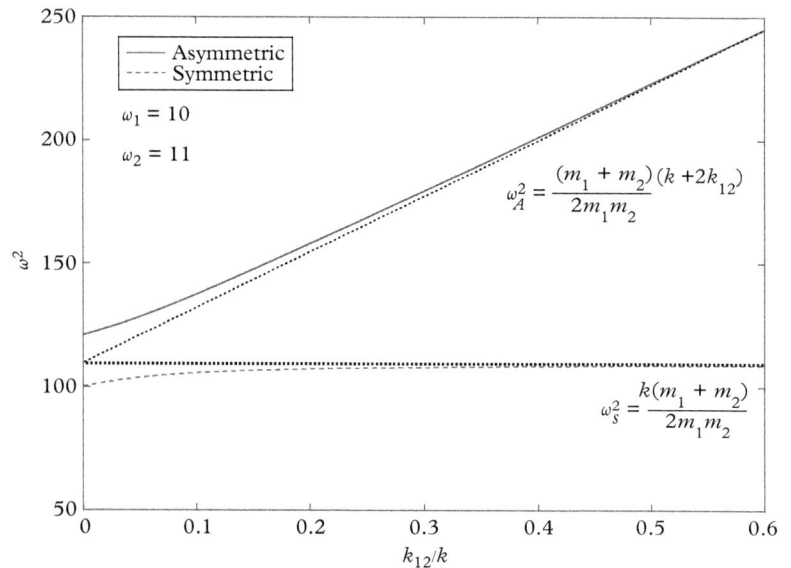

Figure 4.2 *Squared eigenfrequencies ω^2 for two unequal masses with equal spring constants k as a function of coupling k_{12}. At $k_{12} = 0$, $\omega^2 = \omega_1{}^2$ and $\omega_2{}^2$. At large k_{12}, the symmetric mode converges on the compromise frequency.*

equal to the isolated oscillator frequencies. At high coupling, the symmetric mode
has an intermediate frequency (between the two original frequencies) called the
compromise frequency

$$\omega_S^2 = \frac{k\,(m_1 + m_2)}{2m_1 m_2} \tag{4.10}$$

while the asymmetric mode increases with coupling k_{12} and has the asymptotic
dependence

$$\omega_A^2 = \frac{(m_1 + m_2)}{2m_1 m_2}\,(k + 2k_{12}) \tag{4.11}$$

as the two masses oscillate against each other. Coupled linear oscillations abound
in physics and have many analogs in quantum systems and wave mechanics.

4.1.2   Networks of coupled linear oscillators

The general problem of N coupled linear oscillators with heterogeneous couplings
has many interesting aspects. These systems are an analog to the thermal and
dynamical properties of amorphous materials, like glasses. They can show critical
behavior as the density of couplings increases beyond a threshold as a single giant
cluster spans the entire network.

For coupled linear oscillators, it is assumed that they are described by N
generalized coordinates q_k for $k = 1{:}N$ and have N equilibrium positions that
are stationary solutions to Lagrange's equations. The rectilinear coordinates are
functions of the generalized coordinates

$$\begin{aligned} x_\beta^a &= x_\beta^a\left(q^b\right) \\ q^b &= q^b\left(x_\beta^a\right), \end{aligned} \tag{4.12}$$

where x_β^a is the ath coordinate value of the βth oscillator. The kinetic and potential
energies in terms of the generalized coordinates are

$$\begin{aligned} T &= \frac{1}{2} m_{ab}\dot{q}^a \dot{q}^b \\ U &= \frac{1}{2} A_{ab} q^a q^b \end{aligned} \tag{4.13}$$

using the Einstein summation convention. The coefficients of the kinetic energy
term are

$$m_{ab} = \sum_\beta m_\beta \sum_i \frac{\partial x_\beta^i}{\partial q^a}\frac{\partial x_\beta^i}{\partial q^b} \tag{4.14}$$

and the definition of A_{ab} is

$$A_{ab} = \frac{\partial^2 U}{\partial q^a \partial q^b}.\tag{4.15}$$

When the expressions for T and U in Eq. (4.13) are inserted into Lagrange's equation,

$$\frac{\partial L}{\partial q^a} - \frac{d}{dt}\frac{\partial L}{\partial \dot{q}^a} = 0,\tag{4.16}$$

it becomes

$$\frac{\partial U}{\partial q^a} + \frac{d}{dt}\frac{\partial T}{\partial \dot{q}^a} = 0\tag{4.17}$$

and the equations of motion are

$$\sum_a (A_{ab}q^a + m_{ab}\ddot{q}^a) = 0.\tag{4.18}$$

Because we are looking for a stationary solution at a fixed frequency ω, the secular determinant is

$$\left|A_{ab} - \omega^2 m_{ab}\right| = 0,\tag{4.19}$$

leading to the eigenfrequencies and the eigenmodes of the dynamical system.

Example 4.1 Off-diagonal mass tensor

Generalized coordinates are not always selected based on Cartesian coordinates, but may be dictated by symmetry considerations. For a general choice of coordinates, the mass tensor need not be diagonal, leading to kinetic energy terms that are products of different velocities rather than the square of velocities.

As an example, consider the two unequal masses, but with the generalized (normal) coordinate choice

$$\begin{aligned}q_1 &= x_1 - x_2\\ q_2 &= x_1 + x_2\end{aligned}\tag{4.20}$$

to capture the importance of the symmetric mode when both masses move together for $q_1 = 0$. The mass tensor is then

$$m_{ab} = \frac{1}{4}\begin{pmatrix} m_1 + m_2 & m_1 - m_2 \\ m_1 - m_2 & m_1 + m_2 \end{pmatrix} = \frac{1}{2}\begin{pmatrix} \bar{m} & \frac{1}{2}\Delta m \\ \frac{1}{2}\Delta m & \bar{m} \end{pmatrix},\tag{4.21}$$

which is not diagonal in the masses, and the kinetic energy has terms that depend on the products $\dot{q}_a \dot{q}_b$. The potential energy is

$$
\begin{aligned}
U &= \frac{1}{2} k x_1^2 + \frac{1}{2} k x_2^2 + \frac{1}{2} k_{12} (x_1 - x_2)^2 \\
&= \frac{1}{2} \left(k_{12} + \frac{k}{2} \right) q_1^2 + \frac{1}{4} k q_2^2
\end{aligned}
\tag{4.22}
$$

and the potential energy coefficient tensor is

$$
A_{ab} = \frac{1}{2} \left(\begin{array}{cc} k + 2k_{12} & 0 \\ 0 & k \end{array} \right),
\tag{4.23}
$$

which is diagonal for this generalized coordinate choice. The secular determinant from Eq. (4.19) is

$$
\left| \begin{array}{cc} k + 2k_{12} - \omega^2 \bar{m} & -\dfrac{1}{2} \omega^2 \Delta m \\ -\dfrac{1}{2} \omega^2 \Delta m & k - \omega^2 \bar{m} \end{array} \right| = 0.
\tag{4.24}
$$

The choice of generalized coordinates did not simplify the determinant, but it did show the importance of the symmetric mode, and of the unequal masses given by the difference Δm. If the masses were equal, then the secular determinant is already diagonal, whereas Eq. (4.4) was not.

Example 4.2 Linear array of equal masses and springs

Consider a linear chain of N identical masses connected in a circle by N identical springs of spring constant k and separated by a lattice constant α. This model is often used to describe lattice vibrations (known as phonons) in solid state crystals. Because of the translational symmetry, it is easiest to work directly with the equations of motion and to assume propagating solutions. The equations of motion are

$$
m \ddot{q}_n = k (q_{n+1} - q_n) + k (q_{n-1} - q_n)
\tag{4.25}
$$

and the time dependence $e^{-i\omega t}$ converts the equations to

$$
-\omega^2 m q_n = k (q_{n+1} - q_n) + k (q_{n-1} - q_n).
\tag{4.26}
$$

The traveling wave solutions have the form

$$
q_{n+1} = q e^{ik(n+1)a}.
\tag{4.27}
$$

Putting this assumed solution into the equations of motion gives

$$
-\omega^2 m q e^{inka} = kq \left(e^{ik(n+1)a} - e^{inka} \right) + kq \left(e^{ik(n-1)a} - e^{inka} \right).
\tag{4.28}
$$

Canceling out the term $q e^{inka}$ on each side yields

$$
\begin{aligned}
-\omega^2 m &= k \left(e^{ika} - 1 \right) + k \left(e^{-ika} - 1 \right) \\
&= 2k (\cos ka - 1).
\end{aligned}
\tag{4.29}
$$

continued

Example 4.2 *continued*

Therefore, the dispersion equation for traveling waves on the periodic chain is

$$\omega^2 = k\left(e^{ika} - 1\right) + k\left(e^{-ika} - 1\right)$$
$$= \frac{2k}{m}\left(1 - \cos ka\right) \tag{4.30}$$

and with a trigonometric identity becomes

$$\omega(k) = \sqrt{\frac{4k}{m}}\left|\sin\frac{1}{2}ka\right|. \tag{4.31}$$

This dispersion curve is shown in Fig. 4.3 plotted in k-space, which is the Fourier-transform space of the linear lattice. The k-vector takes values between $k = -\pi/a$ to $k = \pi/a$, which is known as the first Brillouin zone. The dispersion at the center of the zone near $k = 0$ (long wavelength waves) is linear, with a phase velocity equal to

$$c_s = \sqrt{\frac{k}{m}}a, \tag{4.32}$$

which is the speed of sound in this lattice. Near the Brillouin zone boundary, the dispersion flattens and the group velocity goes to zero at the boundary, representing a standing wave.

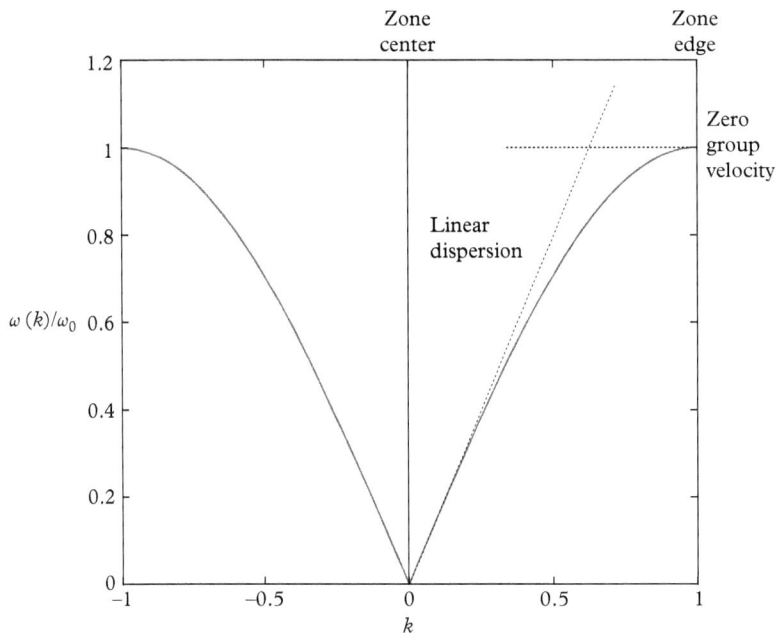

Figure 4.3 *Dispersion curve for a linear chain of N identical masses on springs. The k-vector spans between $k = -\pi/a$ and $k = \pi/a$.*

The coupling of linear systems retains the principle of superposition. The modes may be scrambled, with new normal modes as combinations of the original oscillator motions, but nothing new arises from the mixture. This all changes when the systems become nonlinear, where the oscillators themselves may be nonlinear, or linear oscillators may be coupled nonlinearly. New frequencies can appear, or multiple old frequencies can be entrained to a single frequency as the system synchronizes. Nonlinear synchronization of oscillators is a fundamental physical process that lies at the core of many electronic circuits, as well as in many biological systems like the brain and heart. Some of the simplest models capture the essential features of synchronization that hold true even for these more sophisticated systems.

4.2 Simple models of synchronization

There are many simple models for the synchronization of two oscillators. These can be treated analytically in some cases, and qualitatively in others. The most common examples are: (1) integrate-and-fire oscillators; (2) quasiperiodicity on the torus (action–angle oscillators); and (3) a discrete map called the sine-circle map. These oscillators are no longer the harmonic oscillators of linear physics. The concept of an "oscillator" is more general, often intrinsically nonlinear, such as an integrating circuit that fires (resets) when its voltage passes a threshold, as in an individual neuron.

4.2.1 Integrate-and-fire oscillators

Integrate-and-fire oscillators are free-running linear integrators with a threshold. They integrate linearly in time until they surpass a threshold, and then they reset to zero—they "fire." An example of two uncoupled integrate-and-fire oscillators with different integration rates is shown in Fig. 4.4. There are familiar electronic examples of such oscillators using op-amps and comparator circuits. There are

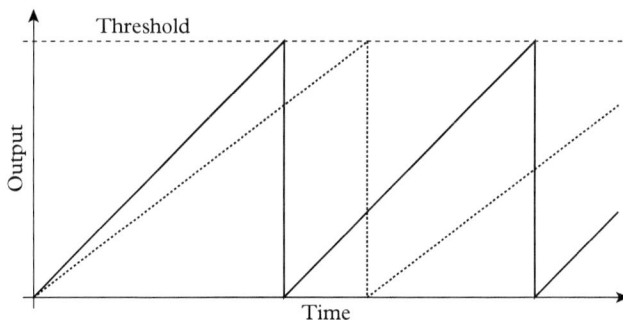

Figure 4.4 *Two different linear integrators that reset at a threshold value. Each is a separate integrate-and-fire oscillator. Because they rise at different rates, the periodicities are different.*

also many biological examples, such as neurons or pacemaker cells in the heart. The existence of a threshold converts such a free-running integrator into a non-linear oscillator, and with the nonlinearity comes the possibility of locking two independent (but coupled) oscillators so that they share the same frequency or phase.

Two oscillators can be coupled in an asymmetric manner in which one oscillator is the master oscillator and the other is called the receiver oscillator. The master oscillator has a natural frequency that is higher than the receiver so that it fires first, driving the second oscillator to reset itself. The asymmetric coupling can take the simple form

$$y_1(t) = \mathrm{mod}\,(\omega_1 t, threshold)$$
$$y_2(t) = \begin{cases} \omega_2 t + g\,|y_1(t) - y_2(t)| & \text{for } y_2 < threshold \\ 0 & \text{for } y_2 > threshold \end{cases} \qquad (4.33)$$

for $\omega_1 > \omega_2$, where $\mathrm{mod}(x, y)$ is the modulus function.[1] In this case, the first oscillator is the master oscillator that has a simple unperturbed integrate-and-fire behavior. The receiver function receives a contribution from the master oscillator with the coupling factor g. Before the master fires, the receiver is noticeably below threshold, because it rises at a slower rate. However, when the master resets to zero, the receiver oscillator gets a large kick upward that puts it over the threshold, and it resets right behind the master. Then they both start integrating again until they trigger again, and so forth, as shown in Fig. 4.5.

It is easy to calculate the necessary coupling strength required to lock two unequal frequencies ω_1 and ω_2. The receiver oscillator must go over the threshold when the master fires and resets to zero. This puts the following condition on the coupling coefficient g just after the master resets to zero

$$\omega_2 t_{th} + g\,|0 - y_2(t_{th})| > threshold$$
$$\omega_2 t_{th} + g\omega_2 t_{th} > \omega_1 t_{th}$$
$$\omega_2(1 + g) > \omega_1 \qquad (4.34)$$
$$g > \frac{\omega_1}{\omega_2} - 1 = \frac{\Delta\omega}{\omega_2}$$

[1] For instance, $\mathrm{mod}(3.72, 2) = 1.72$. This also reads as 3.72 mod 2 equals 1.72.

Figure 4.5 *The frequency locking of two integrate-and-fire oscillators. The master oscillator has the higher frequency. When it resets, it gives a kick to the receiver oscillator that causes it to reset as well.*

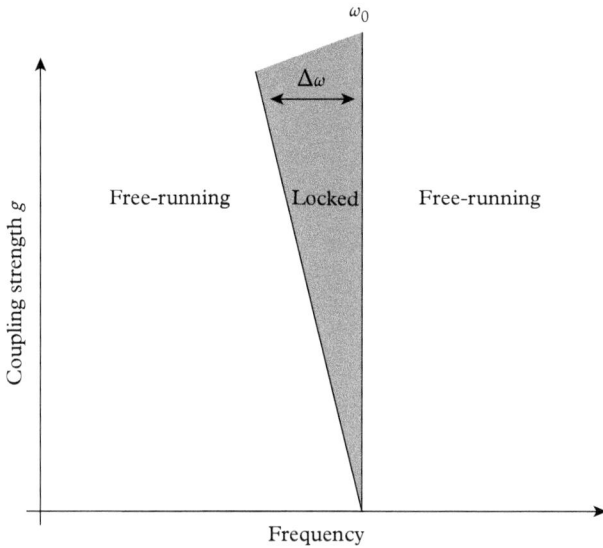

Figure 4.6 *Principle of an Arnold tongue. The range of frequencies that can be locked to a master oscillator increases with increasing coupling.*

which yields the condition for frequency locking

$$g > \frac{\Delta\omega}{\omega_2}. \tag{4.35}$$

Therefore, the coupling strength must be greater than the relative frequency difference between the two oscillators.

This is a general feature of many types of nonlinear oscillator synchronization. It states that for a given value of the coupling g, there is a range of frequencies $\Delta\omega$ that can be phase locked. As the coupling increases, the range of frequencies that can be synchronized gets broader. This relation is depicted graphically in Fig. 4.6 that plots the range of frequency that can be locked as their coupling strength increases. This diagram is called an Arnold Tongue (after V. I. Arnold who studied these features). The surprisingly simple relation stated in Eq. (4.35) is true even for more sophisticated synchronization, as we will see repeatedly in this chapter and the next.

4.2.2 Quasiperiodicity on the torus

In Chapter 2, an integrable Hamiltonian system was described by motion on a hypertorus with the equations of motion

$$
\begin{aligned}
H &= \omega^a \mathcal{J}_a \\
\dot{\mathcal{J}}_a &= -\frac{\partial H}{\partial \theta^a} \\
\dot{\theta}_a &= \omega_a.
\end{aligned}
\tag{4.36}
$$

These dynamics describe an *action–angle oscillator*, or an *autonomous phase oscillator*. For a non-integrable system (but one that reduces to Eq. (4.36) when the non-integrability is small), the equation for the phase angles is

$$\dot{\theta}_a = \omega_a + \varepsilon f\left(\theta_b\right),\tag{4.37}$$

where $f(\theta)$ is periodic in the phase angles. If the phase angles are those of the limit cycles of autonomous oscillators (like a van der Pol oscillator), then Eq. (4.37) describes the mathematical flow of coupled oscillators.

Consider the case of two coupled oscillators that are coupled with a sinusoidal coupling function

$$\begin{aligned}\dot{\theta}_1 &= \omega_1 + g_1 \sin\left(\theta_2 - \theta_1\right)\\\dot{\theta}_2 &= \omega_2 + g_2 \sin\left(\theta_1 - \theta_2\right).\end{aligned}\tag{4.38}$$

In the absence of coupling ($g_1 = 0$ and $g_2 = 0$), the state-space trajectories on a torus (defined by two angles) look like those in Fig. 4.7.

This diagram, and how it appears depending on the ratio of ω_1/ω_2, introduces the concept of quasiperiodicity. Quasiperiodicity occurs when there are ratios of small integer numbers of frequencies present in a time series. Often, the number of frequency components is only two. The ratio of these frequencies can either be rational or irrational numbers. If they are rational, they can be expressed as

$$\frac{\omega_1}{\omega_2} = \frac{p}{q},\tag{4.39}$$

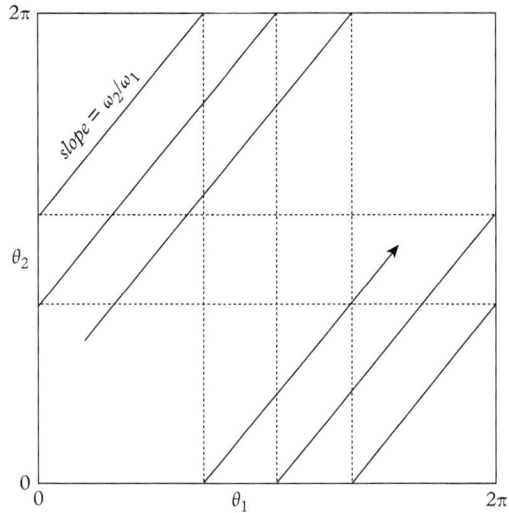

Figure 4.7 *State space on a (flat) torus of two uncoupled oscillators. The slope = ω_2/ω_1. If the frequencies are commensurate (rational ratio), the trajectory is a closed orbit and repeats. If they are incommensurate (irrational ratio), then the trajectory eventually fills the space.*

where p and q have no common factors and are usually small integers. In this case, the trajectories on the flat torus (Fig. 4.7) are closed orbits and repeat after a time

$$T = pT_1 = \frac{2\pi}{\omega_1}p = \frac{2\pi}{\omega_2}q. \qquad (4.40)$$

When the two systems are coupled $(g_1, g_2 > 0)$, the rate of change of the relative phase becomes the measure of synchronization. If the systems are synchronized, then the phase difference remains a constant. The rate of change of the relative phase is

$$\begin{aligned} \dot{\phi} &= \dot{\theta}_1 - \dot{\theta}_2 \\ &= \omega_1 - \omega_2 - (g_1 + g_2)\sin\phi. \end{aligned} \qquad (4.41)$$

This is equivalent to a one-dimensional (1D) oscillator with a phase picture shown in Fig. 4.8 The amplitude of the sine curve is $g_1 + g_2$, and the mean value is $\omega_1 - \omega_2$. The rate of change of the relative phase can only be zero (to define a fixed point) if the amplitude is larger than the mean value

$$|\omega_1 - \omega_2| < g_1 + g_2 \qquad (4.42)$$

for which there is one stable and one unstable node (compare to Fig. 3.1). The system trajectories are attracted to the stable node as the two systems become frequency and phase locked. The constant phase offset is then

$$\sin\phi^* = \frac{\omega_1 - \omega_2}{g_1 + g_2}. \qquad (4.43)$$

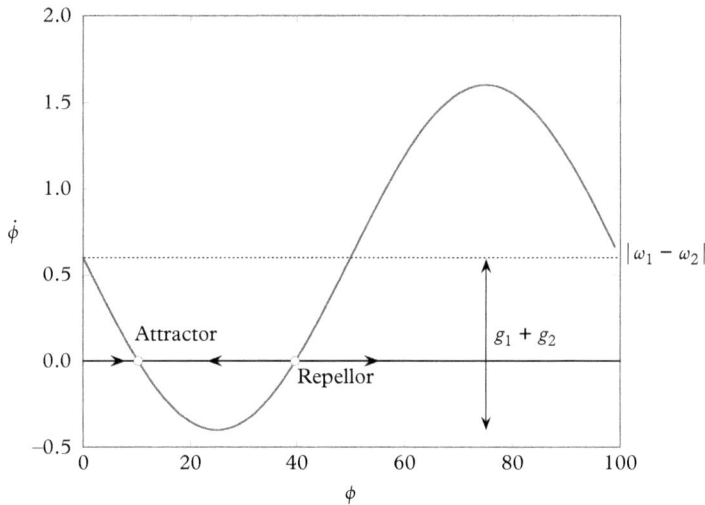

Figure 4.8 *There is one stable node and one unstable node for phase locking on the torus when* $|\omega_1 + \omega_2| < g_1 + g_2$.

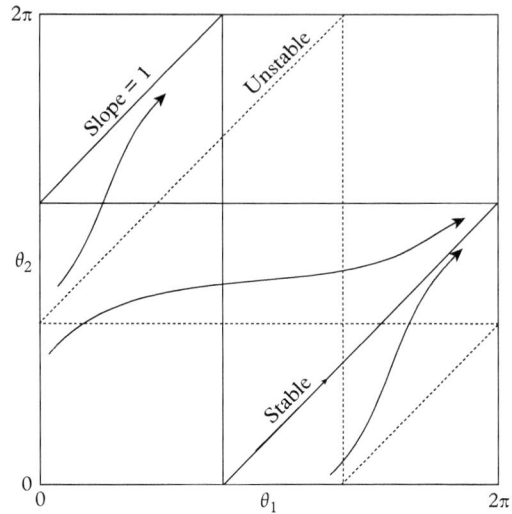

Figure 4.9 *The stable and the unstable limit cycles on the torus have slope = 1 when they are phase locked.*

This situation with a constant relative phase between the two systems is called *phase-locked*. The two systems also share the common frequency

$$\omega^* = \frac{g_2\omega_1 + g_1\omega_2}{g_1 + g_2} \tag{4.44}$$

called the *compromise frequency*. The evolution of the motion on the torus is shown in Fig. 4.9. The trajectories are repelled from the unstable manifold and attracted to the stable manifold.

For phase-locking there is clearly a threshold related to the frequency difference between the autonomous oscillators and the strength of the coupling. As long as the frequency difference is less than the coupling, then the system is phase locked. If the frequency offset is too large, or if the coupling is too small, then the system remains quasiperiodic. This is just what we saw in Eq.(4.35) for the integrate-and-fire oscillators described by the Arnold tongue in Fig. 4.6.

4.2.3 Sine-circle map

In the above example, the coupled-oscillator system was mapped as a simple flow on the flat torus. There is a related discrete map, called the sine-circle map, that captures much of the physics of the coupled oscillators. The discrete map is

$$\theta_{n+1} = f(\theta_n) = \mathrm{mod}\left(\left[\theta_n + \Omega - g\sin(\theta_n)\right], 2\pi\right), \tag{4.45}$$

where the angle is taken mod (2π). In this discrete map the frequency Ω plays the role of the frequency ratio between two oscillators on the torus. Therefore, the 1D sine-circle map captures the physics of the coupled oscillators on the two-dimensional torus. This map has a fixed point when

$$\theta^* = \mod\left(\theta^* + \Omega - g\sin\left(\theta^*\right), 2\pi\right), \tag{4.46}$$

giving

$$\sin\left(\theta^*\right) = \frac{\Omega + N2\pi}{g} \tag{4.47}$$

for any integer N. This is real-valued for

$$g > |\Omega + N2\pi| \tag{4.48}$$

with a stability determined by

$$\frac{df}{d\theta} = 1 - g\cos\theta^*. \tag{4.49}$$

The fixed point is stable when this value lies between -1 and 1.[2] Therefore, the limit cycle is stable when

$$0 < g\cos\theta^* < 2. \tag{4.50}$$

The integer N in Eq. (4.47) provides many possible values of g and Ω that can yield stable limit cycles. It also opens up many routes for complicated iterative values as the system moves in and out of the stability condition Eq. (4.50). In spite of this complicated stability condition, larger coupling allows wider frequency offsets to be synchronized, as we saw before.

Iterations of the sine-circle map are illustrated in Fig. 4.10 for a frequency ratio of $\Omega = 3{:}2$ shown for different values of the coupling g. For $g = 0.8 \times 2\pi$ (Fig. 4.10(a)), the values iterate randomly. For $g = 1.0 \times 2\pi$ (Fig. 4.10(b)), the system has a period-2 cycle with $\theta^* = (\pi/2, 3\pi/2)$. For $g = 1.2 \times 2\pi$ (Fig. 4.10(c)), the system is again chaotic, while for $g = 1.5 \times 2\pi$ (Fig. 4.10(d)) it has a stable fixed point at $\theta^* = \pi/2$. Note that the fixed point lies on the diagonal function $y = x$.

The complicated behavior with increasing g is captured in Fig. 4.11 for a frequency ratio equal to the golden mean $\Omega = 1.618$, shown as successive iterates as g increases from 0 to 2 (units of 2π). There is no frequency locking below $g = 0.1 \times 2\pi$. Above $g = 0.1 \times 2\pi$, some structure appears, until a stable frequency lock occurs between $g = 0.4 \times 2\pi$ and $g = 0.5 \times 2\pi$. The locking threshold occurs at the ratio $\Omega/g = 4$. This is far above the "weak coupling" limit (when frequency locking occurs as a perturbation of uncoupled oscillators), but it shows the rich structure for strong coupling.

The sine-circle map has intriguing properties related to *resonances*. A resonance occurs when two locked frequencies are in ratios of small integers. The system will phase-lock a frequency and a harmonic such as 2:1 or 2:3. This type of frequency locking creates considerable harmonic structures in coupled nonlinear oscillators.

[2] Remember that the sine-circle map is a discrete mapping, and Eq. (4.49) is a Floquet multiplier.

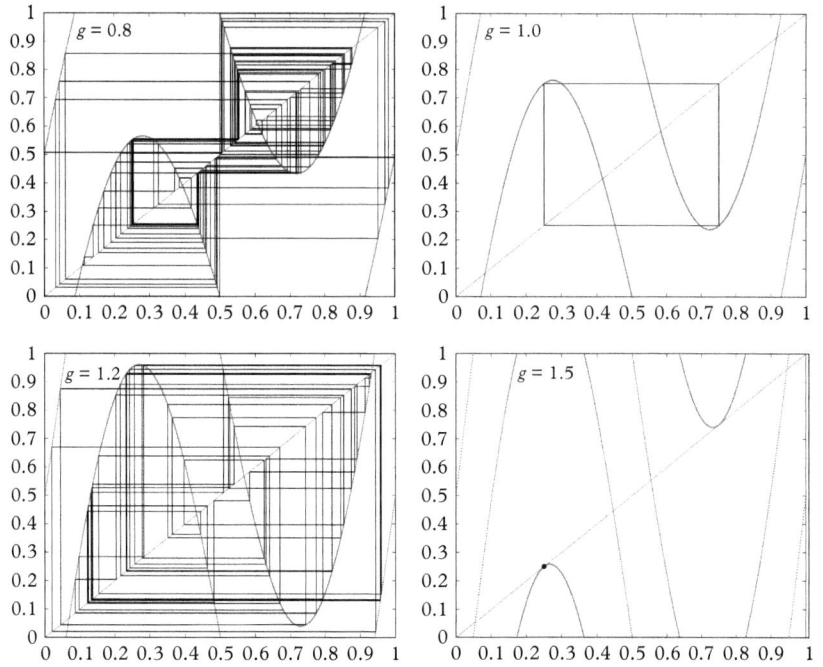

Figure 4.10 *Numerical results of the sine-circle map generated using* $\Omega = 3{:}2$ *with (a)* $g = 0.8$, *(b)* $g = 1$, *(c)* $g = 1.2$, *(d)* $g = 1.5$. *All g values are normalized by* 2π.

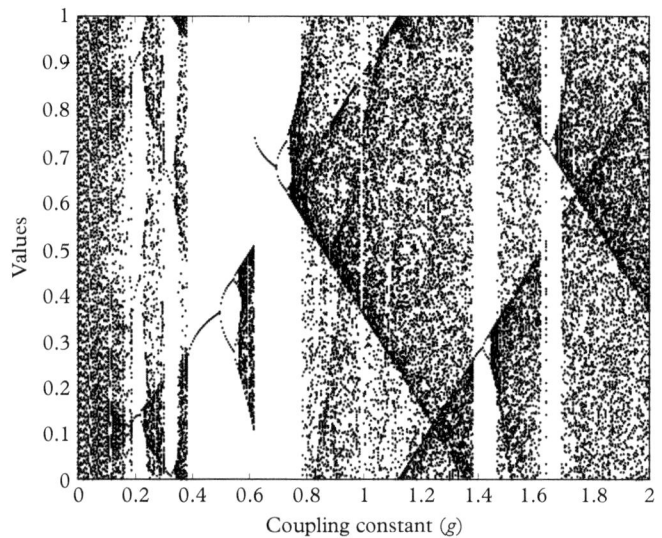

Figure 4.11 *Iterated values for the sine-circle map for* $\Omega = 1.618$ *(the golden mean) shown for* $g = 0$ *to 2 (units of* 2π*).*

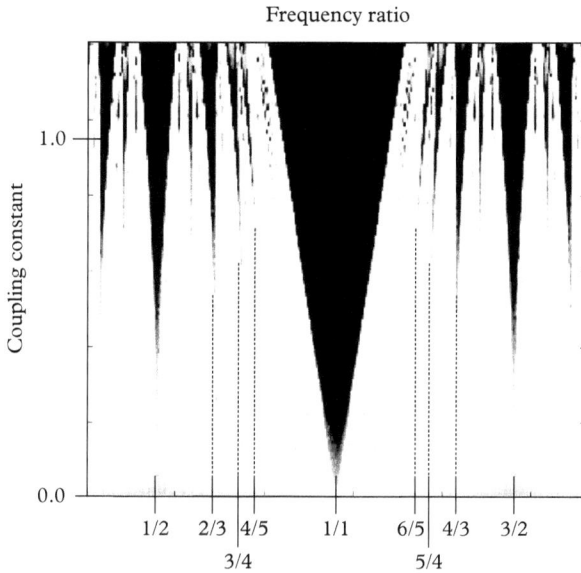

Figure 4.12 *Arnold tongues for the sine-circle map, which captures the behavior of two mutually coupled oscillators. The dark regions are the frequency-locked regions.*

The resonance structure of the sine-circle map is shown in Fig. 4.12 as a plot of the frequency range that is locked for increasing coupling *g*. The gray-scale is based on how tightly the iterates are frequency locked. The figure shows a strong Arnold tongue structure for frequency ratios of small integers. The frequency entrainment range is strongest for the 1:1 resonance, followed by 1/2, 2/3, 3/4, and 4/5, with symmetric tongues for resonances larger than 1:1 at 4/3, 5/4, and 6/5, respectively.

Frequency resonances are partly responsible for the origin of the gaps in the rings of Saturn, as shown in Fig. 4.13. Saturn has many moons, and in the rings there are "shepherd" moons that entrain nearby ring particles in frequency-locked bands with gaps where ring material is swept out. Shepherd moons have been identified for all the major gaps in the rings, corresponding to resonances of frequency ratios of small integers. For instance, the outer edge of the A-ring is in a 7:6 resonance condition with the moon Janus. The ring materials circle Saturn seven times while Janus orbits six times. The boundary between the outer edge of the B-ring and the Cassini division is in a 2:1 resonance condition with the moon Mimas, while the Encke gap is in a 5:3 resonance condition with the same moon.

4.3 External synchronization of an autonomous phase oscillator

Often, a nonlinear autonomous oscillator, such as a van der Pol oscillator or a limit-cycle oscillator, is driven by an external periodic force. If the drive frequency

Figure 4.13 *Details of Saturn's B-ring showing the pronounced gaps of the Cassini division.*
Photograph from <http://photojournal.jpl. nasa.gov/catalog/PIA08955> Image credit: NASA/JPL.

is near the natural frequency of the oscillator, then the oscillator will be entrained by the drive frequency. Just as for the coupled oscillators, stronger coupling to the drive leads to wider frequency ranges over which the oscillator frequency will be locked to the drive.

Consider an autonomous phase oscillator in Eq. (4.36) subject to a sinusoidal external periodic drive that depends on the phase of the oscillator

$$\dot{\theta} = \omega_0 + g \sin (\theta - \omega_d t) , \qquad (4.51)$$

where θ is the angular coordinate of the oscillator, ω_d is the drive angular frequency, and ω_0 is the autonomous frequency of the isolated oscillator. In the absence of the drive, the phase oscillator has the trajectory $\theta = \omega_0 t$. When the oscillator is locked to the external drive then $\dot{\theta} = \omega_d$, and the phase of the oscillator perfectly tracks the phase of the drive such that

$$\omega_d = \omega_0 + g \sin \left(\theta^* - \omega_d t\right) , \qquad (4.52)$$

where $\theta^* - \omega_d t = const$. The condition for frequency locking is set by the condition that $|\sin \theta| < 1$ and hence is set by the condition that

$$|\omega_d - \omega_0| < g. \qquad (4.53)$$

Therefore, if the angular frequency difference between the drive frequency and the natural oscillator frequency is smaller than the coupling constant g, then the system will be locked.

When the frequency offset between the drive frequency and the natural frequency is larger than the coupling constant, then the system is not locked, and the phase of the oscillator drifts relative to the phase of the drive. In this regime, it is possible to approach the dynamics from the point of view of the phase difference between the two systems. This leads to an equation for the "slow phase" ψ defined by

$$\psi = \theta - \omega_d t$$
$$\dot{\psi} = \dot{\theta} - \omega_d. \tag{4.54}$$

When the system is almost frequency-locked, this phase changes slowly. In terms of this slow phase, the driven autonomous oscillator Eq. (4.51) is now

$$\dot{\psi} + \omega_d = \omega_0 + g\sin\psi, \tag{4.55}$$

which is

$$\frac{d\psi}{dt} = -\Delta\omega + g\sin\psi, \tag{4.56}$$

where

$$\Delta\omega = \omega_d - \omega_0 \tag{4.57}$$

is the frequency difference between the drive frequency and the natural frequency of the oscillator. This equation is integrated to yield the period of oscillation of the slow phase as

$$T_\psi = \int_0^{2\pi} \frac{d\psi}{g\sin\psi - \Delta\omega}. \tag{4.58}$$

This period is associated with the beat frequency Ω_ψ between the drive frequency and the altered frequency of the oscillator

$$\Omega_\psi = \frac{2\pi}{T_\psi} = 2\pi \left[\int_0^{2\pi} \frac{d\psi}{g\sin\psi - \Delta\omega}\right]^{-1}. \tag{4.59}$$

To evaluate this beat frequency, average over a single cycle of the slow phase

$$\int_0^{2\pi} \frac{d\psi}{g\sin\psi - \Delta\omega} = \frac{2}{\sqrt{\Delta\omega^2 - g^2}} \tan^{-1}\left[\frac{-\Delta\omega\tan(\psi/2) + g}{\sqrt{\Delta\omega^2 - g^2}}\right]_0^{2\pi}$$
$$= \frac{-2\pi}{\sqrt{\Delta\omega^2 - g^2}} \tag{4.60}$$

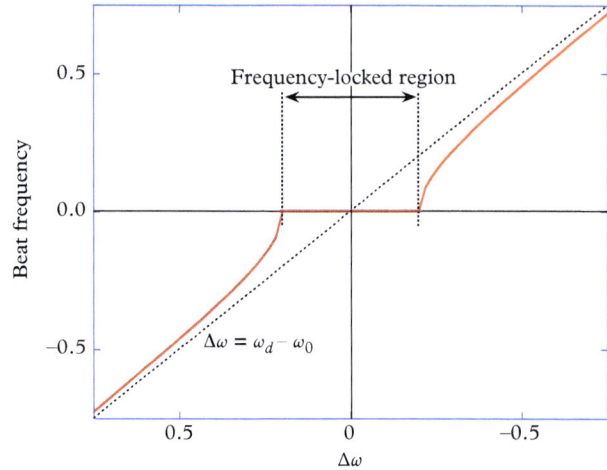

Figure 4.14 *The beat angular frequency as a function of angular frequency detuning between the drive and the autonomous oscillator. The dependence near the locking threshold has a square-root dependence.*

to obtain the important and simple result

$$\text{Beat frequency:} \qquad \Omega_\psi = \sqrt{\Delta\omega^2 - g^2} \quad \text{for} \quad |\Delta\omega| > g \qquad (4.61)$$

Therefore, the frequency shift of the autonomous oscillator varies as a square root away from the frequency-locking threshold. This square-root dependence is a ubiquitous phenomenon exhibited by many externally-driven autonomous oscillators, regardless of the underlying details of the physical system. It is related to universal properties that emerge from critical phenomena in statistical mechanics and in mean field theory. The beat frequency is shown schematically in Fig. 4.14 as a function of frequency detuning, showing the frequency-locked region for small detunings, and the square-root dependence outside the locking region. Prior to frequency locking, the frequency of the autonomous oscillator is pulled towards the external drive frequency, known as frequency entrainment. This is shown numerically in Fig. 4.15 as a function of increasing coupling g for a single oscillator with a frequency 5% larger than the drive frequency. The frequency becomes locked when the coupling exceeds 5%.

4.4   External synchronization of a van der Pol oscillator

The driven van der Pol oscillator is of interest because this system is fundamentally nonlinear and shows limit-cycle behavior even in the absence of a harmonic

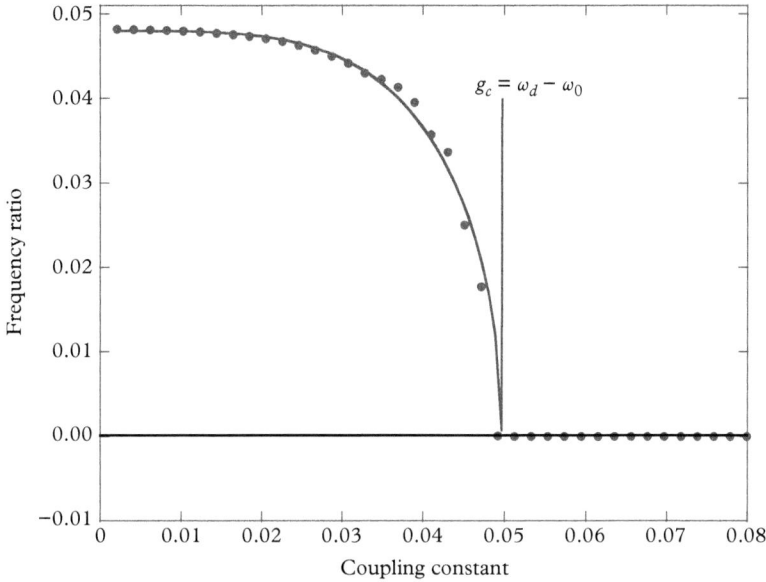

Figure 4.15 *The relative frequency offset of an autonomous oscillator from the drive frequency as a function of coupling. Above the threshold the autonomous phase oscillator is entrained by the drive frequency.*

drive (it is an autonomous oscillator). However, when it is driven harmonically, the self-oscillation is influenced by external periodic forcing that causes the oscillator to oscillate with the same frequency as the external harmonic drive.

Consider a weakly nonlinear van der Pol oscillator:

$$\ddot{x} = 2\mu\dot{x}\left(1 - \beta x^2\right) - \omega_0^2 x. \tag{4.62}$$

In the absence of the gain term in μ, this is an undamped harmonic oscillator with natural angular frequency ω_0. When this is driven by a harmonic forcing function,

$$\ddot{x} - 2\mu\dot{x}\left(1 - \beta x^2\right) + \omega_0^2 x = F \sin \omega_d t, \tag{4.63}$$

the non-autonomous equation is converted to the autonomous flow

$$\begin{aligned} \dot{x} &= y \\ \dot{y} &= 2\mu\dot{x}\left(1 - \beta x^2\right) - \omega_0^2 x + F \sin z \\ \dot{z} &= \omega_d. \end{aligned} \tag{4.64}$$

The frequency offset between the drive frequency and the autonomous frequency (the frequency of the self-oscillating system, *not* of the harmonic oscillator ω_0) is characterized by $\Delta\omega$. For given properties of the oscillator, the parameters to explore for synchronization are the frequency offset $\Delta\omega$ and the magnitude of the force F. The numerical results for the driven system are presented in Fig. 4.16 and Fig. 4.17. The system frequency is pulled toward the external drive frequency as the frequency offset decreases, and then the van der Pol frequency becomes locked to the external frequency. The beat frequency is shown in Fig. 4.17, which shows the classic frequency-locking behavior with the square-root dependence that was shown schematically in Fig. 4.14.

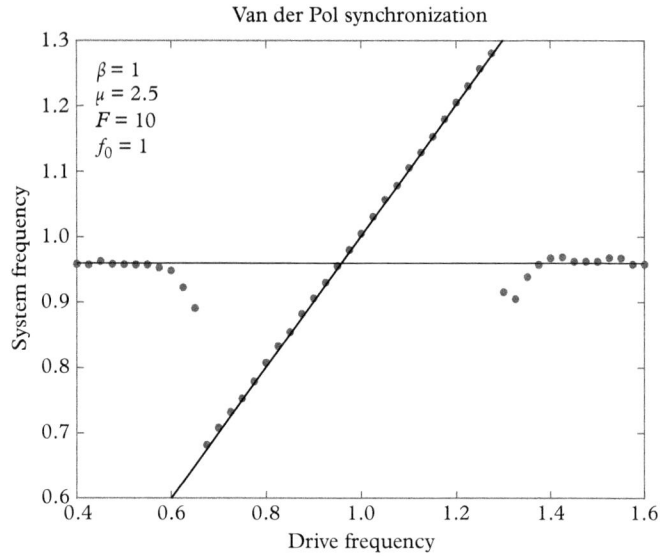

Figure 4.16 *Numerical results for synchronization of a van der Pol oscillator with a harmonic driving force.*

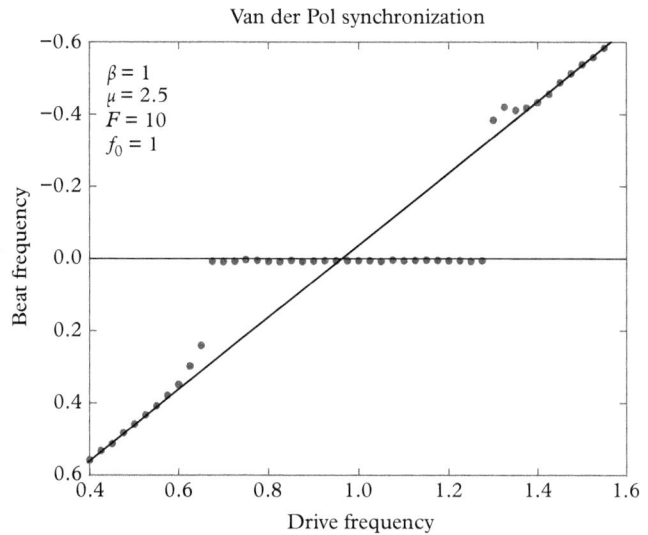

Figure 4.17 *The same data as for Fig. 4.16, but expressed as a beat frequency.*

4.5 Mutual synchronization of two autonomous oscillators

Synchronization was first described and investigated experimentally by Christian Huygens in 1665 when he noticed that two pendulum clocks attached to a wall would become synchronized in the motions of their pendulums, no matter how different their initial conditions were, even if the frequencies were slightly different.

One of the best examples of coupled oscillators are van der Pol oscillators because they are limit-cycle oscillators that behave much as pendulum clocks do. They have a natural frequency, yet are intrinsically nonlinear and amenable to frequency locking. The flow for two coupled van der Pol oscillators is

$$
\begin{aligned}
\dot{x} &= y + g\,(z - x) \\
\dot{y} &= 2\mu\dot{x}\left(1 - \beta x^2\right) - \omega_1^2 x \\
\dot{z} &= w + g\,(x - z) \\
\dot{w} &= 2\mu\dot{z}\left(1 - \beta z^2\right) - \omega_2^2 z.
\end{aligned}
\tag{4.65}
$$

The coupling terms in g are two-way symmetric and linear and occur only between the x and z variables. The oscillators are identical, except that they have different values for their linear autonomous frequencies ω_1 and ω_2. If these frequencies are not too dissimilar, and if the coupling g is strong enough, then the two oscillators will synchronize and oscillate with a common compromise frequency.

Numerical results for the mutual frequency locking are shown in Fig. 4.18. The individual frequencies are shown in Fig. 4.18(a) as a function of the frequency offset for uncoupled oscillators. The two distinct frequencies are pulled and then entrained in the locked regime as the frequency offset decreases. The frequency pulling does not follow a linear behavior, even in the locked regime. However, when the data are re-plotted as a beat frequency, then the system shows the same classic signature as for an externally driven individual oscillator (compare with Fig. 4.14 and Fig. 4.17).

Frequency locking is not necessarily synonymous with phase locking. For instance, coupled chaotic oscillators (like two identical coupled Rössler oscillators) can share the same phase, yet have no well-defined frequency (see Exercise 13). However, for these coupled van der Pol oscillators, both the phase and frequency become locked, as shown numerically in Fig. 4.19. The two oscillators begin out of phase, yet become phase locked after only a few oscillations.

The coupled van der Pol oscillator system is just one example of a wide range of nonlinear systems that exhibit frequency entrainment and phase locking. These are physical systems such as Huygens's wall clocks and the rings of Saturn, or biological systems such as pacemaker cells in the heart and even communities of fireflies that blink in synchrony, among many other examples. The beauty in the analysis of the sine-circle map, simple as it is, is how broadly

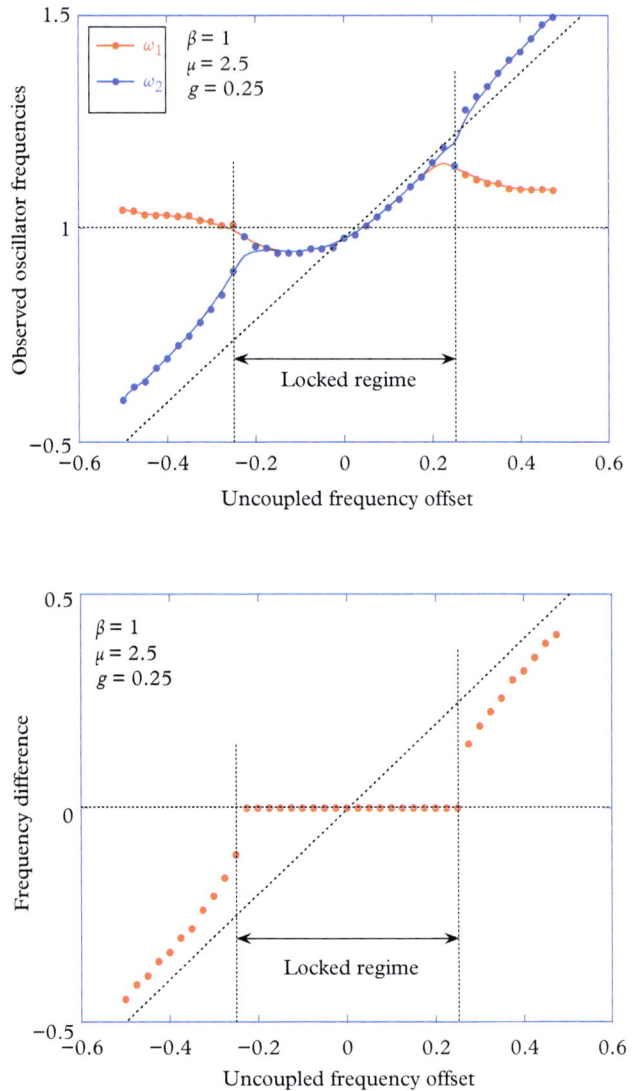

Figure 4.18 *Two coupled van der Pol oscillators. The individual frequencies are shown in (a). The beat frequency is shown in (b). Note that, despite the variable frequencies in (a), the beat frequency in (b) is nearly identical to Fig. 4.16 for an external drive.*

the basic phenomenon is displayed in widely-ranging systems. While true universality is often hard to find across broad systems, the general behavior of complicated systems can be very similar. The basic behavior of coupled van der Pol oscillators is repeated again and again across wide areas of science and engineering.

The logical next step is to go beyond two oscillators and to consider a group of coupled oscillators. The interconnectivity among the members of the

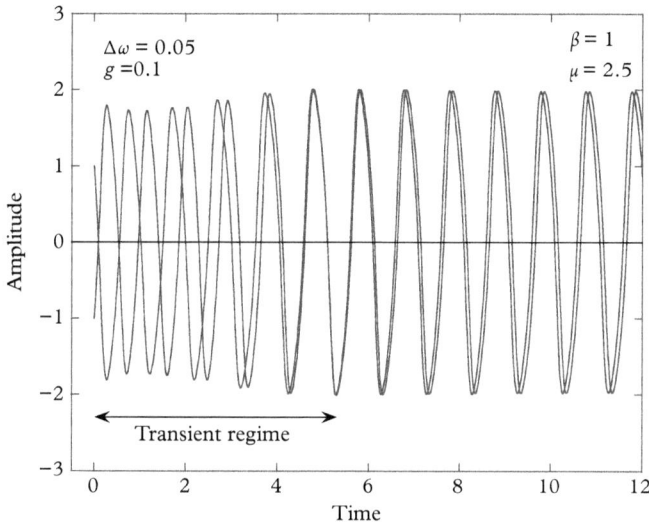

Figure 4.19 *The phase locking of two coupled van der Pol oscillators from an unlocked initial condition.*

group can take on many topological forms, forming different classes of network structures. The behavior of networks of coupled nonlinear systems is the topic of the next chapter.

4.6   Summary

The starting point for understanding many complex systems begins with the coupling of two semi-autonomous systems. The behavior of the coupled pair provides insight into the behavior that emerges for larger numbers of coupled systems. In the case of linear coupled oscillators, the result is a weighted mixture of the individual motions, and the principle of superposition applies in which the whole quite simply is the sum of the parts. In the case of nonlinear coupling, on the other hand, superposition breaks down. For example, with the nonlinear coupling of two autonomous oscillators, *synchronization* is possible. Simple asymmetric coupling of *integrate-and-fire oscillators* captures the essence of frequency locking in which the ability to synchronize the system compares the frequency offset between the two oscillators against the coupling strength. Quasiperiodicity on the torus (*action-angle oscillators*) with nonlinear coupling demonstrates phase locking, while the *sine-circle map* is a discrete map that displays multiple *Arnold tongues* for frequency locking *resonances*. External synchronization of a phase oscillator can be analyzed in terms of the *"slow" phase* difference, resulting in a *beat frequency* and *frequency entrainment* that are functions of the coupling strength. *Mutual synchronization* of two unequal van der Pol oscillators displays the same square-root frequency entrainment as external synchronization of a single oscillator.

General condition for frequency locking

The critical dimensionless coupling strength required to couple two unequal frequencies is

$$g > \frac{\Delta\omega}{\bar{\omega}}, \tag{4.35}$$

where $\bar{\omega}$ is the mean frequency and $\Delta\omega$ is the frequency difference.

Sine-circle map

The discrete sine-circle map is

$$\theta_{n+1} = \mathrm{mod}\left([\theta_n + \Omega - g\sin(\theta_n)], 2\pi\right), \tag{4.45}$$

where the angle is taken mod (2π).

External synchronization of a phase oscillator

This is one of the simplest models for synchronization

$$\dot{\theta} = \omega_1 + g\sin(\omega_d t - \theta), \tag{4.51}$$

where θ represents the phase offset between the oscillator and the drive, ω_d is the drive frequency, and ω_1 is the autonomous frequency.

Beat frequency

When the coupling between an external drive and a single phase oscillator is not strong enough, the beat frequency is given by

$$\Omega_\psi = \sqrt{\Delta\omega^2 - g^2} \quad \text{for } \Delta\omega > g \tag{4.61}$$

and is zero when $\Delta\omega < g$.

4.7 Bibliography

V. I. Arnold, *Mathematical Methods of Classical Mechanics*, 2nd ed. (Springer-Verlag, New York, 1989).
A mind-expanding text on classical mechanics that uses geometric proofs more often than analytical proofs.

L. Glass and M. Mackey, From Clocks to Chaos (Princeton University Press, 1988).
This is an easy level book for general audiences on aspects of biological rhythms from the viewpoint of nonlinear dynamics.

R. Hilborn, *Chaos and Nonlinear Dynamics: An Introduction for Scientists and Engineers* (Oxford University Press, 2001).
This is a good undergraduate textbook on nonlinear dynamics. It makes a good companion to Strogatz, as they complement each other well.

A. Pikovsky, M. Rosenblum and J. Kurths, *Synchronization: A Universal Concept in Nonlinear Science* (Cambridge University Press, 2001).
A highly technical mathematical text on synchronization, but it has several chapters on experimental results and phenomenological approaches that are accessible to the general reader.

Steven H. Strogatz, *Nonlinear Dynamics and Chaos* (Westview Press, 1994).
This is one of the most readable and accessible books at the introductory undergraduate level. The book has a great collection of homework problems.

A. T. Winfree, The Geometry of Biological Time (Springer-Verlag, New York, 1990).
A classic of biological oscillations and biorhythms by the founder of the field.

4.8 Homework exercises

Analytic problems

1. **Coupled oscillators:** Find the secular determinant in ω^4 for a periodic linear cycle of masses and springs composed of two unequal masses (lattice with a basis).
2. **Global coupling:** Calculate the eigenfrequency spectrum of N identical harmonic oscillators of equal frequencies $\omega^2 = k/m$ that are coupled globally to all other oscillators with spring coupling constants κ. Interpret the eigenvalue results physically. What are the eigenmodes?
3. **Integrate-and-fire oscillator:** For a single integrate-and-fire oscillator, derive the frequency-locking condition required on g and $\Delta\omega$ for the case when the upper threshold is sinusoidally modulated

$$y_1(t) = \text{mod}(\omega_1 t, \text{threshold}) \qquad Threshold = 1 + g \sin \omega_0 t$$

4. **Sudden Perturbation:** Consider the phase oscillator

$$\dot{r} = r\left(c - r^2\right) \qquad \dot{\theta} = 2\pi$$

for $c > 0$. Let the system run in steady state. What is the period T_0 of the oscillation. At time t_0 and phase $\theta_0(t_0)$ apply a jump in the x-direction Δx (remember that $r = \sqrt{x^2 + y^2}$). What is the time T it takes to return to the phase $\theta_0(T)$? If the system is repetitively perturbed at phase θ_0 with period T, it will follow the new period.

5. **Slow phase dynamics:** Consider a driven phase oscillator

$$\dot{\phi} = \omega_1 + g \sin(\phi - \omega_0 t).$$

For $\psi = \phi - \omega_0 t$ there is a "slow" phase dynamics given by $\dot{\psi} = -\Delta\omega + g \sin\psi$ where $\Delta\omega = \omega_1 - \omega_0$. Find the fixed point ψ^*. Show that the beat frequency is

$$\Omega_\psi = \frac{2\pi}{T_\psi} = 2\pi \left[\int_{\psi^*-\pi}^{\psi^*+\pi} \frac{d\psi}{g \sin(\psi - \psi^*) - \Delta\omega} \right]^{-1}.$$

6. **Complex amplitude equation:** The complex amplitude equation (also known as the Landau–Stuart or the lambda–omega equation) is an autonomous flow. The equation is

$$\frac{dA}{dt} = (1 + i\eta) A - (1 + i\alpha) |A|^2 A.$$

Show that this can be written in polar coordinates

$$\frac{dR}{dt} = R\left(1 - R^2\right)$$
$$\frac{d\theta}{dt} = \eta - \alpha R^2$$

and that the solution is a limit cycle.

7. **Sine-circle map:** Derive the discrete sine-circle map Eq. (4.45) from the coupled phase oscillators on the torus Eq. (4.38). Show that the frequency offset $\omega_1 - \omega_2$ in the continuous model is captured by the frequency ratio $\Omega = \omega_1/\omega_2$ in the discrete map.

8. **Limit cycle:** For the numerical results and parameters in Fig. 4.19, analytically derive the relaxation time to the phase-locked state for the van der Pol limit cycle.

Computational projects[3]

9. **Driven oscillator:** For a single externally-driven phase oscillator (Eq. (4.51) for a fixed g, numerically simulate the frequency as the drive frequency ω_0 is swept through ω_1. Pick a wide-enough frequency range to capture the frequency entrainment and locking.

10. **Sine-circle map:** For the sine-circle map at $\Omega = 0.50$ and then again at $\Omega = 0.51$, create a bifurcation plot for g increasing from 0 to 2. Why are the results so sensitive to Ω at small g?

[3] Matlab programming examples are found at www.works.bepress.com/ddnolte.

11. **Stability:** Consider the phase oscillator

$$\dot{r} = r\left(c + 2r^2 - r^4\right) \quad \dot{\theta} = 2\pi$$

for $-1 < c < 0$. Find the two stable solutions. One is oscillatory, while one is not (it is a steady state).

(a) Set the initial conditions such that the stable state is oscillatory. Explore what perturbations do to the oscillations by giving a jump in radius Δr of varying size and at varying times. For some choice of perturbation parameters, the oscillations can be quenched and the system goes into steady state. In physiology, a healthy oscillation (beating heart) can be quenched by a perturbation (electric shock).

(b) Set the initial conditions such that the system is in its steady state. Apply a perturbation to make the system settle in the oscillatory state. In physiology, a perturbation (electric shock) can restart healthy oscillation (beating heart).

12. **Coupled van der Pol oscillators:** For two coupled van der Pol oscillators, put the coupling on the velocity terms instead of the position variables. Study the phase-locking as a function of frequency offset for a fixed coupling g. What differences do you note between the two types of coupling (positional vs. velocity)?

13. **Coupled chaos:** Consider two identical Rössler chaotic oscillators that are coupled with a strength g

$$\begin{aligned}
\dot{x}_1 &= -0.97x_1 - z_1 + g\left(x_2 - x_1\right) \\
\dot{y}_1 &= 0.97x_1 + ay_1 + g\left(y_2 - y_1\right) \\
\dot{z}_1 &= x_1\left(z_1 - 8.5\right) + 0.4 + g\left(z_2 - z_1\right)
\end{aligned}$$

$$\begin{aligned}
\dot{x}_2 &= -0.97x_2 - z_2 + g\left(x_1 - x_2\right) \\
\dot{y}_2 &= 0.97x_2 + ay_2 + g\left(y_1 - y_2\right) \\
\dot{z}_2 &= x_2\left(z_2 - 8.5\right) + 0.4 + g\left(z_1 - z_2\right)
\end{aligned}$$

(a) For the parameter $a = 0.15$, the phase of the individual oscillators is well-defined to be $\phi_i = \arctan\left(y_i/x_i\right)$. Study the phase synchronization of this coupled chaotic system as a function of coupling strength g. For instance, define the "order parameter" to be the phase difference between the two oscillators and find when it becomes bounded as a function of time.

(b) For the parameter $a = 0.25$, the phase is no longer coherent. Can you define an "order parameter" for the phase incoherent regime that shows a synchronization transition?

(c) Make the two oscillators non-identical. How do the phase-coherent and phase-incoherent regimes behave now? How does synchronization depend on the non-identicality?

(For a network study, see C. S. Zhou and J. Kurths, "Hierarchical synchronization in complex networks with heterogeneous degrees," *Chaos*, vol. 16, 015104, 2006.)

5 Network Dynamics

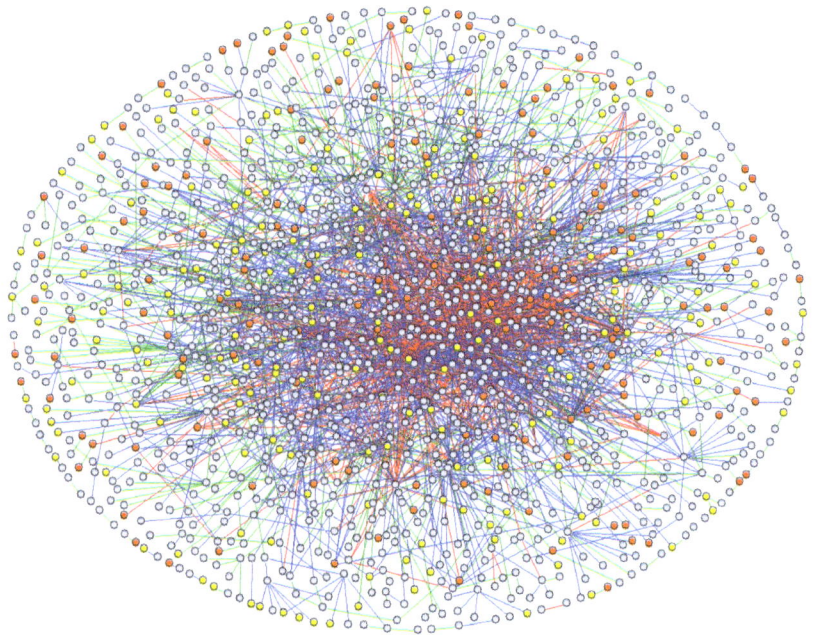

The human protein interactome.

We live in a connected world. We interact and communicate within vast social networks using email and text messaging and phone calls. The World Wide Web contains nodes and links that branch out and connect across the Earth. Ecosystems contain countless species that live off of each other in symbiosis, or in competition as predator and prey. Within your body, thousands of proteins interact with thousands of others doing the work that keeps you alive. All of these systems represent networks of interacting elements. They are dynamic networks, with nodes and links forming and disappearing over time. The nodes themselves may be dynamical systems, with their own characteristic frequencies, coupled to other nodes and subject to synchronization. The synchronization of simple oscillators on networks displays richly complex phenomena. Conversely, synchronization of chaotic oscillators can show surprisingly coherent behavior. This chapter introduces common network structures, studies the diffusion of viruses across these networks, and finishes with an exploration of synchronization on networks of coupled dynamical systems.

5.1   Network structures

Networks come in many shapes and sizes, and the global dynamical properties can depend on what shape they have. Actually, shape is not quite the right description, rather, it is the network topology that determines how its properties evolve dynamically. Topology defines how the nodes of a network are connected or linked. For instance, all the nodes may be connected to all others in a network topology called a "complete graph." Or each node may be connected in a linear sequence that closes on itself in a network topology called a "linear cycle." The connectivity patterns are very different between a complete graph and a linear cycle. Networks are often defined in terms of statistics of nodes and links. For instance, the number of links that are attached to a specific node is known as the degree of a node, and one way to describe a network can be in terms of the average degree across the net. Many other types of statistical descriptions are useful as well, including measures of connectivity and clustering.

5.1.1   Types of graphs

The word "graph" is the mathematical term for a network. A graph is a set of nodes and the links that connect them. Links can be undirected, or they can be directed from one node to another. A directed link starts on one node and ends on another node. They are drawn as lines with arrows. An undirected link is two-way, and is drawn without arrows as in Fig. 5.1.[1]

Regular graphs have definite (non-random) connectivities. One example of a regular graph is the complete graph in which every node is connected to every other. Other examples of regular graphs include linear chains, cycles, trees, and lattices.

Random graphs have random connections among the nodes. There are many possible network topologies for random graphs. The three most common are: (1) Erdös–Rényi (ER) graphs[2], (2) small-world (SW) networks, and (3) scale-free (SF) networks. The ER graphs have N nodes that are randomly connected by M links. Small-world networks are characterized by many local connections, with a few long-distance connections. Scale-free networks are characterized by a few highly connected hubs, a larger number of moderately-connected hubs, and many lightly-connected nodes.

Examples of these network topologies are found in economics, sociology, computer science, biology, materials science, statistical mechanics, evolutionary dynamics, and telecommunications, among others. Perhaps the most famous examples of networks are the world-wide-web, biological neural networks and protein interaction networks in cellular biology.

[1] This chapter confines itself to undirected graphs.

[2] Paul Erdös (1913–1996) was a Hungarian mathematician who was possibly more prolific than Leonhard Euler. Of the many fields and topics he explored, it was combinatorics that lead him to study the statistical properties of random graphs, which he published in 1959 with Alfréd Rényi.

Regular graphs

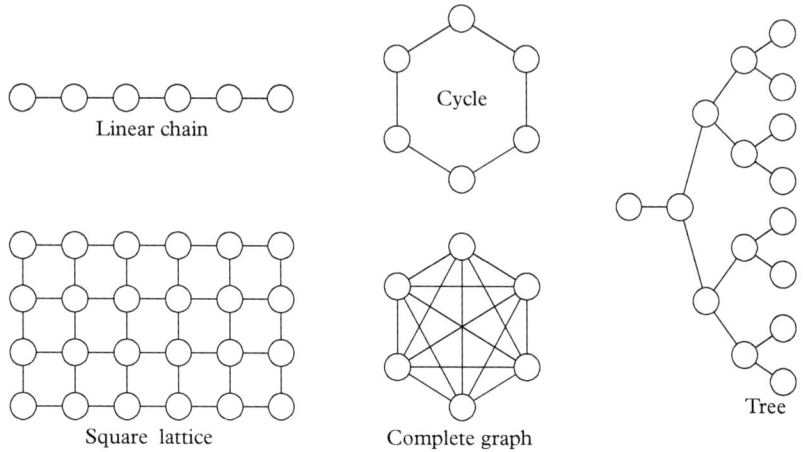

Figure 5.1 *Regular undirected graphs with nodes (circles) and links (lines).*

5.1.2   Statistical properties of networks

There are many statistical measures of the topology of networks that capture mean values as well as the statistical fluctuations. The physical behavior of dynamical networks may be distinguished depending on the different values of these measures.

5.1.2.1   Degree and moments

The degree of a node is the number of links attached to the node. The degree distribution probability is given by

$$p_k = \frac{N_k}{N},\tag{5.1}$$

where N_k is the number of nodes of degree k and N is the total number of nodes. The average degree of a network is

$$\langle k \rangle = \sum_{j=0}^{N} j p_j,\tag{5.2}$$

with higher moments given by

$$\langle k^m \rangle = \sum_{j=0}^{N} j^m p_j.\tag{5.3}$$

Not all moments are defined within a given network topology. For instance $\langle k^2 \rangle$ diverges for scale-free networks.

5.1.2.2 *Adjacency matrix*

The adjacency matrix for a graph with N nodes is an N-by-N matrix with elements

$$
\begin{aligned}
A_{ij} &= 1 \quad i \text{ and } j \text{ connected} \\
A_{ij} &= 0 \quad i \text{ and } j \text{ not connected} \\
A_{ii} &= 0 \quad \text{zero diagonal}
\end{aligned}
\tag{5.4}
$$

The degree of the ith node is given by

$$
k_i = \sum_j A_{ij}
\tag{5.5}
$$

and the average degree of the network is

$$
\langle k \rangle = \frac{1}{N} \sum_i k_i = \frac{1}{N} \sum_{ij} A_{ij}.
\tag{5.6}
$$

The adjacency matrix is symmetric and has N real eigenvalues. The eigenvalue spectrum of the adjacency matrix provides a "fingerprint" of the network topology. The spectral density of the network is defined by the eigenvalues λ_i as

$$
\rho(\lambda) = \frac{1}{N} \sum_j \delta\left(\lambda - \lambda_j\right),
\tag{5.7}
$$

which provides a convenient way to find the moments of the spectral density by

$$
\begin{aligned}
\langle \lambda^m \rangle &= \int d\lambda \, \lambda^m \rho(\lambda) = \frac{1}{N} \sum_{j=1}^{N} \left(\lambda_j\right)^m \\
&= \tfrac{1}{N} Tr\left(A^m\right).
\end{aligned}
\tag{5.8}
$$

The quantity $\langle \lambda^m \rangle$ can be interpreted as the number of closed loops of length m in the network.

5.1.2.3 *Graph Laplacian*

The graph Laplacian of a network is an operator that is analogous to the Laplacian operator of differential calculus. It is defined as

$$
L_{ij} = \left(\sum_a A_{ia}\right)\delta_{ij} - A_{ij} =
\begin{cases}
k_i & i = j \\
-1 & i \text{ and } j \text{ connected} \\
0 & \textit{otherwise.}
\end{cases}
\tag{5.9}
$$

The eigenvalues of the graph Laplacian provide another fingerprint of the network topology. For instance, the eigenvalues of the graph Laplacian are ordered as

$$
0 = \lambda_0 \leq \lambda_1 \leq \ldots \leq \lambda_{N-1} \leq N,
\tag{5.10}
$$

where the degeneracy of the eigenvalue λ_0 is the number of disconnected sub-graphs in the network. For the special case of a complete graph (every node is connected to every other), then

$$\lambda_0 = 0 \qquad \lambda_i = N. \tag{5.11}$$

On the other hand, for densely connected graphs (but not complete), the eigenvalue spectrum will have many eigenvalues close to N.

5.1.2.4 *Distance matrix*

The distance between two nodes in a graph is the smallest number of links that connects the nodes. The distance matrix is defined as the N-by-N symmetric matrix of internode distances. The shortest path between two nodes on a network is known as the geodesic path, in analogy with geodesics in metric spaces. The algorithm to obtain the distance matrix uses what is known as a breadth-first search. A node is selected, and then all nearest neighbor nodes are found that are one link away. These are tabulated as distance 1. Then *their* nearest neighbors are found and tabulated as distance 2, unless they were already assigned a prior distance. This is iterated until no new nodes are accessed. Remaining inaccessible nodes (disconnected clusters) are assigned a negative distance. An example of a distance matrix is shown in Fig. 5.2 for a random graph with 100 nodes and an average degree of 5. The maximum distance between any two nodes is equal to 6, which is defined as the network "diameter."

Distances on networks tend to be surprisingly small. This is known as the "small World" effect, also known as "six degrees of separation." The popular

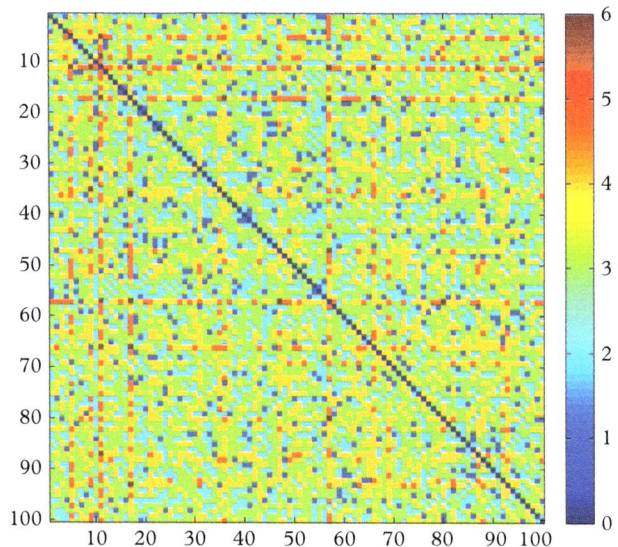

Figure 5.2 *Distance matrix for a 100-node random graph with an average degree of 5. The maximum distance between any two nodes, the network diameter, is equal to 6 links.*

description of this effect is that anyone in the United States is only six acquaintances away from the President of the United States. In other words, you know someone who knows someone who knows someone who knows someone who knows the President. The reason for this small number, in spite of nearly 400 million Americans, is that the average number of vertices that are s steps away from a randomly chosen vertex scales as $\exp(s/\langle k \rangle)$. Therefore, the average distance between nodes of a random network scales logarithmically as

$$\ell = \ell_0 + \frac{\ln N}{\ln \langle k \rangle}, \tag{5.12}$$

where ℓ_0 is a constant of order unity (in the limit of large N) that depends on the network topology. The logarithmic scaling is approximately valid for random graphs and depends on the network topology. Although it fails for some regular graphs (such as lattices), it holds true even for regular tree networks, and it is generally a good rule of thumb. For the example in Fig. 5.2, the average distance is $\ln(100)/\ln(5) \approx 3$, which is equal to the mean value of the distance matrix. The network diameter in this example is equal to 6.

5.2   Random network topologies

Random graphs are most like the networks encountered in the real world. Three of the most common are discussed here: Erdös–Rényi (ER), small-world (SW), and scale-free (SF) networks.

5.2.1   ER graphs

An ER (Erdös–Rényi) graph is constructed by taking N nodes and connecting them with an average $Nz/2$ links between randomly chosen nodes. The average coordination z of the graph is a number between zero and $N-1$. The connection probability p is related to z by

$$p = \frac{Nz}{2} \frac{2}{N(N-1)} = \frac{z}{N-1}. \tag{5.13}$$

The degree distribution for an ER graph is given by the permutations of all the possible configurations that use k connections among $N-1$ nodes. These permutations are quantified by the binomial coefficient, giving the degree distribution

$$p_k = \binom{N-1}{k} p^k (1-p)^{N-1-k}, \tag{5.14}$$

where the binomial coefficient is

$$\binom{n}{k} = \frac{n(n-1)\ldots(n-k+1)}{k(k-1)\ldots 1}. \tag{5.15}$$

For large N, this can be expanded using

$$\binom{N-1}{k} = \frac{(N-1)!}{k!\,(N-1-k)!} \simeq \frac{(N-1)^k}{k!} \tag{5.16}$$

and

$$\lim_{N\to\infty} \left(1 - \frac{x}{N}\right)^N = e^{-x} \tag{5.17}$$

to give

$$p_k = e^{-pN}\frac{(pN)^k}{k!} = e^{-z}\frac{z^k}{k!}, \tag{5.18}$$

where $z = pN$ is the average number of connections per node. This is a Poisson distribution with the mean value

$$\langle k \rangle = \sum_{k=0}^{\infty} k e^{-z}\frac{z^k}{k!} = z e^{-z}\sum_{k=1}^{\infty}\frac{z^{k-1}}{(k-1)!} = z. \tag{5.19}$$

The mean squared value for a large ER graph is

$$\langle k^2 \rangle = \sum_{k=0}^{\infty} k^2 e^{-z}\frac{z^k}{k!} = z e^{-z}\sum_{k=1}^{\infty} k\frac{z^{k-1}}{(k-1)!} = z\,(z+1) \approx \langle k \rangle^2, \tag{5.20}$$

But this is a special case that holds for the ER graph and is not satisfied for general random graphs. An example of an ER graph is shown in Fig. 5.3 for $N = 64$ and $p = 0.125$. The connectivity diagram in (Fig. 5.3(a) shows the nodes and links, and the distance matrix is in Fig. 5.3(b). The diameter of the network is equal to 4, and the mean degree is 7.9.

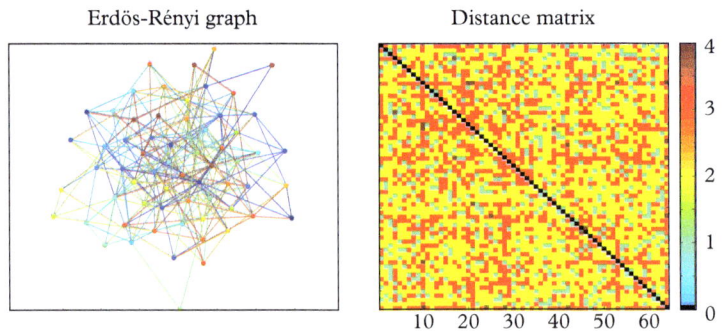

Erdös-Rényi graph   Distance matrix

Figure 5.3 *(a) Erdös–Rényi random graph connectivity diagram for N = 64 and p = 0.125. (b) Distance matrix. This graph has a diameter equal to 4 links. The average degree is* $\langle k \rangle$ *= 7.9, and there are 252 edges.*

5.2.2   Small-world networks

Small-world (SW) networks provide a continuous bridge between regular graphs and completely random (ER) graphs. This connection was established by Watts and Strogatz in 1998 (Watts, 2002). Small-world networks have the property of local clusters of short-range connections, with a few long-range connections that jump between local clusters. The existence of local clusters is a natural organizational principle in networks. These are cliques (using a social term), and like all cliques, they include a few individuals that belong to other cliques. A parameter p ranges from 0 to 1 to bring the Strogatz–Watts network continuously from a regular graph to an SW graph. This parameter is a rewiring probability.

The Strogatz–Watts algorithm is

1. Define a regular linear lattice with an average degree k and cyclic boundary conditions.
2. Pick each end of a link and move it to a randomly selected node with probability p, or keep it attached to the same node with probability $(1 - p)$.
3. Continue until each end of each link has been processed.

This procedure, for mid-range values of the rewiring probability, produces clusters of locally-connected nodes that are linked by a few long-distance links to other local clusters. This "small-world network" has the important property that the clustering is high, while the average distance between two nodes is small. These networks have groups of highly interconnected nodes, in cliques that tend to separate the network into distinct subgroups. Yet every node is only a few links away from any other node. The rewiring of a small graph is shown in Fig. 5.4, and a small-world graph with $N = 64$, degree $k = 8$, and $p = 0.1$ is shown in Fig. 5.5 with its distance matrix.

5.2.3   Scale-free networks

Many real-world networks grow dynamically, like new web pages popping up on the World Wide Web. These new nodes often are not connected randomly

$k = 4$

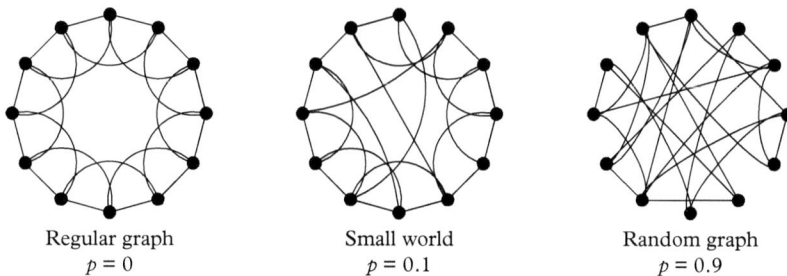

| Regular graph | Small world | Random graph |
| $p = 0$ | $p = 0.1$ | $p = 0.9$ |

Figure 5.4 *Small-world networks of degree 4 with rewiring probability p varying from 0 to 1.*

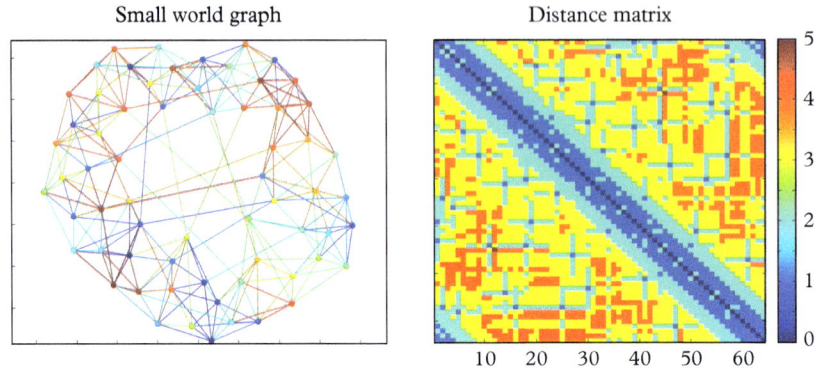

Small world graph   Distance matrix

Figure 5.5 *(a) Small-world (SW) connectivity diagram for N = 64 nodes, 8 links per node, and a rewiring probability of 10%. (b) Distance matrix. This graph has a diameter equal to 5 links. The average degree is ⟨k⟩ = 7.9, and there are 253 edges.*

to existing nodes, but instead tend to preferentially make links to nodes that already have many links. This produces a "rich-get-richer" phenomenon in terms of web page hits, as early nodes become hubs with high degree (very popular), attaching to later hubs of lower degree (less popular), attaching to hubs of yet lower degree, and eventually to individuals with degree of one (loners). This type of network topology is called a "scale-free" network. The special property of scale-free networks is that their degree distribution obeys a power-law function

$$p_k \sim k^{-\gamma}, \tag{5.21}$$

in which γ tends to be between 2 and 3 in real-world examples.

The scale-free algorithm has a single iterative step:

> To a number m_0 of existing nodes, add a new node and connect it with a fixed number $m < m_0$ links to the existing nodes with a preferential attachment probability to the ith node given by
>
> $$P_i = \frac{k_i}{\sum_j k_j}$$
>
> Then repeat for t time steps.

For sufficiently large networks, this yields a scale-free network with an exponent $\gamma = 3$.

An example of a scale-free network is shown in Fig. 5.6 for $N = 64$ nodes and 4 links added per new node. The highest degree nodes are near the center of the network, and the low-degree nodes are on the outside. The average degree is 7.9, but the maximum degree is 26 towards the center of the figure and corresponding with the small distances at the upper left corner of the distance matrix. The network diameter is equal to 4.

Scale free graph   Distance matrix

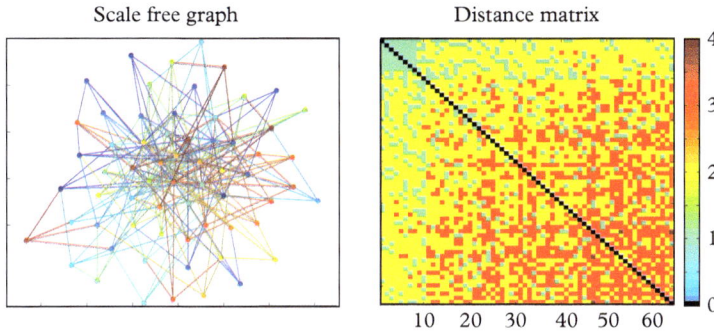

Figure 5.6 *(a) Scale-free (SF) network connectivity diagram for N = 64 nodes and 4 links per new node. (b) Distance matrix. This graph has a diameter equal to 4 links. The average degree is ⟨k⟩ = 7.9, and there are 252 edges.*

5.3 Diffusion and epidemics on networks

The principal property of networks is their connectivity which facilitates the movement of conditions, such as states, information, commodities, disease or whatever can be "transmitted" from one node to another. The movement of some quantity over time is known as transport, and it is a dynamic process. For physical transport in many real-world systems, spatial distances are important, but in this chapter we only consider topological connectivity with "distanceless" links. For instance, diffusion will be a random walk on a network. Diffusion across networks is responsible for the epidemic spread of viruses or disease, of "viral" YouTube videos, of rumors (both good and bad), memes and ideas. Therefore, the dynamical properties of diffusion on networks is an important and current topic of interest, and shares much in common with evolutionary dynamics (Chapter 7), that occurs within networks of interacting species.

5.3.1 Percolation on networks

The propagation of a state or condition on a network requires most of the nodes of a graph to be connected to a single giant cluster. For instance, the goal for the internet is to have every node connected within a single network. On the other hand, to prevent the propagation of viruses, one might like the network to have tenuous links that could be vaccinated to break apart the network into smaller isolated clusters that would be protected, as behind a firewall. There is often a single giant cluster that contains most of the nodes of the network, and the network is said to *percolate*. The word "percolation" comes from the notion of information, or some other condition, percolating across a network from node to node.

Not all networks have a giant component (percolation cluster), so it is important to find what conditions are necessary for a giant component to exist. It is also important to find what fraction of nodes belongs to the giant component, and how changes in the network properties affect the size of the giant component. For instance, there is usually a percolation threshold, at some finite value of a network parameter, at which the size of the giant component vanishes. Different types of networks have different percolation thresholds, and each case has to be investigated individually.

The random ER graph is one type of graph for which the percolation threshold can be studied analytically. The fraction of nodes S in the giant component is a function of the average degree $\langle k \rangle$. S is obtained by the solution to

$$S = 1 - \exp\left(-\langle k \rangle S\right), \tag{5.22}$$

which is a transcendental equation that can be solved graphically or numerically for the fraction S as a function of $\langle k \rangle$. For a graph with increasing average degree (by adding links to an existing sub-percolating graph), S rises from zero at the percolation threshold when

$$\frac{d}{dS}\left(1 - \exp\left(-\langle k \rangle S\right)\right) = 1, \tag{5.23}$$

for which

$$\langle k \rangle \exp\left(-\langle k \rangle S\right) = 1. \tag{5.24}$$

At the threshold $S = 0$, so the threshold for the random graph is

$$\langle k \rangle = 1. \tag{5.25}$$

This means that when the average degree exceeds unity in an ER graph, there is a giant component that contains a finite fraction of the nodes of the network. The size of the giant component as a function of $\langle k \rangle$ is shown in Fig. 5.7. For average degree less than unity, there are many small clusters, but no single cluster contains most of the nodes. As the average degree crosses $\langle k \rangle = 1$, the size grows linearly above threshold. The slope of the curve is discontinuous at the threshold, but the fraction is continuous. This percolation transition is known as a continuous phase transition, or as a second-order phase transition.[3] An important condition for a well-defined percolation transition to exist is to have a system that is very large, with a large number N of nodes. For smaller networks, the threshold is more smeared out. The strict existence of a threshold only takes place in the limit as N goes to infinity. But for practical purposes, a network with very large N is sufficient to display a relatively sharp transition. As a rule of thumb, one would want $N > 1000$ for a sharp transition.

It often happens that a starting network has a giant component, but then some nodes (and their links) are removed. A natural question is whether the giant component falls apart (goes through a percolation transition), and how the size of the giant component depends on the fraction of deleted nodes. This question is important for real-world networks like computer networks. The resilience of a network to attack or to partial failure defines the robustness of the network.

Consider a random ER graph with an average degree $\langle k \rangle$ and average squared degree $\langle k^2 \rangle$. The nodes are all originally in place. Then individual nodes are

[3] Another famous example of a continuous phase transition is the superconductivity transition.

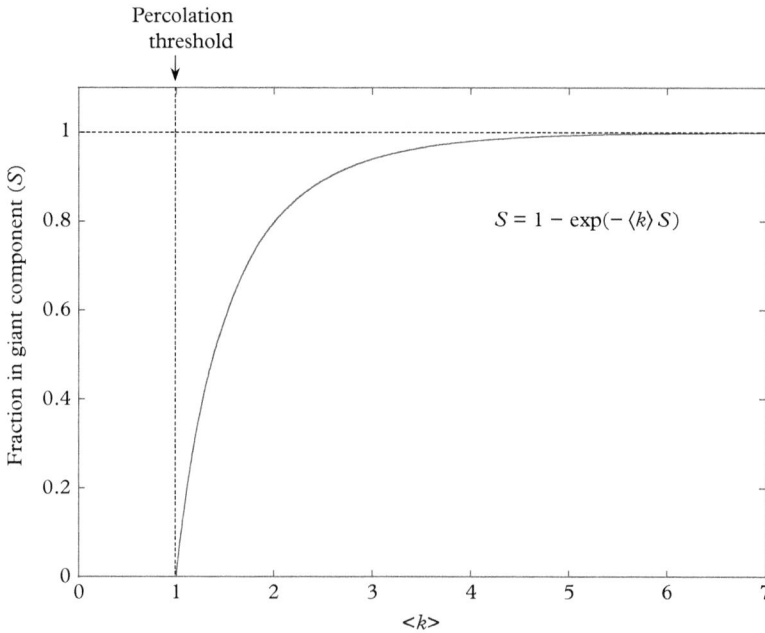

Figure 5.7 *Size of the giant component vs. average degree $\langle k \rangle$ for a random graph. There is a giant component for $\langle k \rangle > 1$.*

deleted randomly, as well as their associated links. The fraction of nodes of the total that remain is called the occupation probability ϕ. As the occupation probability decreases, the fraction of nodes S that belong to the giant component decreases. Eventually, so many nodes have been removed, that the giant component vanishes at a threshold value of the occupation probability ϕ_c. For a random graph with randomly deleted nodes, this threshold is

$$\phi_c = \frac{\langle k \rangle}{\langle k^2 \rangle - \langle k \rangle}, \qquad (5.26)$$

which, for a Poisson degree distribution, is

$$\phi_c = \frac{1}{\langle k \rangle}. \qquad (5.27)$$

On the other hand, for a power-law degree distribution with a power between 2 and 3, the squared mean $\langle k^2 \rangle$ diverges to infinity and hence $\phi_c = 0$. This means that power-law distributions are extremely robust. Virtually all of the nodes must be removed to destroy the giant component. The size of the giant cluster is plotted in Fig. 5.8 for an exponential ($p_k \propto e^{-\lambda k}$) and for a power-law ($p_k \propto k^{-\alpha}$) degree distribution. There is a clear percolation threshold for the exponential case, but not for the power-law case.

There are many types of degree distributions, many types of network topologies, and many different ways nodes can be removed. Thresholds and sizes need

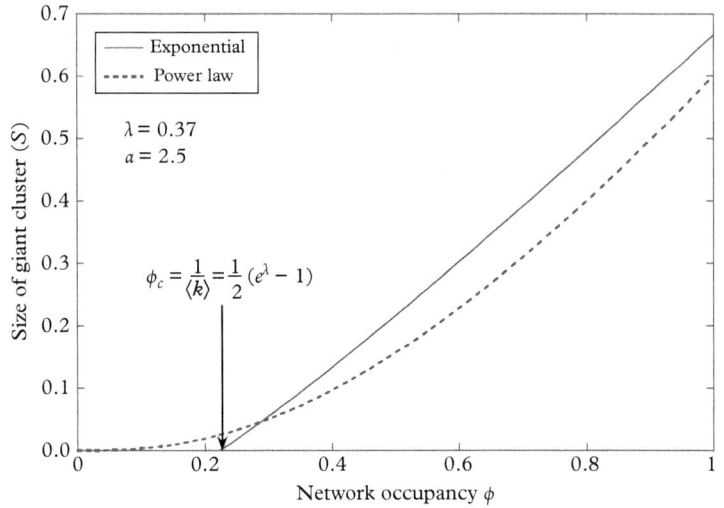

Figure 5.8 *Size of the giant component S vs. occupation probability ϕ. An exponential degree probability distribution with $\lambda = 0.37$ is compared to a power-law degree distribution with $\alpha = 2.5$.*

to be calculated for each case. For instance, while being robust against *random* attacks, power-law networks can be badly vulnerable if the attack is directed preferentially at the highest-degree nodes. Conversely, epidemic spread can be effectively stopped by vaccinating these high-degree nodes. These topics are beyond the scope of this textbook. For detailed discussions, see Newman (2010).

5.3.2   Diffusion

Diffusion on networks is a combination of physical diffusion and percolation. For a percolation analysis, the state of a node is all or nothing. For diffusion, on the other hand, the interest is on continuous values of concentrations or other continuous properties on a node, and how these values are transported through different network structures.

The change of a property or concentration c_i of the ith node is related to the adjacent values of that property on neighboring nodes. The rate of change is given by

$$\frac{dc_i}{dt} = \beta \sum_j A_{ij} \left(c_j - c_i \right), \tag{5.28}$$

where A_{ij} is the adjacency matrix and β is the diffusion coefficient. This becomes

$$\frac{dc_i}{dt} = \beta \sum_j A_{ij} c_j - \beta c_i \sum_j A_{ij} = \beta \sum_j A_{ij} c_j - \beta c_i k_i$$

$$= \beta \sum_j \left(A_{ij} - \delta_{ij} k_i \right) c_j, \tag{5.29}$$

where use has been made of the degree of the ith node

$$\sum_j A_{ij} = k_i. \tag{5.30}$$

In matrix form this is

$$\frac{d\mathbf{c}}{dt} = \beta\,(\mathbf{A} - \mathbf{D})\,\mathbf{c}, \tag{5.31}$$

where the diagonal matrix **D** is

$$\mathbf{D} = \begin{pmatrix} k_1 & 0 & 0 & 0 \\ 0 & k_2 & 0 & 0 \\ 0 & 0 & k_3 & 0 \\ 0 & 0 & 0 & \dots \end{pmatrix}. \tag{5.32}$$

A matrix **L** is defined as

$$\mathbf{L} = \mathbf{D} - \mathbf{A}, \tag{5.33}$$

which is recognized as the graph Laplacian of Eq. (5.9). Using the graph Laplacian, the dynamical flow for diffusion on a network becomes

Diffusion on a network:
$$\frac{d\mathbf{c}}{dt} + \beta\mathbf{L}\mathbf{c} = 0, \tag{5.34}$$

where L plays a role analogous to that of the Laplacian $(-\nabla^2)$ in continuous systems. This is the reason why L is called the graph Laplacian, and it appears in many instances of dynamic properties of networks.

The solution to Eq. (5.34) is obtained as a linear combination of eigenvectors of the Laplacian

$$\mathbf{c}\,(t) = \sum_i a_i\,(t)\,\mathbf{v}_i. \tag{5.35}$$

The eigenvectors of the Laplacian satisfy $\mathbf{L}\mathbf{v}_i = \lambda_i\mathbf{v}_i$, which allows Eq. (5.34) to be written as

$$\sum_i \left(\frac{da_i}{dt} + \beta\lambda_i a_i\right)\mathbf{v}_i = 0. \tag{5.36}$$

The solutions are

$$a_i\,(t) = a_i\,(0)\,\mathrm{e}^{-\beta\lambda_i t}, \tag{5.37}$$

which decay exponentially from the initial values. This is not to say that all concentrations at all nodes decay. For instance, if the initial concentration is localized on one part of the network, then the concentration everywhere else is zero. But these zero concentrations are because of the cancellation of the coefficients of the

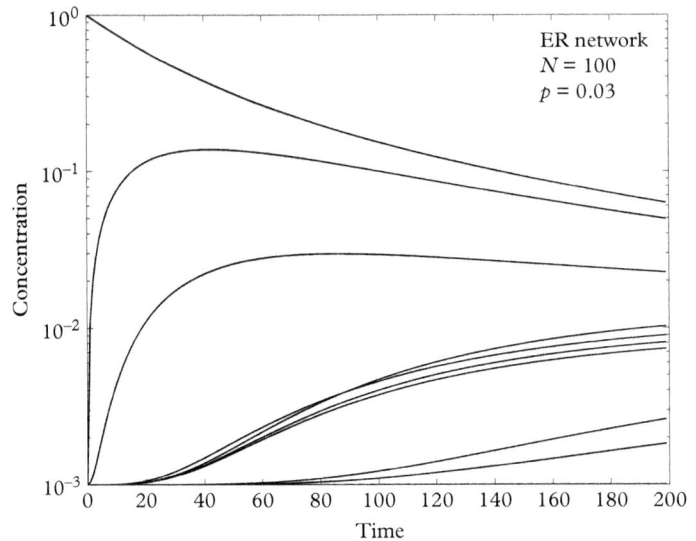

Figure 5.9 *Concentrations on selected nodes of a ER network with N = 100 and link probability = 0.03. The beginning concentration is isolated to a single node and diffuses to its neighbors. In the long-time limit, the concentration asymptotes to 0.01.*

eigenvectors in Eq. (5.35) in the initial condition. As the system evolves, and the coefficients decay, this cancellation is removed and the concentration increases on most of the nodes. This is just what diffusion means: an initial concentration diffuses into the network until there is a uniform concentration everywhere.

An example of diffusion is shown in Fig. 5.9 for an ER network with $p = 0.03$ and $N = 100$. The concentrations of 10 selected nodes are shown in the figure, including the single initial node that was given a concentration equal to unity. The original node's concentration decays exponentially. Some of the neighboring nodes increase initially, then decrease to reach steady state, while other farther neighbors rise asymptotically to the steady state. At long times, all concentrations eventually become equal as the system reaches equilibrium.

5.3.3   Epidemics on networks

One of the consequences of connected networks is the possibility for the spread of an infectious agent. Among computers, the infectious agent can be a digital virus that is spread through email or through downloads. In forests, the infectious agent can be a beetle or a forest fire that spreads from nearest neighbors and sweeps across the landscape. Among humans, the infectious agent can be a virus or a bacterium that is passed by physical contact. Alternatively, the infectious agent can be a thought, or a phrase (a meme) that sweeps across a population. All of these instances involve dynamics as the states of numerous individuals within the network become the multiple variables of a dynamical flow. The rate of change of the individual states, and the rate of spread, are directly controlled by the interconnectivity of the specific networks.

Table 5.1 *Three common infection models*

SI	Susceptible-infected
SIS	Susceptible-infected-susceptible
SIR	Susceptible-infected-removed

In the study of epidemics, there are three simple models of disease spread that capture many of the features of real-world situations. These are known as the SI, SIS, and SIR models (Table 5.1). SI stands for susceptible-infected, SIS stands for susceptible-infected-susceptible, and SIR stands for susceptible-infectious-removed. In the SI model, an individual can either be susceptible (not yet infected) or infected. Once infected, the individual remains infected and can spread the disease to other individuals who are susceptible. Obviously, the end state of the SI model is complete infection within any connected cluster of individuals. In the SIS model, an individual is either susceptible or infected, but can recover after being infected to become susceptible again. In the SIR model there are three states: susceptible, infectious, and removed. In the SIR model, after being infected an individual recovers, is no longer infectious and has acquired immunity to the disease. The end states of the SIS and SIR models depend on the character of the infection and on the connectivity properties of the network of interactions. In these models, a disease may cause an epidemic, or may die out. Even in the SIS model, in which individuals always become susceptible again, an epidemic may not spread if the recovery rate is fast enough.

As a first approach to understand the qualitative differences among the SI/SIS/SIR epidemic models, it can be assumed (to lowest-order) that the network has a homogeneous connectivity in which all individuals interact with $\langle k \rangle$ others. This is not always a good assumption for real networks, because it ignores the actual heterogeneous connectivity, but it does show how epidemic thresholds can be reached. The dynamic equations and fixed points for these homogeneous models are given in Table 5.2. The diffusion coefficient is β, the decay rate is μ, $i(t)$ is the infected population, and $r(t)$ is the recovered population. The individual nodes have disappeared in these homogeneous models, and have been replaced by averages.

The fixed points for each model have different saturation levels. Aside from the trivial solution of no initial infection, $i^* = 0$, the SI model evolves to complete saturation while the SIS model evolves to a steady-state nonzero level of infection equal to $(1 - \mu/\langle k \rangle \beta)$. In the SIR model everyone eventually is removed (either by death or by immunity).

The stability analysis around the fixed point $i^* = 0$ shows a very important aspect of epidemic spread. The Lyapunov exponents for the three models are

$$
\begin{aligned}
\lambda_{SI} &= \langle k \rangle \beta \\
\lambda_{SIS} &= \langle k \rangle \beta - \mu \\
\lambda_{SIR} &= \langle k \rangle \beta - \mu.
\end{aligned}
\tag{5.38}
$$

Table 5.2 *Homogeneous infection models*

SI: $$\frac{di(t)}{dt} = \langle k \rangle \beta i(t)[1 - i(t)]$$	$i^* = [0, 1]$
SIS: $$\frac{di(t)}{dt} = -\mu i(t) + \langle k \rangle \beta i(t)[1 - i(t)]$$	$i^* = \left[0, \left(1 - \frac{\mu}{\langle k \rangle \beta} \right) \right]$
SIR: $$\frac{di(t)}{dt} = -\mu i(t) + \langle k \rangle \beta i(t)[1 - r(t) - i(t)]$$ $$\frac{dr(t)}{dt} = \mu i(t)$$ $$s(t) = 1 - i(t)$$	$[i^*, r^*] = [0, 0], [0, 1]$

For the SI model, the fixed point at $i^* = 0$ is unstable and leads to exponential growth. On the other hand, for the SIS and the SIR models, there is a distinct threshold for the spread of the epidemic. If the recovery rate $\mu > \langle k \rangle \beta$, then the fixed point at $i^* = 0$ is stable and the infection never spreads. This is because individuals recover faster than they infect other susceptible individuals. However, once the threshold is exceeded, the epidemic spreads exponentially until it saturates to the final steady state.

The existence of an epidemic threshold is a crucial concept with widespread ramifications, and is also a key to preventing epidemics through immunization. If a fraction g of individuals are immunized, then the rate of infection becomes $\beta(1 - g)$. This leads to an immunization threshold given by

$$g_c = 1 - \frac{\mu}{\langle k \rangle \beta}. \tag{5.39}$$

Once the fraction of immunized individuals is larger than the threshold, the infection cannot spread exponentially. Furthermore, by sustaining the immunization, the disease can eventually die out and disappear forever, even if not everyone in the network is immunized. Only a critical fraction needs to be immunized to eradicate the disease (if it does not lie dormant). This is how the World Health Organization eradicated smallpox from the human population. These approximate conclusions apply to networks with relatively high degrees and high homogeneity. For heterogeneous networks, such as scale-free networks and small-world networks, the quantitative spread of disease depends on the details of the degree distributions and the on eigenvalues and eigenvectors of the adjacency matrix.

Diffusion on a network represents a dynamical process with a multidimensional dynamical flow equation (Eq. (5.34)) whose solution depends on the network topology through the adjacency matrix and the graph Laplacian. Critical phenomena, such as percolation thresholds, arise in the static structural properties of the network connectivity, but dynamic thresholds may also occur, such as an immunization threshold. In this sense, dynamics on networks becomes a probe of the network properties. A different type of dynamical process on networks is the synchronization of individual oscillators coupled through the network links. Therefore, synchronization also acts as a probe of network topology through the adjacency matrix and the graph Laplacian, and these dynamics can display additional types of critical phenomena such as a global synchronization threshold.

5.4   Linear synchronization of identical oscillators

Networks of interacting elements are often distinguished by synchronized behavior of some or all of the nodes. The synchronization properties of networks are studied by populating the nodes of a network with a dynamical system, and then coupling these according to the network topology. One particularly simple form of synchronization occurs when all the nodes have the same time evolution. The question is, what properties do the node dynamics need to have to allow such a solution? This section explores the problem of networks of identical (possibly chaotic) oscillators with *linear* coupling, and then in the next section turns to *nonlinear* coupling among non-identical oscillators.

To begin, consider a multi-variable flow in which each variable corresponds to the state of a node, and each node has the same one-variable dynamics:

$$\frac{d\phi_i}{dt} = F(\phi_i) \tag{5.40}$$

for N nodes whose states have the values ϕ_i. When these nodes are interconnected by links into a network, their dynamics may depend on the states of their neighbors to which they are coupled. If the coupling is linear, the time evolution becomes

$$\frac{d\phi_i}{dt} = F(\phi_i) + g \sum_{j=1}^{N} C_{ij} H(\phi_j), \tag{5.41}$$

where g is the linear coupling coefficient, C_{ij} is the coupling matrix, and $H(\phi_j)$ is a response function. If we make the further restriction that the coupling is proportional to the difference between the outputs, then the evolution equations become

$$\frac{d\phi_i}{dt} = F(\phi_i) + g \sum_{j=1}^{N} L_{ij} H(\phi_j), \tag{5.42}$$

where L_{ij} is the graph Laplacian. The graph Laplacian has the values $L_{ij} = 0$ if the nodes are not connected, $L_{ij} = -1$ if they are connected, and $L_{ii} = k_i$ is the degree of the ith node.

A trivial synchronized solution to the linearly-coupled equation is when all oscillators have identical time evolution $s(t)$. In this case the coupling term vanishes and each oscillator oscillates with the same phase. For chaotic oscillators, such as the Rössler or Lorenz oscillators, keeping the same phase is not allowed because of the sensitivity to initial conditions, and the relative phase would slowly drift among the many coupled chaotic oscillators in the absence of coupling. However, in the presence of coupling, the key question is whether perturbations to the time evolution of $s(t)$ grow or decay in time.

To answer this question, consider a perturbation such that

$$\phi = s + \xi \tag{5.43}$$

and the oscillator and response functions are expanded as

$$\begin{aligned} F(\phi_i) &\approx F(s) + \xi_i F'(s) \\ H(\phi_i) &\approx H(s) + \xi_i H'(s). \end{aligned} \tag{5.44}$$

The evolution of the perturbation is therefore given by

$$\frac{d\xi_i}{dt} = F'(s)\,\xi_i + g \sum_{j=1}^{N} L_{ij} H'(s)\,\xi_i. \tag{5.45}$$

These equations represent N coupled differential equations that are coupled through the graph Laplacian L_{ij}. This equation can be re-expressed in terms of the eigenvalues and eigenvectors of the graph Laplacian as

$$\frac{dv_i}{dt} = \left[F'(s) + g\lambda_i H'(s) \right] v_i, \tag{5.46}$$

where the λ_i are the eigenvalues of the graph Laplacian and the v_i are the eigenvectors. For the system to be synchronized, the term in brackets must be negative for all eigenvalues λ_i and for all values of the free-running solution $s(t)$. This makes it possible to define a master function

$$\Lambda(g) = \max_{s(t),\lambda_i} \left[F'(s) + g\lambda_i H'(s) \right], \tag{5.47}$$

which is maximized over the free-running trajectory and over all the eigenvalues of the graph Laplacian. Whenever this quantity is negative, for given network topologies and appropriate values of g, then the system is synchronizable.

Many physical systems have a general structure to the master function in which $\Lambda(g)$ is positive for small coupling, negative for intermediate coupling strength, and positive for large coupling strength. In this case, there are threshold values for g, shown in Fig. 5.10. Such systems are subject to "over-coupling," in which synchronization is lost if coupling strengths get too large. Therefore, for global synchronization, the coupling g must be between $g_{\min}$ and $g_{\max}$ and

$$\frac{\lambda_{\max}}{\lambda_2} < \frac{g_{\max}}{g_{\min}}, \tag{5.48}$$

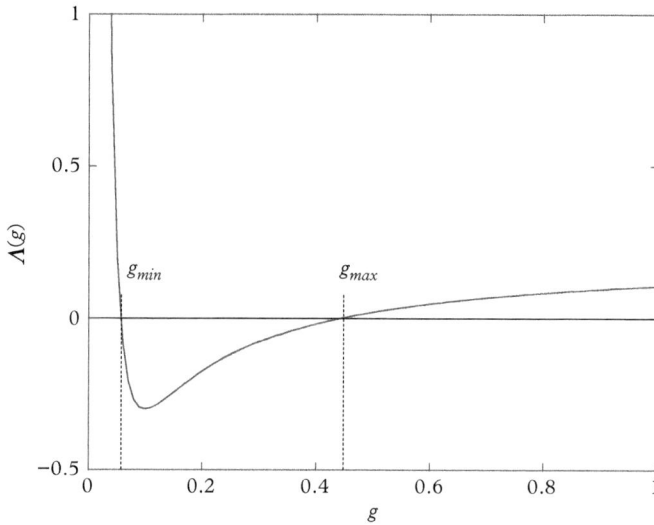

Figure 5.10 *Conceptual graph of $\Lambda(g)$ as a function of coupling g for a broad class of dynamical networks. When the eigenvalue is negative, the global synchronized state is stable. In this example, the system globally synchronizes for intermediate values of coupling, but synchronization is destroyed for too weak or too strong coupling.*

where $\lambda_{\max}$ is the largest eigenvalue of the graph Laplacian and λ_2 is the smallest nonzero eigenvalue. (The graph Laplacian is guaranteed to have one eigenvalue equal to zero because each row sums to zero.)

This relation is an important result that holds for many dynamical systems. It states that, as the ratio $\lambda_{\max}/\lambda_2$ decreases, the synchronizability of a network increases. In other words, networks with very narrow ranges for their eigenvalues are more easily synchronized than networks with a wide range of eigenvalues. Although the details may vary from system-to-system, and synchronization is not always guaranteed, this relation provides good estimates of synchronization, and is good at predicting whether changes in a network will cause it to be more easily or less easily synchronized.

An example is given in Fig. 5.11 for 50 coupled Rössler chaotic oscillators on an ER graph with $N = 50$ and $p = 0.1$. The network has a diameter equal to 5. This case is shown for a weak coupling coefficient $g = 0.002$ that plots the y-value of the 33rd node in the network against the x-value of the 13th node—nodes which are one network diameter apart. Although the relative phase between the two nodes evolves monotonically (governed by the choice of oscillator parameters), the trajectory is chaotic. When the coupling is increased to $g = 0.5$, the nodes synchronize across the network. This is shown in Fig. 5.12, which plots the y-value of the 33rd oscillator against the x-value of the 13th oscillator. In this case, the chaotic oscillators are synchronized.

This section provided an analytical approach (Eq. (5.47)) to the prediction of synchronizability of a network of identical nonlinear oscillators that are linearly coupled. Analytical results are sometimes difficult to find in complex network dynamics, and one often must resort to computer simulations in specific cases. Analytic results are also possible, even in the case of nonlinear coupling among

Figure 5.11 *N = 50 Rössler oscillators on an ER graph with p = 0.1 (a = 0.15, b = 0.4, c = 8.5) with coupling g = 0.002. The graph on the left is the (x_1, x_2) strange attractor of the 13th oscillator. The graph on the right is x_1 of the 13th oscillator plotted against x_2 of the 33rd oscillator. There is no apparent structure in the figure on the right (no synchronization).*

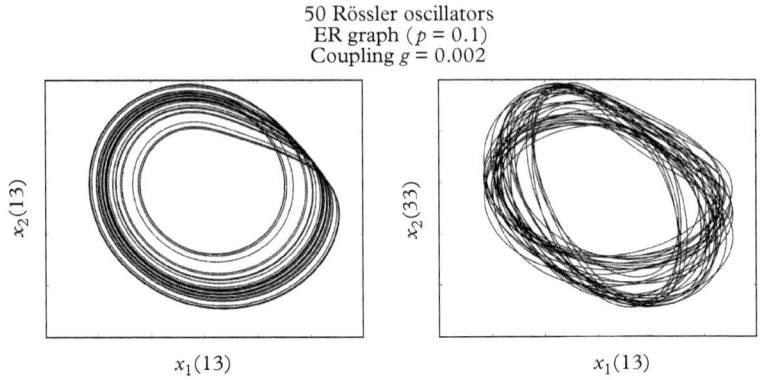

50 Rössler oscillators
ER graph ($p = 0.1$)
Coupling $g = 0.002$

Figure 5.12 *N = 50 Rössler oscillators on the ER graph of Fig. 5.11 with coupling g = 0.25. The graph on the left is the (x_1, x_2) strange attractor of the 13th oscillator. The graph on the right is x_1 of the 13th oscillator plotted against x_2 of the 33rd oscillator. Despite the large distance separating oscillator 13 and 33 (network diameter = 5), the strange attractors are identical.*

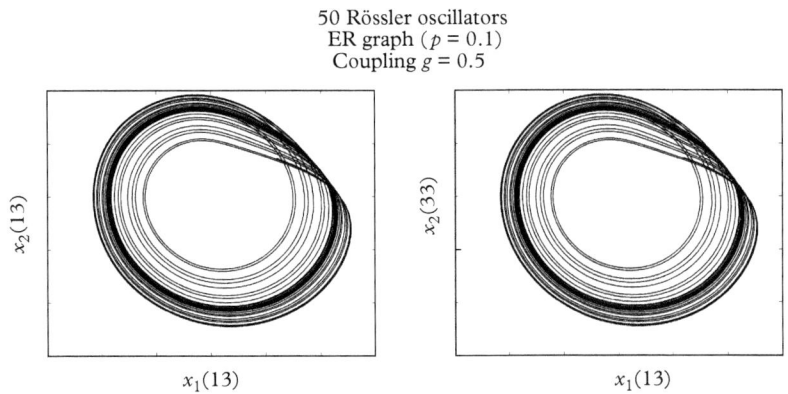

50 Rössler oscillators
ER graph ($p = 0.1$)
Coupling $g = 0.5$

non-identical oscillators, as long as the coupling is on a complete graph—meaning that every node is coupled to every other. This is the case for the Kuramoto model of coupled phase oscillators.

5.5 Nonlinear synchronization of coupled phase oscillators on regular graphs

The synchronization of non-identical nonlinearly-coupled phase oscillators (each oscillator has a different frequency) on regular graphs can demonstrate the existence of global synchronization thresholds that represent a phase transition. In particular, the complete graph displays a Kuramoto transition to global synchronization which is analyzed using mean field theory.

5.5.1   Kuramoto model of coupled phase oscillators on a complete graph

An isolated phase oscillator (also known as a Poincaré oscillator, see Eq. 2.177), has the simplest possible flow given by

$$\frac{d\phi_k}{dt} = \omega_k \qquad (5.49)$$

in which the kth oscillator has a natural frequency ω_k. A population of N phase oscillators can be coupled globally described by the flow

$$\frac{d\phi_k}{dt} = \omega_k + g\frac{1}{N}\sum_{j=1}^{N}\sin\left(\phi_j - \phi_k\right) \qquad (5.50)$$

in which the coupling constant g is the same for all pairs of oscillators.

To gain an analytical understanding of the dynamics of this system we use an approach called a mean-field theory. In mean-field theory, each oscillator is considered to interact with the average field (mean field) of all the other oscillators. This approach extracts the average behavior of the network as a function of the coupling strength and of the range in natural frequencies $\Delta\omega = \text{std}\,(\omega_k)$ where std stands for standard deviation of the distribution of ω_k values. The mean field is a single complex-valued field given by

$$Ke^{i\Theta} = \frac{1}{N}\sum_{k=1}^{N} e^{i\phi_k} \qquad (5.51)$$

with mean amplitude K and mean phase Θ. The population dynamics are rewritten in terms of the mean field values K and Θ as

$$\frac{d\phi_k}{dt} = \omega_k + gK\sin\left(\Theta - \phi_k\right). \qquad (5.52)$$

The mean-field values have the properties

$$\begin{aligned}
\Theta &= \bar{\omega}t \\
K &= const. \\
\psi_k &= \phi_k - \bar{\omega}t,
\end{aligned} \qquad (5.53)$$

where the last line is the slow phase that was encountered previously in Chapter 4. The evolution becomes

$$\begin{aligned}
\frac{d\psi_k}{dt} &= (\omega_k - \bar{\omega}) - gK\sin\psi_k \\
&= \omega_k' - gK\sin\psi_k.
\end{aligned} \qquad (5.54)$$

From this equation, we see that we have a multi-oscillator system analogous to a collection of driven oscillators in which the mean field is the driving force. Each oscillator is driven by the mean field of the full collection. The oscillators that are entrained by the mean field contribute to the mean field.

There is a synchronous solution defined by

$$\psi_k = \sin^{-1}\left(\frac{\omega_k - \bar{\omega}}{gK}\right) \tag{5.55}$$

for $gK > |\omega_k - \bar{\omega}|$, and the number of oscillators that are synchronized is

$$n_s(\psi) = \rho(\omega)\left|\frac{d\omega}{d\psi}\right| = gK\rho(\bar{\omega} + gK\sin\psi)\cos\psi, \tag{5.56}$$

where $\rho(\omega)$ is the probability distribution of initial isolated frequencies. The mean field is caused by the entrained oscillators

$$\begin{aligned} Ke^{i\bar{\omega}t} &= \int_{-\pi}^{\pi} e^{i\psi + i\bar{\omega}t} n_s(\psi)\,d\psi \\ &= gKe^{i\bar{\omega}t} \int_{-\pi/2}^{\pi/2} \cos^2\psi\,\rho(\bar{\omega} + gK\sin\psi)\,d\psi, \end{aligned} \tag{5.57}$$

which is a self-consistency equation for K. This can be solved for a Lorentzian distribution of starting frequencies

$$\rho(\omega) = \frac{\gamma}{\pi\left[(\omega - \bar{\omega})^2 + \gamma^2\right]} \tag{5.58}$$

with the solution

$$g_c = \frac{2}{\pi\rho(\bar{\omega})} = 2\gamma$$
$$K = \sqrt{\frac{g - g_c}{g}}. \tag{5.59}$$

The most important feature of this solution is the existence of a threshold g_c and the square root dependence of the mean field K on $g - g_c$ above threshold.[4] As the coupling strength increases, at first there is no global synchronization as all oscillators oscillate at their natural frequencies. But when g increases above a threshold that is related to the width of the distribution of natural frequencies, then subclusters synchronize. The size of the subclusters increases as the square root of g as it increases above the threshold. Far above the threshold, the full system synchronizes to the single common frequency $\bar{\omega}$. The transition can be relatively sudden, as shown in Fig. 5.13 for $N = 256$ phase oscillators.

5.5.2 Synchronization and topology

Complete graphs in dynamical systems are an exception rather than a rule. In many physical systems, nearest neighbors tend to interact strongly because

[4] This square-root dependence is common for mean-field solutions, for instance for magnetism in solid state physics.

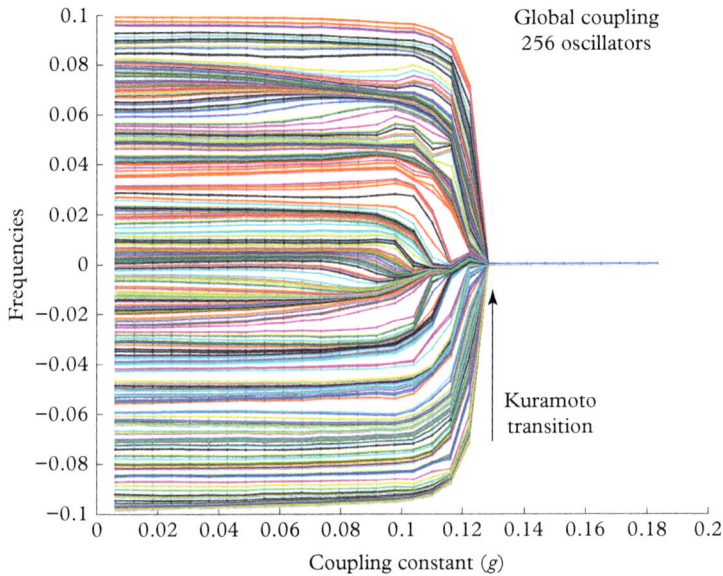

Figure 5.13 *Kuramoto synchronization transition as a function of coupling g for N = 256 oscillators uniformly distributed in frequency between −0.1 and 0.1. The graph has global (all-to-all) coupling. The global entrainment transition is sharp, although small groups of oscillators with similar frequencies frequency-lock at lower couplings.*

of their close spatial proximity, while farther neighbors interact weakly, if at all. Therefore, in most physical cases, local connectivity is favored over global connectivity. As an example, spins in magnetic systems tend to interact strongly locally, as do the network of pacemaker cells in the heart.

Perhaps one of the simplest topologies in physical networks are lattices. Lattices are common in solid state physics, but also apply to more abstract systems, such as percolating systems. The simplest lattice networks have only nearest-neighbor connections. In some sense, a lattice is at one extreme of local connectivity, while global connectivity is at the other extreme. Therefore, we may expect a different behavior for a lattice than for global connectivity.

Consider a 16-by-16 square lattice of non-identical phase oscillators. The degree of the graph is equal to 4. The frequencies as a function of coupling strength are shown in Fig. 5.14 with a gradual coalescence of the frequencies around the average frequency. This gradual synchronization is different from the sudden global synchronization in the Kuramoto model in Fig. 5.13. To understand this, we can plot a 16-by-16 lattice in terms of common (entrained) frequencies. This is shown in Fig. 5.15. For weak coupling (on the left) all the oscillators have different frequencies. As the coupling increases, clusters form and grow with common frequencies. At strong coupling, a single large cluster of synchronized oscillators spans the lattice and eventually entrains all the remaining oscillators.

Networks can have effective "dimensions" that influence how easily they can achieve global synchronization. Two other examples are a linear cycle (1D) and a lattice (2D). The 2D lattice exhibits more entrainment at low coupling coefficients, but less entrainment at high coupling compared to a complete graph

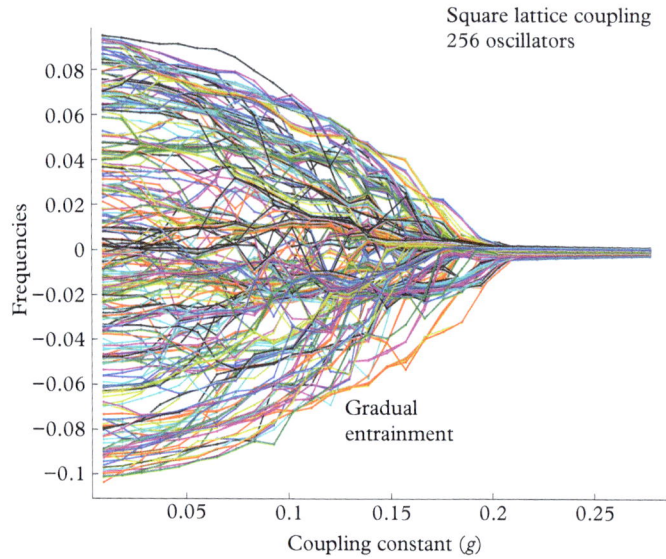

Figure 5.14 *Synchronization of 256 phase oscillators on a square lattice with increasing coupling. Entrainment proceeds more gradually than for global coupling, and it takes larger coupling to achieve full entrainment.*

Figure 5.15 *A sequence of a 16 × 16 square lattice for low coupling (left), intermediate coupling (center), and strong coupling (right). These lattices show increasingly larger clusters of entrained frequencies.*

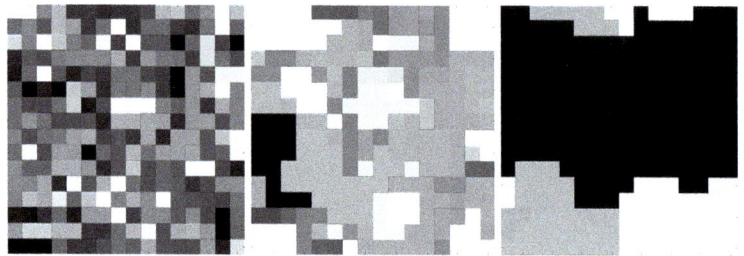

(global coupling). This has the overall effect of "smearing out" the entrainment transition. This wider transition is directly related to the dimensionality of the 2D lattice.

Because networks are composed of nodes and links, there is no intrinsic embedding dimension for networks. While some networks, like a 1D cycle or the 2D lattice, have a clear dimensionality related to the topology of their interconnections, other networks are harder to define dimensionally. For instance, a tree network does not seem to have a clear dimensionality because of its hierarchy of branches. However, a tree topology is a close analog to a 1D string of nodes—namely, if a single link is cut, then an entire subgraph disconnects from the global network. In this sense, a tree is quasi-one-dimensional.

A comparison of frequency entrainment on three different network topologies is shown in Fig. 5.16 for $N = 256$ nodes. Each line represents a separate oscillator. The initial frequencies are distributed uniformly from -0.1 to 0.1 relative to the mean frequency. The Kuramoto model of the complete graph is shown at the

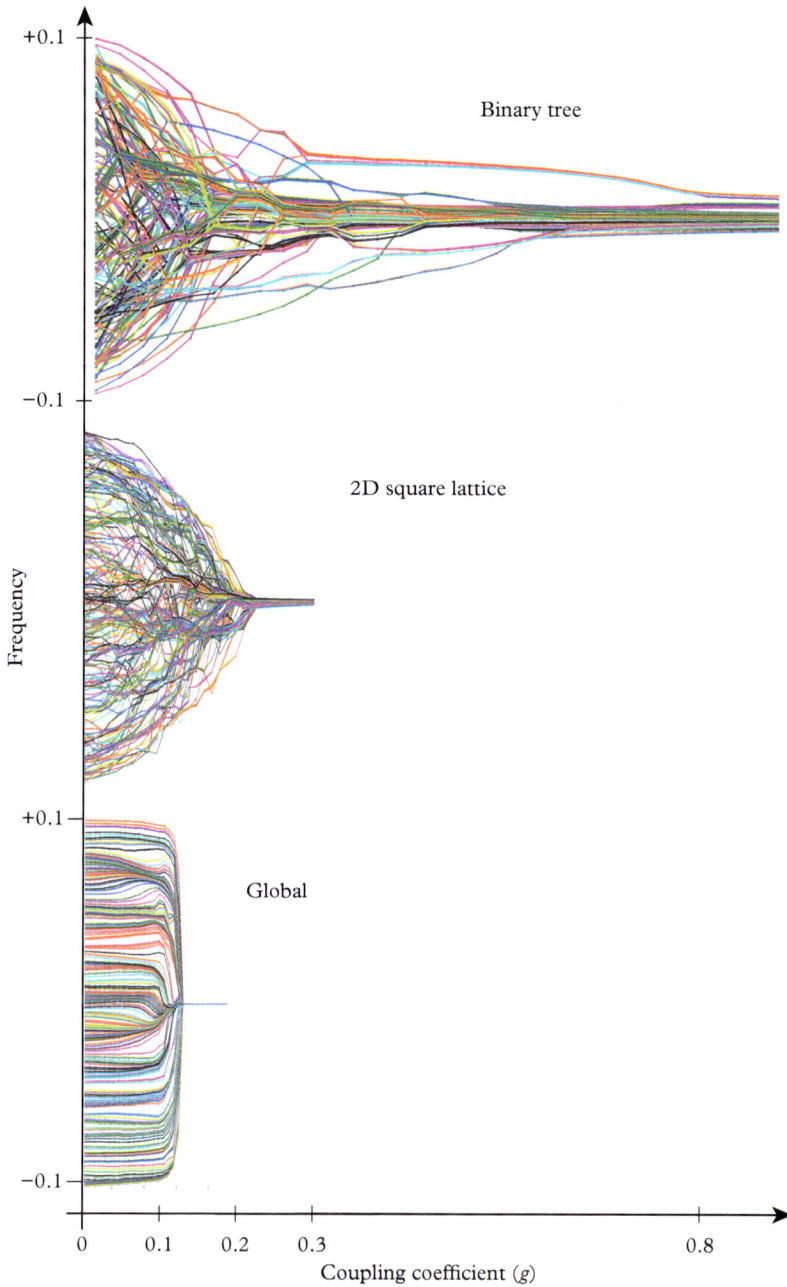

Figure 5.16 *Plots of frequency vs. coupling coefficient for a binary tree, a 2D lattice, and a complete graph. The rate of entrainment is strikingly different depending on the effective dimensionality of the graphs. (A binary tree is effectively 1D.)*

bottom and has the sharpest transition to global synchronization near $g_c = 0.1$. The 2D lattice still has a global synchronization threshold, but the transition is more gradual. On the other hand, the binary tree has very indistinct transitions and requires very high coupling to induce full entrainment of all the nodes (which has not happened yet in the figure with g up to 1). However, for both the binary tree and the 2D lattice the initial synchronization of some of the oscillators occurs near $g = 0.1$.

The synchronization of oscillators on networks provides a versatile laboratory for the study of critical phenomena. The critical behavior of dynamical processes (such as global synchronization) can differ from topological structure (percolation thresholds), especially in their scaling properties near the critical transition. Network dimensionality and topology play central roles in the transitions, but the nature of the oscillators (linear or nonlinear) and their coupling (linear or nonlinear) also come to bear on the existence of thresholds, the scaling behavior near threshold, and the dependence on network size. These are current topics of advanced research in physics and mathematics, but the basic tools and methods needed to approach these problems have been provided here in this chapter.

5.6   Summary

Dynamical processes on interconnected networks have become a central part of modern life through the growth in importance of communication and social networks. *Network topology* differs from the topology of metric spaces, and even from the topology of dynamical spaces like state-space. A separate language of *nodes* and *links*, *degree* and *moments*, *adjacency matrix* and *distance matrix*, among others, are defined and used to capture the wide range of different types and properties of network topologies. *Regular graphs* and *random graphs* have fundamentally different connectivities that play a role in dynamic processes such as *diffusion* and *synchronization* on a network. Three common random graphs are the *Erdös–Rényi* (ER) graph, the *small-world* (SW) graph, and the *scale-free* (SF) graph. Random graphs give rise to *critical phenomena* based on static connectivity properties, such as the *percolation threshold*, but also exhibit dynamical thresholds for the diffusion of states across networks and the synchronization of oscillators. The *vaccination threshold* for diseases propagating on networks, and the *global synchronization* transition in the *Kuramoto model* are examples of dynamical processes that can be used to probe network topologies.

Adjacency matrix

The adjacency matrix for a graph with N nodes is an N-by-N matrix with elements

$$
\begin{aligned}
A_{ij} &= 1 \quad i \text{ and } j \text{ connected} \\
A_{ij} &= 0 \quad i \text{ and } j \text{ not connected} \\
A_{ii} &= 0 \quad \text{zero diagonal}
\end{aligned}
\tag{5.4}
$$

Graph Laplacian

The graph Laplacian of a network is an operator that is defined as

$$L_{ij} = \left(\sum_j A_{ij} \right) \delta_{ij} - A_{ij} = \begin{cases} k_i & i = j \\ -1 & i \text{ and } j \text{ connected} \\ 0 & \text{otherwise} \end{cases} \qquad (5.9)$$

Network diameter

For a random graph, the network diameter is defined as the largest distance between two nodes. It scales with number of nodes N and average degree $\langle k \rangle$ as

$$\ell = \ell_0 + \frac{\ln N}{\ln \langle k \rangle}, \qquad (5.12)$$

where ℓ_0 is of order unity.

Synchronization range

The range of coupling values for synchronization depends on the largest and smallest eigenvalues of the graph Laplacian:

$$\frac{\lambda_{\max}}{\lambda_2} < \frac{g_{\max}}{g_{\min}}. \qquad (5.48)$$

Kuramoto model

The Kuramoto model assumes a complete graph of identical coupled phase oscillators. The flow equation is

$$\frac{d\phi_k}{dt} = \omega_k + g \frac{1}{N} \sum_{j=1}^{N} \sin \left(\phi_j - \phi_k \right). \qquad (5.50)$$

The Kuramoto model has a mean field solution that shows a sharp synchronization transition.

5.7   Bibliography

M. Barahona and L. M. Pecora, Synchronization in small-world systems, *Physical Review Letters*, vol. 89, 054101, 2002.

A. Barrat, M. Barthelemy and A. Vespignani, *Dynamical Processes on Complex Networks* (Cambridge University Press, 2008).
   A high-level text with excellent intermediate introductory material.

L. Glass and M. C. Mackey, *From Clocks to Chaos* (Princeton University Press, 1988).
An enjoyable popular description of synchronization phenomena.

M. E. J. Newman, *Networks: An Introduction* (Oxford University Press, 2010).
This has quickly become the standard introductory text of network theory.

A. Pikovsky, M. Rosenblum and J. Kurths, *Synchronization: A Universal Concept in Nonlinear Science* (Cambridge University Press, 2001).
An excellent text that is both introductory and comprehensive on the topic of synchronization.

D. J. Watts, *Small Worlds: The Dynamics of Networks Between Order and Randomness* (Princeton University Press, 2003).
A nontechnical book for a general audience.

A. T. Winfree, *The Geometry of Biological Time* (Springer, 2001).
Full of gems with many examples of synchronization from biology.

5.8 Homework exercises

Analytic problems

1. **Eigenvalues**: Show (prove) for a complete graph that Λ_{ij} has $\lambda_i = N$ except for $i = 1$.

2. **Kuramoto model**: For the Kuramoto model, explicitly do the integration of

$$K = \int_{-\pi/2}^{\pi/2} gK\cos^2\psi\rho\,(\bar{\omega} + gK\sin\psi)\,d\psi$$

using $\rho(\omega) = \dfrac{\gamma}{\pi\left[(\omega-\bar{\omega})^2+\gamma^2\right]}$ to find g_c and to confirm the square-root dependence at threshold.

3. **Average distance**: Find the average distance between nodes for the Strogatz–Watts model as a function of rewiring probability p. Choose a range of 3 to 4 orders of magnitude on p.

4. **Clustering coefficient**: Define the clustering coefficient C_i is the average fraction of pairs of neighbors of a node i that are neighbors of each other (shared edges):

$$C_i = \frac{e_i}{k_i\,(k_i - 1)/2},$$

where e_i is the number of edges shared among the k_i neighbors of node i. The number of shared edges is given by

$$e_i = \frac{1}{2}\sum_{jk} A_{ij}A_{jk}A_{ki},$$

where A_{jk} is the adjacency matrix. What is the average clustering coefficient of the Strogatz–Watts model as a function of rewiring probability p. Choose a range of 3 to 4 orders of magnitude on p

Computational projects[5]

5. **Graph Laplacian**: Numerically calculate $\frac{\lambda_{max}}{\lambda_2}$ (from the graph Laplacian) for small-world graphs as a function of the rewiring probability. Choose $N = 100$, $k = 4$, and $p = 0.1$. Write a program that begins with the adjacency matrix of a regular graph $k = 4$, then loop through the columns to rewire each link with a probability p (be careful to only work on the upper or lower diagonal of the matrix and keep it symmetric). Then construct the Laplacian and use the Matlab function eig(L). (You can type "help eig" at the >> prompt to learn how to use the output of the function.)

6. **Spectral width**: How does the width $\Delta\lambda_i$ of eigenvalues of the Laplacian depend on the rewiring probability for small-world graphs? After eliminating the lowest eigenvalue, you can simply take the standard deviation of the remaining eigenvalues, and track how that changes as a function of p.

7. **Synchronized chaos**: Track the range of g for which identical Rössler oscillators are synchronized for the nets of Figs. 5.3, 5.5 and 5.6 as a function of the rewiring probability.

8. **Synchronized chaos**: Find parameters for the identical Rössler oscillators for which the network is only just barely synchronizable ($\frac{\alpha_2}{\alpha_1} \simeq \frac{\lambda_{max}}{\lambda_2}$).

9. **Entrainment**: Study synchronization on random graphs. Plot the entrainment probability of Poincaré phase oscillators as a function of coupling constant for ER, SF, and SW graphs $N = 100$ for the same average degree.

[5] Matlab programming examples are found at www.works.bepress.com/ddnolte.

Part 3

Complex Systems

Modern life abounds in complex systems—too many to describe here. However, some have such profound influence on our lives that they rise to the top of the list and deserve special mention. These include the nature of neural circuitry (the origins of intelligence), the evolution of new forms and the stability of ecosystems in the face of competition and selection (the origins of species and the ascent of man), and the dynamics of economies (the origins of the wealth of nations and the welfare of their peoples). This section introduces these three topics, and applies many of the tools and techniques that were developed in the previous chapters.

Neurodynamics
and Neural Networks

6

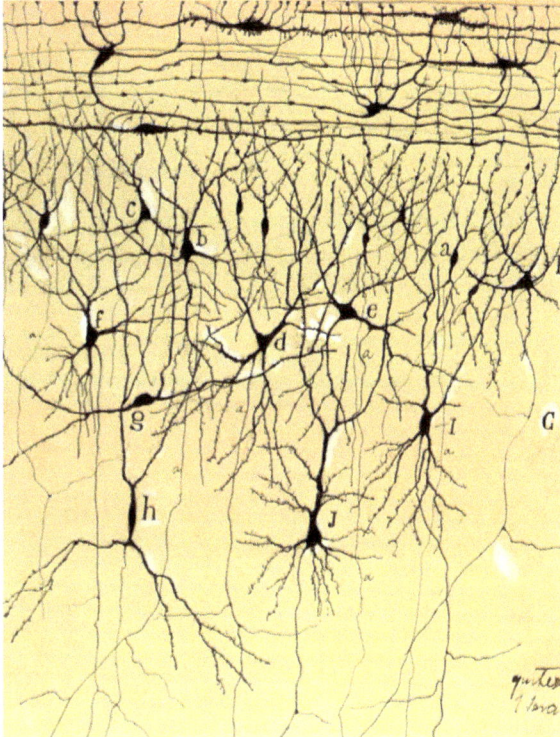

Santiago Ramón y Cajal
Capas 1ª y 2ª de la corteza olfativa de la circunvolución del hipocampo del niño, n 1901
℗ Herederos de Ramón y Cajal

The brain, for us, is perhaps the most important dynamic network we can study. The brain is composed of nearly 100 billion neurons organized into functional units with high connectivity. The sophistication of the information processing capabilities of the brain is unsurpassed, partly because of the immense dimensionality of the network. However, much simpler collections of far fewer neurons also exhibit surprisingly sophisticated processing behavior.

This chapter begins with the dynamic properties of a single neuron, which are already complex, displaying bursts and bifurcations and limit cycles. Then it looks at the simplest forms of neural networks, and analyzes a recurrent network known as the Hopfield network. The overall behavior of the Hopfield network is understood in terms of fixed points with characteristic Lyapunov exponents distributed across multiple basins of attraction that represent fundamental memories as well as false memories.

6.1   Neuron structure and function

The single neuron is an information processing unit—it receives multiple inputs, sums them and compares them with a threshold value. If the inputs exceed the threshold, then the nerve sends a signal to other neurons. The inputs are received on the neuron dendrites (see Fig. 6.1), and can be either excitatory or inhibitory. Once the collective signal exceeds the threshold, the nerve body fires an electrical pulse at the axon hillock. The pulse propagates down the axon to the diverging axon terminals that transmit to the dendrites of many other downstream neurons. The signal that propagates down the axon has a specified amplitude that is independent of the strength of the input signals and propagates without attenuation because it is re-amplified along its path at the many nodes of Ranvier between the junctions of the Schwann cells. The strength of a neural signal is encoded in the frequency of the neural pulses rather than by the amplitude of the pulses.

An example of an action potential pulse is shown in Fig. 6.2, which represents the voltage at a node of Ranvier along an axon. The resting potential inside the neuron is at −65 mV. This negative value is a thermodynamic equilibrium value determined by the permeability of the neuron membrane to sodium and potassium ions. When the neuron potential is at the resting potential, the neuron is said to be *polarized*. As the membrane voltage increases, it passes a threshold and sustains a rapid run-away effect as the entire region of the neuron depolarizes, and the voltage overshoots zero potential to positive values, followed by a relaxation that can undershoot the resting potential. Eventually, the resting potential is re-established. The action potential spike is then transmitted down the axon to the next node of Ranvier, where the process repeats. The runaway spike of the action potential is caused by positive feedback among ion channels in the

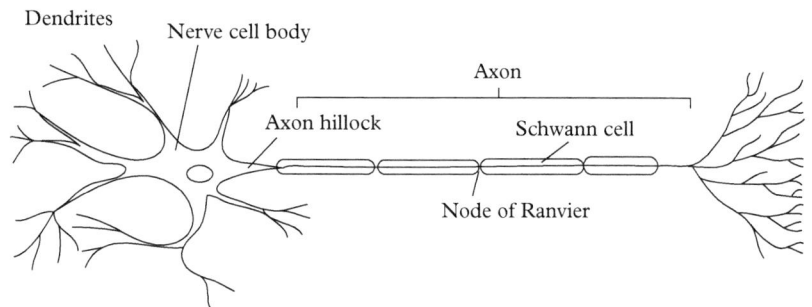

Figure 6.1 *Neuron physiological structure. The neuron receives signals on its dendrites. If the sum of the signals exceeds a threshold, the axon hillock initiates a nerve pulse, called an action potential, that propagates down the axon to the axon terminals that transmit to the dendrites of many other neurons.*

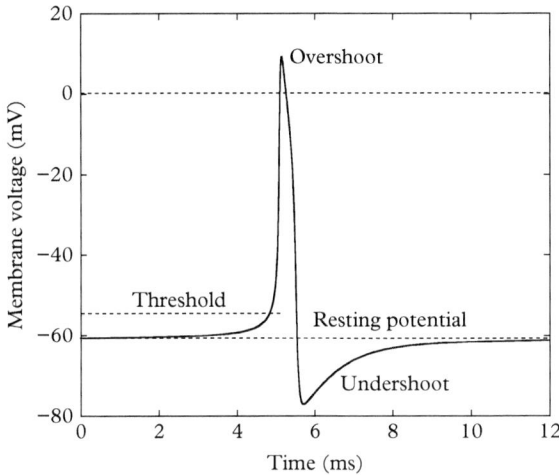

Figure 6.2 *Action potential spike of a neuron. The equilibrium resting value is typically −65 mV. When the potential rises above a threshold value, the neuron potential spikes and then recovers within a few milliseconds.*

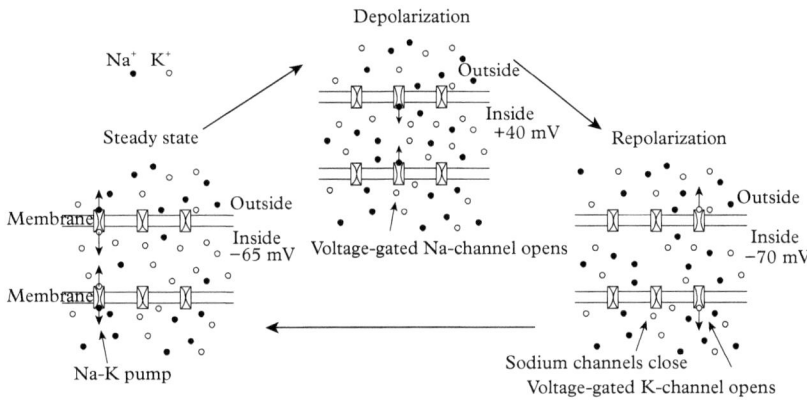

Figure 6.3 *Biochemical and physiological process of action potential generation. The Na–K pump maintains the resting potential prior to an action potential pulse (left). During the upswing of the action potential (middle), voltage-gated Na channels open. Recovery occurs through equalization of the internal ionic charge when the Na channels close and the voltage-gated K channels open (right) to return to steady state and be prepared for the next action potential.*

neuron membrane. These ion channels are voltage gated, meaning that they open depending on the voltage difference across the membrane.

The physiological and biochemical origin of the resting potential and the action potential is illustrated in Fig. 6.3 that shows sodium and potassium ion channels in the cell membrane. In the resting state, a sodium–potassium pump maintains a steady-state polarization with an internal potential around −65 mV. The pump is actively driven by the energy of ATP hydrolysis. In addition to the sodium-potassium pump, there are voltage-gated ion channels for sodium and potassium, separately. At the resting potential, the voltage-gated channels are closed. However, as the potential increases (becomes less negative) caused by an external stimulus, some of the voltage-gated sodium channels open, allowing sodium ions outside the membrane to flow inside. This influx of positive ions causes the internal potential to further increase. As the potential rises, more and more of the voltage-gated sodium channels open, in a positive-feedback that causes a runaway effect in which all of the sodium channels open, and the potential rises above 0 mV. In this depolarized state, the potassium voltage-gated ion channels now open to allow potassium ions to flow out of the neuron, causing the potential to fall by reducing the number of positive ions inside the membrane. After the action potential has passed, the sodium–potassium pump re-establishes the steady state condition and the neuron is ready to generate another action potential.

The role of voltage-gated ion channels in the generation and propagation of action potentials was first uncovered by Hodgkin and Huxley (1952) in studies of the squid giant axon. They identified three major current contributions to depolarization that included the sodium and potassium channels, but also a third leakage current that depends mainly on Cl⁻ ions. They further discovered that potassium channels outnumbered sodium channels (accounting for the strong repolarization and even hyperpolarization caused by the out flux of potassium) and that sodium channels had an inactivation gate that partially balances activation gates (accounting for the need to pass a certain threshold to set off the positive-feedback runaway). Hodgkin and Huxley devised an equivalent circuit model for the neuron that included the effects of voltage-dependent permeabilities with activation and inactivation gates. The Hodgkin and Huxley model equation is

$$C\dot{V} = I - g_K n^4 \left(V - E_K\right) - g_{Na} m^3 h \left(V - E_{Na}\right) - g_L \left(V - E_L\right), \qquad (6.1)$$

where C is the effective capacitance, V is the action potential, I is a bias current, g_K, g_{Na}, and g_L are the conductances, E_K, E_{Na}, and E_L are the equilibrium potentials, $n(V)$ is the voltage-dependent potassium activation variable, $m(V)$ is the voltage-dependent sodium activation variable (there are three sodium activation channels and four potassium channels, which determines the exponents), and $h(V)$ is the voltage-dependent sodium inactivation variable. The model is completed by specifying three additional equations for $\dot{n}$, $\dot{m}$, and $\dot{h}$ that are combinations of decaying or asymptotic time-dependent exponentials.

The Hodgkin and Huxley model is a four-dimensional nonlinear flow that can display an astounding variety of behaviors that include spikes and bursts, limit cycles and chaos, homoclinic orbits and saddle bifurcations, and many more. The field of neural dynamics provides many excellent examples of nonlinear dynamics. In the following sections, some of the simpler variations on the model are explored.

6.2   Neuron dynamics

The dynamic behavior of a neuron can be simplified by dividing it into segments. The body of the neuron is a separate segment from the axon and dendrites. These, in turn, can be divided into chains of dynamic segments that interact in sequence to propagate action potentials. Each segment can be described as an autonomous nonlinear system.

6.2.1 Fitzhugh–Nagumo model

One early simplification of the neural dynamics is the Fitzhugh–Nagumo model of the neuron. Under appropriate conditions, it is a simple third-order limit-cycle oscillator like the van der Pol oscillator. The Fitzhugh–Nagumo model has only two dynamical variables: the membrane potential V and the activation variable n that is proportional to the number of activated membrane channels. Values for the model that closely match experimental observations lead to the equations

$$\dot{V} = (V + 60)\,(a - V/75 - 0.8)\,(V/75 - 0.2) - 25n + B$$
$$\dot{n} = 0.03\,(V/75 + 0.8) - 0.02n, \tag{6.2}$$

where B is a bias current and a is a control parameter that determines the size of the limit cycle.[1] The two nullclines ($\dot{V} = 0$, $\dot{n} = 0$) and a few stream lines are shown in Fig. 6.4 for several values of the control parameter a. The control parameter plays the role of the integrated stimulus of the neuron with larger negative values of a corresponding to stronger stimulus. This model can show stable fixed-point behavior (Figs. 6.4a and 6.4d) or limit cycles (Figs. 6.4c and 6.4d) depending on the value of the control parameter. When the stimulus is weak (larger positive values of a), the system is in a stable fixed point (resting potential). As the stimulus increases, (larger negative values of a), the system converts from the resting potential to a limit cycle (spiking). The transition to the limit

[1] The coefficients in the Fitzhugh–Nagumo equations have implicit dimensions to allow voltages to be expressed in millivolts. See E. M. Izhikevich, *Dynamical Systems in Neuroscience: The Geometry of Excitability and Bursting* (MIT University Press, 2007).

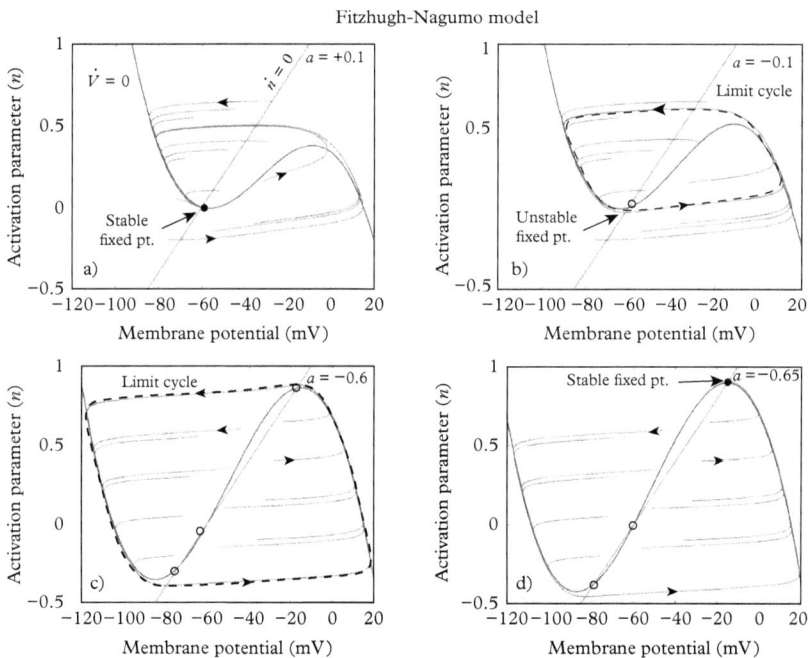

Figure 6.4 *Examples of the Fitzhugh–Nagumo model. The horizontal axis is the membrane potential V, and the vertical axis is the activation parameter n. The nullclines are plotted for n (straight line) and V (cubic curve) for I = 0. In (a) for a = +0.1 there is a stable fixed point (resting potential), while in (b) for a = −0.1 and (c) for a = −0.6 there are limit cycles representing neural spiking. In (d) when a = −0.67, the system converts back to a stable fixed point.*

cycle is seen graphically in Fig. 6.4 when the nullclines ($\dot{V} = 0$ and $\dot{n} = 0$) intersect at the fixed point in the region where the $\dot{V}$ nullcline has positive slope, and a stability analysis shows the fixed point is unstable. This model captures the qualitative transition from resting to spiking that mimics the behavior of neurons. The Fitzhugh–Nagumo model in Eq. (6.2) captures the essential nonlinearity of Eq. (6.1) caused by the voltage-gated channels and allows a simple stability analysis, but it cannot represent the detailed dynamics of neural action potentials.

6.2.2   The NaK model

To capture subtler behavior of neurons, a more realistic model called the NaK model includes voltage-dependent variables, such as relaxation times and excitation parameters. This model is more closely linked to the physical processes of ion currents through neuron membranes and the dependence of these currents on membrane potentials. The dynamics can still be described in a two-dimensional phase plane, but the nullclines take on forms that lead to interesting behavior such as bifurcations and homoclinic orbits.

The starting point of the NaK model is the Hodgkin–Huxley model of voltage-gated currents across the membrane. The voltage-gated currents take the form

$$I_j = g_j h_j (V) \left(V - E_j\right), \tag{6.3}$$

where g_j is a channel conductance of the jth ion, E_j is the equilibrium potential, and $h_j(V)$ is an activation variable related to the fraction of channels that are open. The activation variable is a function of membrane potential, which makes the voltage-gated current nonlinear.

In the NaK model, there are four assumed currents: a bias current I, a leakage current (L), a sodium (Na) current, and a potassium (K) current. The neuron is assumed to have capacitance C and membrane potential V that are driven by the currents as

$$\begin{aligned}
C\dot{V} &= I - g_L (V - E_L) - \sum_{j=1}^{2} g_j h_j (V) \left(V - E_j\right) \\
&= I - g_L (V - E_L) - g_{Na} m (V) (V - E_{Na}) - g_K n (V) (V - E_K),
\end{aligned} \tag{6.4}$$

where there are two activation variables $m(V)$ for the sodium channel and $n(V)$ for the potassium channel. In the NaK model, it is usually assumed that the sodium current is instantaneous, without any lag time. The NaK model is then a two-dimensional flow in V and n through

$$\begin{aligned}
C\dot{V} &= I - g_L (V - E_L) - g_{Na} m_{\infty} (V) (V - E_{Na}) - g_K n (V) (V - E_K) \\
\dot{n} &= \frac{n_{\infty} (V) - n (V)}{\tau (V)},
\end{aligned} \tag{6.5}$$

where

$$m_\infty(V) = \frac{1}{1 + \exp\left((V_m - V)/k_m\right)}$$

$$n_\infty(V) = \frac{1}{1 + \exp\left((V_n - V)/k_n\right)} \qquad (6.6)$$

$$\tau(V) = \tau^0 + \Delta\tau \exp\left(-(V_{\max} - V)^2/2\Delta V^2\right).$$

The NaK model is a two-dimensional phase-plane model that can show complex dynamics, depending on the bias current I (which is integrated by the neuron dendrite over the inputs from many other contacting neurons). With sufficient current input, the dynamics of the target neuron can change dramatically, projecting its own stimulus downstream to other neurons in a vast network.

Despite the apparent simplicity of the NaK model, it shows rich dynamics, in particular bistability and bifurcations. It is therefore fruitful to explore its parameter space to understand some of the different types of behavior it is capable of.

6.2.3   Bifurcations, bistability, and homoclinic orbits

Neurodynamics provides many examples of nonlinear bifurcations, bistability, and homoclinic orbits. The Fitzhugh–Nagumo and NaK models are relatively simple, with V and n nullclines that intersect at only a few fixed points. However, a wealth of different types of behavior are possible, depending on the choice of parameters in the models. Many of these types of behavior have been observed experimentally among physiological nerve cells.

The Fitzhugh–Nagumo model, shown in Fig. 6.4, displays a bifurcation as the control parameter decreases from $a = +1$. The stable fixed point converts discontinuously into a limit cycle (for the parameters used in Fig. 6.4) at the critical value $a_c = -1.0$. As the control parameter decreases further, the area of the limit cycle increases until the limit cycle converts discontinuously to a new fixed point at $a_c = 0.65$. The conversions occur when the n-nullcline crosses the local minimum and then the local maximum of the V-nullcline, respectively. This type of bifurcation is called an Andronov–Hopf bifurcation, when a stable fixed point gives rise to a stable limit cycle as a control parameter is varied.

Bistability is a particularly important property for neurodynamics. This allows a "prepared" neuron that is not currently spiking to receive a stimulus that causes it to spike continuously, even after the stimulus has been removed. The NaK model can display bistability, which is understood by considering Fig. 6.5. For the parameters in the figure, there are two fixed points (a saddle and a stable node) as well as a limit cycle. If the neuron is in the resting state, it is at the stable node. However, a transient pulse that is applied and then removed quickly can cause the system to cross the separatrix into the region where the dynamics relax onto the limit cycle and the system persists in a continuous oscillation, as shown in Fig. 6.6.

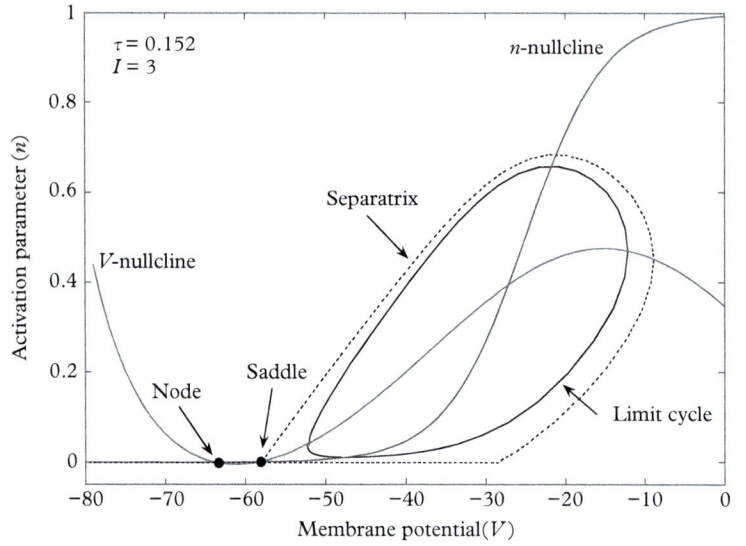

Figure 6.5 *Bistability between a node and a limit cycle. A current spike can take the system from the resting potential (the node) across the separatrix to the spiking state (the limit cycle).*

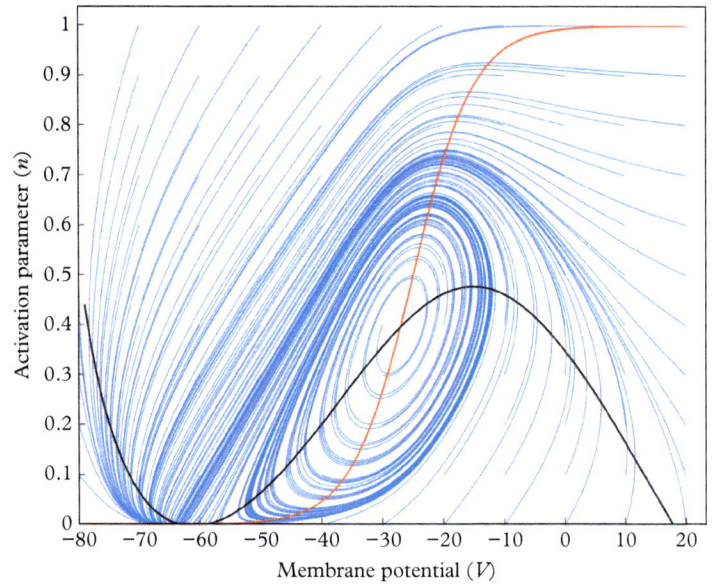

Figure 6.6 *Stream lines for the model that shows bistability.*

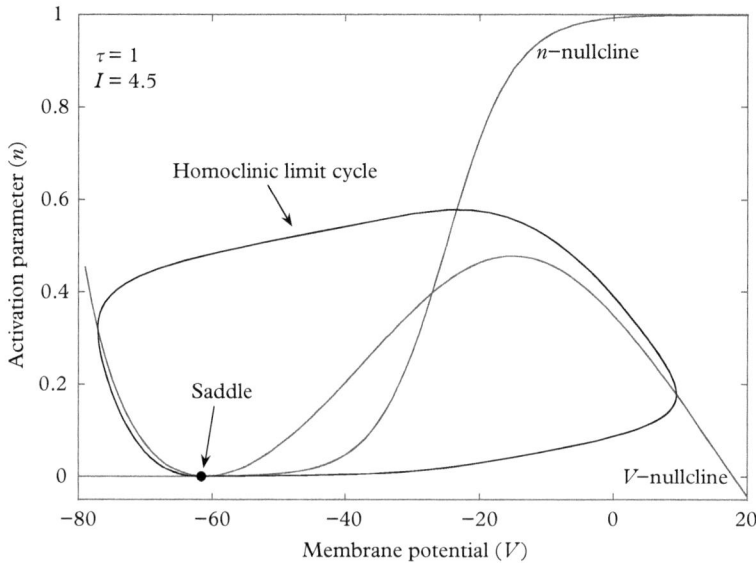

Figure 6.7 *A homoclinic orbit in the NaK model connected to the saddle point.*

The NaK model can display other types of bifurcations as well as bistability. The phase diagram of the NaK model is shown in Fig. 6.7 for the conditions when the stable node and the saddle point collide (merge) into a single saddle point. The limit cycle becomes a homoclinic orbit with infinite period. This type of bifurcation is called a saddle bifurcation (also known as a fold or a tangent bifurcation). The membrane potential for a trajectory near the homoclinic orbit is shown in Fig. 6.8. The rapid traversal of the orbit is followed by a very slow approach to the saddle, giving the membrane potential a characteristic spiking appearance.

6.3   Network nodes: artificial neurons

Biological neurons provide excellent examples of the complexities of neural behavior, but they are too complex as models for artificial neurons in artificial neural networks. Therefore, it is helpful to construct idealized neuron behavior, either computationally, or in hardware implementations.

A simple model of an artificial neuron is shown in Fig. 6.9. It has N inputs, attached with neural synaptic weights w_k^a from the ath neuron to the summation junction of the kth neuron. The output voltage v_k of the summation junction is

$$v_k = R_k \sum_{a=1}^{N} y_a w_k^a - b_k, \tag{6.7}$$

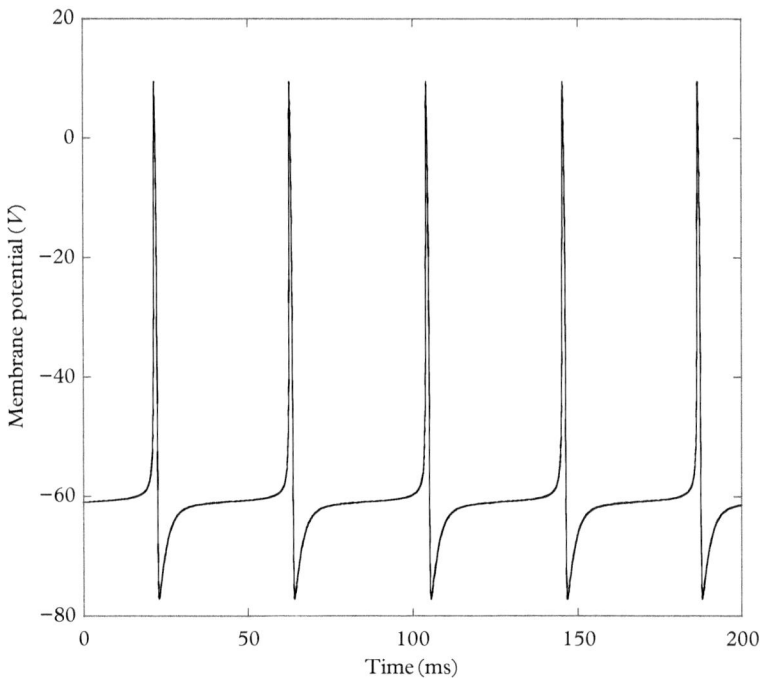

Figure 6.8 *Membrane potential as a function of time for an orbit near the homoclinic orbit.*

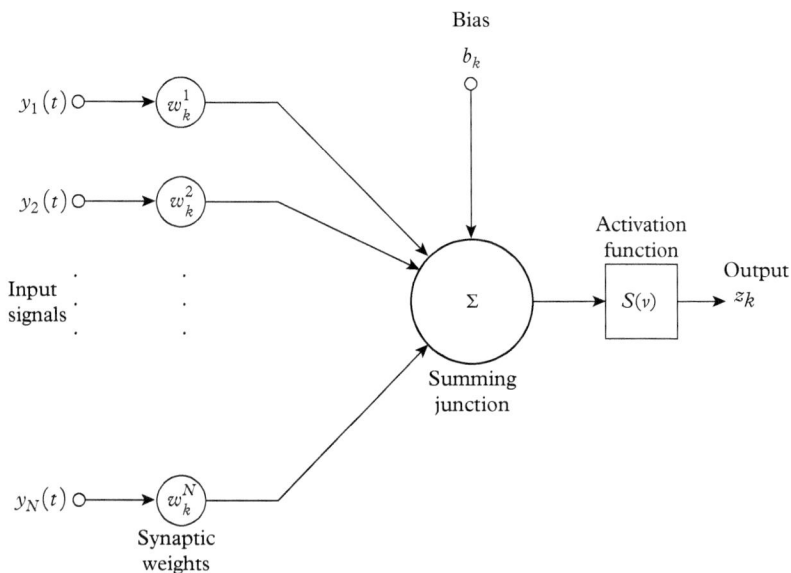

Figure 6.9 *Model of an artificial neuron. There are m inputs that are connected to the summing junction with weights w_k^m where they are summed with a bias b_k to yield v_k that is passed through a saturating nonlinear function $S(v_k)$ to give the neuron output.*

where the y_a are the current input signals,[2] b_k is the neuron bias voltage, and R_k is a characteristic resistance.[3] This value is passed through a nonlinear activation function $S(v_k)$ to yield the output voltage value z_k:

$$z_k = S(v_k) \tag{6.8}$$

The saturating function is the nonlinear element that is essential for the neuron to change its state. It is common to use a sigmoidal function that has two limiting values for very small and very large arguments. Examples of the saturating nonlinear activation function are given in Table 6.1. These are antisymmetric functions that asymptote to ± 1, and the parameter g is called the gain.

The sigmoidal functions are the most common functions used in artificial neurons. These are smooth functions that vary between the two saturation limits and are differentiable, which is important for use in algorithms that select weighting values w_k^m for the network. The two sigmoidal functions are plotted in Fig. 6.10, both with gain $g = 1$. The atan function has a slower asymptote.

The neurodynamics of a network sometimes are modeled by adopting a physical representation for a node that has the essential features of a neuron. This model neuron is shown in Fig. 6.11 with capacitors and resistors, a summation junction and the nonlinear activation function.[4] The sum of the currents at the input node, following Kirchhoff's rule, is

$$C_k \frac{dv_k}{dt} = -\frac{v_k(t)}{R_k} + \sum_{a=1}^{N} y_a(t) w_k^a + I_k \tag{6.9}$$

[2] Subscripts denote row vectors, the superscript of the neural weight matrix is the row index, and the subscript is the column index. The choice to use row vectors in this chapter is arbitrary, and all of the equations could be transposed to operate on column vectors.

[3] Many equations in this chapter have parameters with repeated indices, but do *not* imply the Einstein summation convention. All summations are written out explicitly.

[4] See S. Haykin, *Neural Networks: A Comprehensive Foundation* (Prentice Hall, 1999).

Table 6.1 *Neural response functions*

The Heaviside function:

$$S(v_k) = \begin{cases} 1 & v_k \geq 1 \\ -1 & v_k < 1 \end{cases}$$

Piecewise linear:

$$S(v_k) = \begin{cases} 1 & gv_k \geq 1 \\ gv_k & -1 < gv_k < 1 \\ -1 & gv_k < -1 \end{cases}$$

Sigmoid functions:

Logistic (Fermi) function:

$$S(v_k) = \left[1 - \frac{2}{1 + \exp(2gv_k)} \right] = \tanh(gv_k)$$

Inverse tangent function:

$$S(v_k) = \frac{2}{\pi} \operatorname{atan}\left(\frac{\pi}{2} gv_k \right)$$

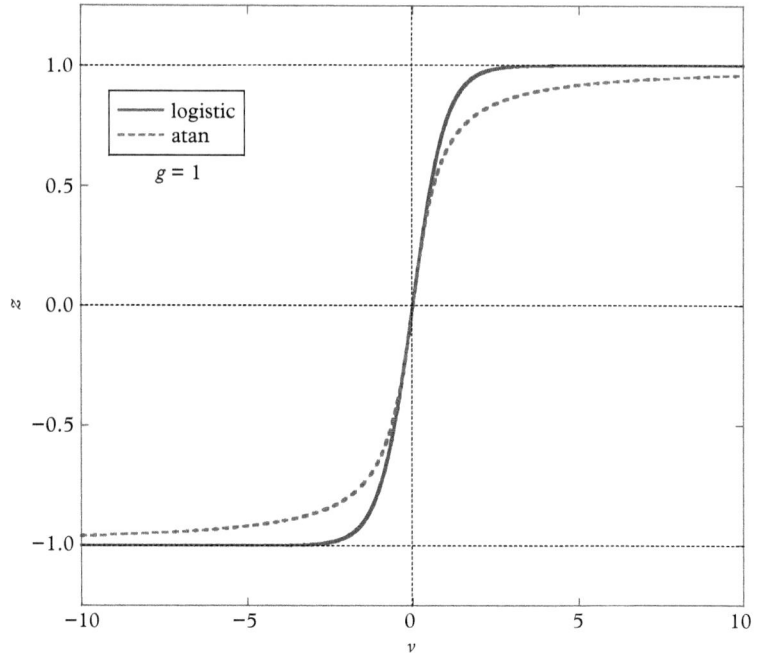

Figure 6.10 *Two common choices for antisymmetric sigmoid functions with a gain g = 1: the logistic function (also known as the Fermi function or tanh function) and the atan function.*

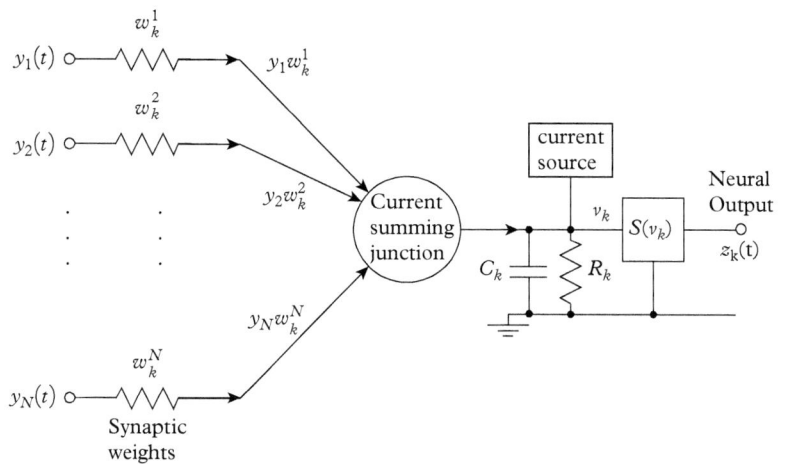

Figure 6.11 *Physical (hardware) model of an artificial neuron with resistors, capacitors, a summing junction, and a nonlinear transfer function.*

Neurodynamics and Neural Networks **223**

and the output voltage of the neuron is

$$z_k(t) = S(v_k(t)). \tag{6.10}$$

The input signals y_k are voltages, and the resistors provide the matrix elements w_k^j. The RC element gives a smooth response through the RC time constant.

6.4 Neural network architectures

The variety of neural network architectures is vast because there are so many possible ways that neurons can be wired together. There are two general classifications that pertain to the way that the neural network learns, which is either supervised or unsupervised. In supervised learning the networks are trained by presenting selected inputs with the "correct" outputs. In unsupervised learning the networks are allowed to self-organize. The human mind learns primarily through unsupervised learning as we interact with our environment as babies and our brains wire themselves in response to those interactions. On the other hand, there are many machine tasks in which specific inputs are expected to give specific outputs, and supervised learning may be most appropriate.

Within the class of neural networks with supervised learning, there are generally three classifications of neural network architectures. These are (1) single-layer feed forward, (2) multilayer feed forward networks without loops, and (3) recurrent networks with feedback loops. There can also be mixtures, such as multilayer networks that may have loops. However, the three main types have straightforward training approaches, which will be described in this section.

6.4.1 Single-layer feed forward (perceptron)

Single-layer feed forward neural networks are the simplest architectures, often referred to as single-layer *perceptrons*. They consist of a single layer of neurons that map multiple inputs onto multiple outputs. They often are used for low-level pattern detection and feature recognition applications, much like the way the retina pre-processes information into specific features before transmission to the visual cortex of the brain. A schematic of a single-layer perceptron is shown in Fig. 6.12. Input neurons are attached to the inputs of a single neural processing layer where the signals are summed according to the weights w_k^j and are output through the nonlinear transfer function. The output of a neuron is

$$z_k = S\left(\sum_{j=1}^{N} y_j w_k^j - b_k\right), \tag{6.11}$$

where the y_j are the inputs, w_k^j is the weight for the jth input to the kth output neuron, and b_k is the threshold of the kth output neuron. The transfer function $S(v_k)$ is one of the sigmoidal functions of Table 6.1.

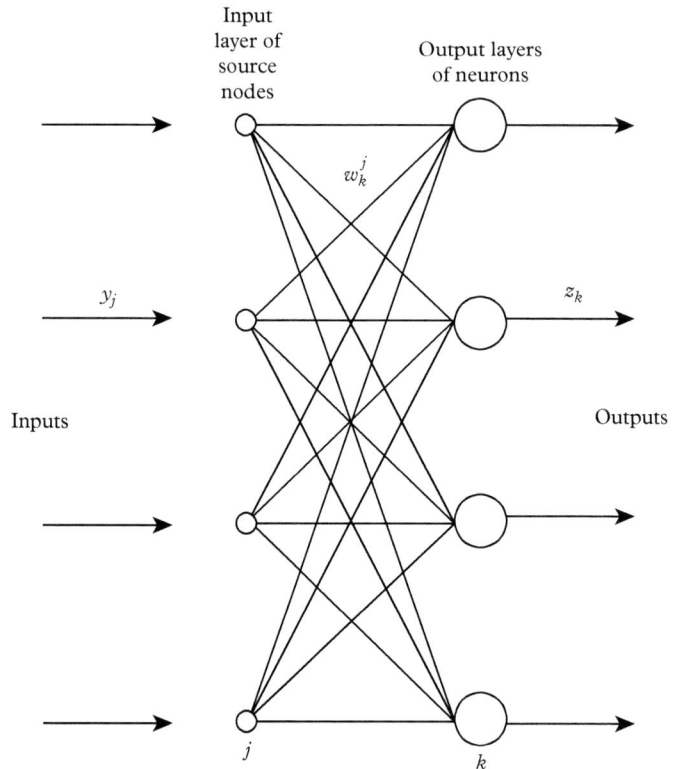

Figure 6.12 *Single-layer feed forward network (single-layer perceptron). Inputs are directed to a single layer of processing neurons with chosen weights to accomplish a supervised task.*

Single-layer perceptrons execute simple neural processing, such as edge detection, which is one of the functions that neurons perform in the neuronal layer of the retina of the eye. The retina is composed of three layers of neurons: a photoreceptor layer that acts as the inputs in Fig. 6.12, a layer of interconnected wiring that acts as the weights w_k^j, and a layer of ganglion cells that collect the signals and send them up the optic nerve to the visual cortex of the brain. One of the simplest neural architectures in the retina is the center-surround architecture. A ganglion cell receives signals from a central photoreceptor that has a positive weight, and signals from surrounding photoreceptors that have negative weight, so that when the excitation is uniform (either bright or dark), the signals cancel out and the ganglion cell does not fire. The equivalent neural diagram is shown in Fig. 6.13. Three input neurons feed forward to the output neuron (ganglion cell) with weights $[-w, 2w, -w]$. The signals for a uniform excitation cancel out. This sensor configuration responds most strongly to directional curvature in the input field, and the weights perform a Laplacian differential operation.

Simple perceptrons, like the center-surround design, can be "wired" easily to perform simple tasks. However, when there are more inputs and more output

Center−surround architecture

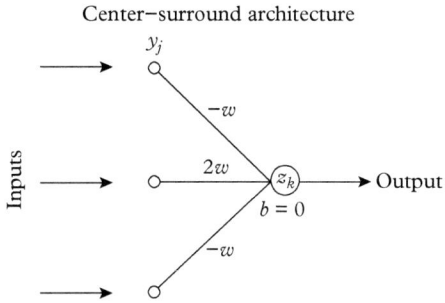

Figure 6.13 *Center-surround archi-tecture of a ganglion cell in the retina. With a simple weighting value w = 1, this acts as an edge-detection neural circuit.*

neurons, even simple perceptrons need to be "programmed," which means the weights w_k^j need to be assigned to accomplish the desired task. There is no closed form for the selection of the weights for a given training set of inputs and desired outputs. This is because the training sets are usually not a one-to-one map-ping. In other words, the problem is ill-posed. But this is precisely the situation in which neural networks excel. Neural networks work best when the problems have roughly correct solutions, but in which ambiguities remain. Our own neural networks—our brains—are especially good at finding good solutions to ambigu-ous problems. Likewise, the simple single-layer perceptron also can perform well under ambiguous conditions, but only after the best compromise is made during the assignment of the weights.

To find the desired weights and thresholds, a training set is used, and an it-erative procedure is established that allows the weights and thresholds to relax to good compromise values. There is a set of μ selected inputs Y_j^μ for which specific outputs Z_k^μ are known. This gives the mapping

$$Y_j^\mu \rightarrow Z_k^\mu \tag{6.12}$$

or, in other words, for a perfectly trained perceptron, the weights w_k^j and thresholds b_k would be chosen such that

$$Z_k^\mu = S\left(\sum_{j=1}^m Y_j^\mu w_k^j - b_k\right). \tag{6.13}$$

However, at the initiation of the perceptron, the weights and thresholds are set randomly to initial values. When the perceptron is presented with an input from the training set, the actual output z_k^μ differs from the desired training output Z_k^μ, and a deviation signal is constructed as

$$D = \frac{1}{2}\sum_{\mu,k}\left(Z_k^\mu - z_k^\mu\right)^2$$

$$= \frac{1}{2}\sum_{\mu,k}\left(Z_k^\mu - S\left(\sum_{j=1}^m Y_j^\mu w_k^j - b_k\right)\right)^2. \tag{6.14}$$

The deviation changes as the weights and thresholds are changed. This is expressed as

$$
\frac{\partial D}{\partial w_j^k} = -\sum_\mu \left[Z_k^\mu - S\left(v_k^\mu\right) \right] S'\left(v_k^\mu\right) \frac{\partial v_k^\mu}{\partial w_j^k} = -\sum_\mu \Delta_k^\mu Y_j^\mu
$$

$$
\frac{\partial D}{\partial b_k} = -\sum_\mu \left[Z_k^\mu - S\left(v_k^\mu\right) \right] S'\left(v_k^\mu\right) \frac{\partial v_k^\mu}{\partial b_k} = \sum_\mu \Delta_k^\mu,
\tag{6.15}
$$

where the delta is

$$
\Delta_k^\mu = \left[Z_k^\mu - S\left(v_k^\mu\right) \right] S'\left(v_k^\mu\right)
\tag{6.16}
$$

and the prime denotes the derivative of the transfer function, which is why this approach requires the transfer function to be smoothly differentiable. In the iterative adjustments of the weights and thresholds, the deviation is decreased by adding small adjustments to the values. These adjustments are

$$
\delta w_j^k = \varepsilon \sum_\mu \Delta_k^\mu Y_j^\mu
$$

$$
\delta b_k = -\varepsilon \sum_\mu \Delta_k^\mu,
\tag{6.17}
$$

where ε is a small value that prevents the adjustments from overshooting as the process is iterated. This is known as the *delta rule* for perceptron training.

In practical implementations of the delta rule, the weights are adjusted sequentially as each training example is presented, rather than performing the sum over μ. A complete presentation of the training set is known as an epoch, and many epochs are used to allow the values to converge on their compromise values. Within each epoch, the sequence of training patterns is chosen randomly to avoid oscillating limit cycles. Training is faster when the sigmoidal transfer function φ is chosen to be antisymmetric, as in Table 6.1.

Single-layer perceptrons have a limited space of valid problems that they can solve. For instance, in a famous example, a single-layer perceptron *cannot* solve the simple XOR Boolean function.[5] To extend the capabilities of feed-forward neural networks, the next step is to add a layer of "hidden" neurons between the input and the output.

6.4.2   Multilayer feed-forward networks

Multilayer feed-forward networks can display complicated behavior and can extract global properties of patterns. There are still some limitations on the types of problems they can solve, but the range for a two-layer perceptron is already much broader than for a single-layer. An example of a two-layer perceptron with a single "hidden" neuron layer is shown in Fig. 6.14.

The two-layer perceptron that performs the 2-bit Boolean function XOR is shown in Fig. 6.15. There are two input values, two hidden neurons, and a single

[5] Marvin Minsky of MIT (with Seymor Papert, 1969) noted that perceptrons cannot implement the XOR gate, and proved that locally-connected perceptrons cannot solve certain classes of problems. Universal computability requires neural networks with either multiple layers of locally-connected neurons or networks with non-local connections (see the recurrent Hopfield model in Section 7.5).

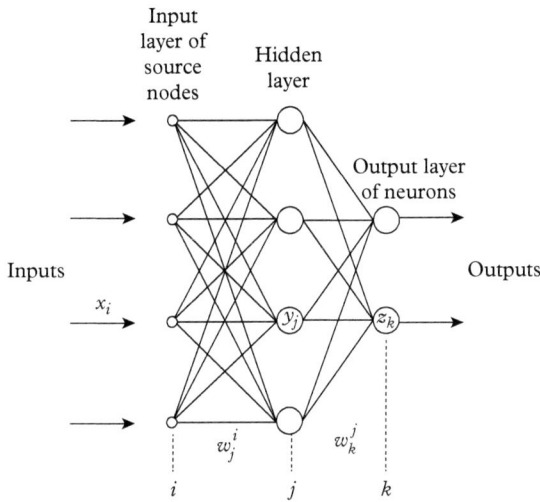

Figure 6.14 *A two-layer feed forward network with a single layer of* hidden *neurons.*

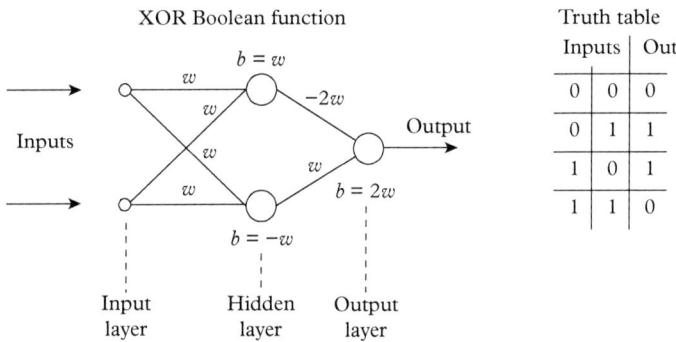

Figure 6.15 *Synaptic weights and thresholds for the XOR Boolean function. The truth table is in the limit* $w \to \infty$.

output neuron. The synaptic weight matrix between the inputs and the hidden layer is symmetric and has a weight value w, and two thresholds in the hidden layer are set to w for the upper node and $-w$ for the lower node. The synaptic weight matrix between the hidden layer and the output is $[-2w, w]$, and the threshold on the output neuron is set to $2w$. The weight w can be set at a moderate value (e.g. $w = 5$) for testing, and then can be set to a large value to ensure that the Boolean output is ± 1.

The key to training a network with hidden neurons is to extend the delta rule to two layers. This is accomplished through a method known as error back-propagation. As in the single-layer perceptron, the two-layer perceptron is trained with a training set composed of μ examples

$$X_i^\mu \to y_j^\mu \to Z_k^\mu, \tag{6.18}$$

where X_i are the training states, y_j are the hidden states, and Z_k are the desired output states. For the three-layer network, the indices i, j, and k correspond to input, hidden, and output layers, respectively. There are now two weight matrices and two threshold vectors that need to be chosen to solve the problem. The weight matrix for the output layer is adjusted in the same way as the delta rule for the single-layer perceptron

$$\delta w_j^k = \varepsilon \sum_\mu \Delta_k^\mu y_j^\mu$$

$$\delta b_k = -\varepsilon \sum_\mu \Delta_k^\mu, \qquad\qquad (6.19)$$

where

$$\Delta_k^\mu = \left[Z_k^\mu - S \left(v_k^\mu \right) \right] S' \left(v_k^\mu \right) \qquad\qquad (6.20)$$

as before. The adjustments of the weight matrix of the hidden layer are obtained in a similar way as for the single-layer network, giving

$$\begin{aligned}
\delta w_i^j &= -\varepsilon \frac{\partial D}{\partial w_i^j} = \varepsilon \sum_{\mu,k} \left[Z_k^\mu - S \left(v_k^\mu \right) \right] S' \left(v_k^\mu \right) \frac{\partial v_k^\mu}{\partial y_j} \frac{\partial y_j}{\partial w_i^j} \\
&= \varepsilon \sum_{\mu,k} \Delta_k^\mu w_j^k S' \left(v_j^\mu \right) \frac{\partial y_j}{\partial w_i^j} \qquad\qquad (6.21) \\
&= \varepsilon \sum_\mu \bar{\Delta}_j^\mu x_i^\mu
\end{aligned}$$

$$\begin{aligned}
\delta b_j &= -\varepsilon \frac{\partial D}{\partial b_j} = \varepsilon \sum_{\mu,k} \left[Z_k^\mu - S \left(v_k^\mu \right) \right] S' \left(v_k^\mu \right) \frac{\partial v_k^\mu}{\partial y_j} \frac{\partial y_j}{\partial b_j} \\
&= \varepsilon \sum_{\mu,k} \Delta_k^\mu w_j^k S' \left(v_j^\mu \right) \frac{\partial y_j}{\partial b_j} \qquad\qquad (6.22) \\
&= -\varepsilon \sum_\mu \bar{\Delta}_j^\mu
\end{aligned}$$

where

$$\bar{\Delta}_j^\mu = \left(\sum_k \Delta_k^\mu w_j^k \right) S' \left(v_j^\mu \right). \qquad\qquad (6.23)$$

The delta rule for the hidden layer has the same form as for the output layer, except that the deviation has been propagated back from the output layer. This is why the delta rule for the two-layer perceptron is called *error back propagation*.

Boolean functions are among the common applications of multilayer perceptrons. The sigmoid function has two clear asymptotes at +1 and −1 that can represent 1 and 0, respectively. An example of a 4-bit AND training session is shown in Fig. 6.16. The network has four inputs, five neurons in the hidden

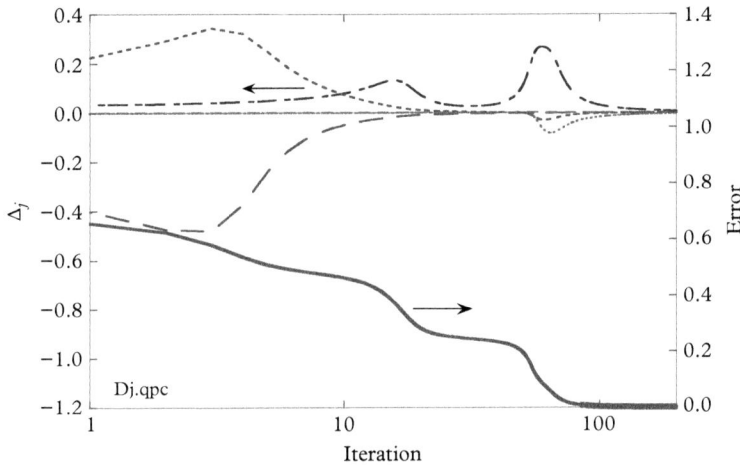

Figure 6.16 *Training a 4-bit AND with one layer of five hidden neurons. The individual errors Δ_j on the five hidden neurons are shown versus iteration number. The output error is on the single output neron.*

layer, and a single output neuron. The training session uses eight patterns out of the total possible 16 inputs among 4 bits, and in this example takes 200 iterations (known as epochs) of training. The five back-propagated errors on the hidden neurons are plotted in Fig. 6.16, as well as the error on the single output neuron. The largest adjustments of the neural weights and thresholds tend to occur in sudden steps, as the neuron values rearrange. The total error on the training set eventually falls below 0.001. After the training, the network is applied on all 16 possible inputs to test if the neural network correctly generalized to the cases not in the training set. If not, then the training is restarted. It is important to note that many training sessions (beginning with randomly assigned weights) failed to converge on the correct function, and even when the training set converged correctly, the network failed to generalize for the remaining input cases. However, testing a neural network is fast and simple, and if a network fails after training, then it is simply trained again until it gives the correct performance with the correct generalizations.

The two-layer perceptron can solve problems that are linearly separable,[6] such as the XOR or a multi-bit AND. But there are still simple examples of problems that cannot be solved. For instance, a simple problem might be identifying multi-bit patterns that are balanced, e.g. [0 1 0 1] and [1 1 0 0] have equal numbers of 1 and 0, while [1 1 1 0] does not. This problem is not linearly separable, and cannot be solved with a single hidden layer. However, adding a second hidden layer can extend the range of valid problems even farther. The art of selecting feed-forward architectures that are best suited to solving certain types of problems is a major field that is beyond the scope of this chapter. Nonetheless, there is one more type of neural network architecture that we will study that is well suited to problems associated with content-addressable memory. These are recurrent networks.

---

[6] In linearly separable problems, points in the classification space can be separated by lines in two dimensions, planes in three dimensions, or hyperplanes in higher dimensions.

6.5 Hopfield neural network

The third general class of neural network architecture is composed of feedback networks. Networks that use feedback have loops and hence are recurrent. One example is shown in Fig. 6.17. The network typically evolves from an initial state (input) to an attractor (fundamental memory). The recalled memory is in some sense the "closest" memory associated with the input. For this reason, these networks work on the principle of associative memory. Partial or faulty inputs recall the full fundamental memory. There are a wide variety of feedback architectures, but one particularly powerful architecture is also very simple, known as the Hopfield network.

The Hopfield network is a recurrent network with no hidden layers and no self-feedback. In this specific model the synaptic weights are symmetric, and the neuron states are updated asynchronously to prevent oscillation. An example is shown in Fig. 6.18 where the outputs are fed back into the input neurons. The delay units ensure "causal" operation consisting of successive distinct iterations to update each neuron state. The network is operated by assigning initial states of the neurons (a possibly corrupted input pattern), and then iterating the network until the neuron states settle down into steady states (the pattern that is "recalled" from memory). The network is "trained" by assigning the w_{kj} feedback weights according to a set of M fundamental memories. In the Hopfield network, the training is performed once, rather than iteratively as in the delta rule of the feed forward networks. The iterative stage for the Hopfield network allows the system dynamics to relax onto an attractive fixed point (memory).

For convenience, the nonlinear activation function in the following examples is set to be the tanh function

$$x = S_a(v) = \tanh\left(\frac{g_a v}{2}\right) \tag{6.24}$$

with the inverse that is also a smooth function

$$v = S_a^{-1}(x) = -\frac{1}{g_a}\ln\left(\frac{1-x}{1+x}\right). \tag{6.25}$$

Feedback (recurrent) network

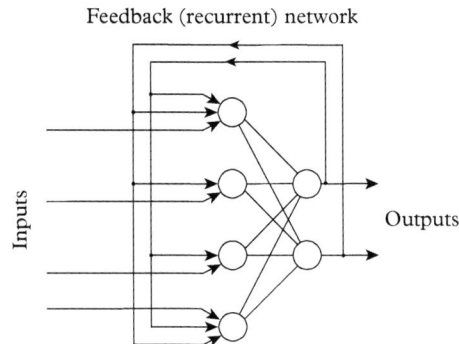

Figure 6.17 *Recurrent network with feedback loops.*

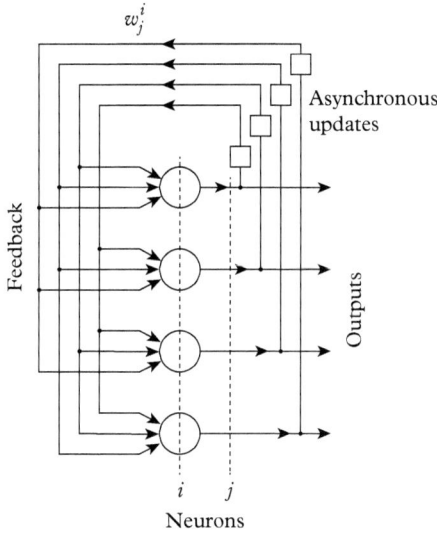

Figure 6.18 *Recurrent network for the Hopfield model. To train the network, the weights w_j^i are assigned according to the fundamental memories. To operate the network, the neuron states are initialized to an input pattern, and then the network is iterated until it settles into the steady state associated with the "recalled" pattern.*

A key question about a dynamical network is whether it will converge on a stable fixed point. This is partially answered by Lyapunov's theorem, which sets the conditions required for all Lyapunov exponents to be negative:

For a state vector $x^a(t)$ and an equilibrium state $\bar{x}$, the equilibrium state $\bar{x}$ is stable if, in a small neighborhood of $\bar{x}$, there exists a smooth positive definite function $V(x^a(t))$ such that

$$\frac{dV(x^a(t))}{dt} \leq 0 \; \forall x^a \text{ in the neighborhood.}$$

This theorem is general for any smooth positive-definite function V, no matter what detailed form it takes. For the Hopfield model, adopting the physical representation of resistors, capacitors and currents (see Fig. 6.11), an energy function serves in the role of the positive definite function. The energy function is

$$E = -\frac{1}{2}\sum_{a=1}^{N}\sum_{b=1}^{N}w_a^b x^a x_b + \sum_{b=1}^{N}\frac{1}{R_b}\int_0^{x_b}S_b^{-1}(x)\,dx - \sum_{b=1}^{N}I_b x_b. \tag{6.26}$$

The energy describes an energy landscape with numerous local minima, where each minimum corresponds to a fundamental memory of the network.[7]

Differentiating the energy E with respect to time gives

$$\frac{dE}{dt} = -\sum_{b=1}^{N}\left(\sum_{a=1}^{N}w_a^b x^a - \frac{v_b}{R_b} + I_b\right)\frac{dx_b}{dt}. \tag{6.27}$$

[7] This energy function is analogous to the spin-Hamiltonian of quantum magnets in spin-glass states when the components of the matrix w_j^i have both signs and compete.

The quantity inside the parentheses is $C_b dv_b/dt$. The energy differential simplifies to

$$\frac{dE}{dt} = -\sum_{b=1}^{N} C_b \left(\frac{dv_b}{dt} \right) \frac{dx_b}{dt}. \tag{6.28}$$

Using the expression for the inverse nonlinear activation function gives

$$\frac{dE}{dt} = -\sum_{b=1}^{N} C_b \left(\frac{d}{dt} S_b^{-1}(x_b) \right) \frac{dx_b}{dt}$$
$$= -\sum_{b=1}^{N} C_b \left(\frac{dx_b}{dt} \right)^2 \left(\frac{d}{dx_b} S_b^{-1}(x_b) \right). \tag{6.29}$$

The right hand side of this expression is strictly negative when the response function $S_b(v)$ is monotonic increasing (sigmoid function), and when it has a smooth inverse. Therefore,

$$\frac{dE}{dt} \leq 0, \tag{6.30}$$

the energy is bounded, and the attracting fixed point to the dynamical system is stable for this physical neuron model. The energy function of a Hopfield network is a continuously decreasing function of time. The system executes a trajectory in a state space that approaches the fixed point at the energy minimum. This fixed point is a recalled memory.

6.6 Content-addressable (associative) memory

A content-addressable memory is one where a partial or a corrupted version of a pattern recalls the full ideal pattern from memory. To make the state description simpler, the Hopfield model is considered in the high-gain limit that leads to binary discrete output states of each neuron. Under high gain, the control parameter $g_i \rightarrow \infty$ in Eq. (6.25), and the simple energy expression is

$$E = -\frac{1}{2} \sum_{a=1}^{N} \sum_{b=1}^{N} w_a^b x^a x_b \quad \text{for } x_b = \pm 1. \tag{6.31}$$

The state of the network is

$$x_b = [x_1, x_2, ..., x_N] \quad \text{for } x_b = \pm 1. \tag{6.32}$$

This state induces a local field

$$v_a = \sum_{b=1}^{N} w_a^b x_b - b_a, \tag{6.33}$$

where

$$x_a = \text{sgn}\,[v_a] \tag{6.34}$$

and where b_a is a local bias, possibly different for each neuron. The biases can be zero for each neuron, or they can be chosen to improve the performance of the recall. In the following examples, the bias is set to zero.

There are two phases in the operation of the content-addressable memory: (1) storage and (2) retrieval. The storage phase selects the ideal patterns that act as fundamental memories and uses these to define the appropriate elements of w_a^b. The retrieval phase presents a partial or corrupted version of the pattern to the network as the network iterates until it converges on an ideal pattern. If the corruption is not too severe, the resulting ideal pattern is usually the correct one.

6.6.1   Storage phase

In the storage phase, M fundamental memories are used to define the weight matrix. For instance, consider M fundamental memories $\vec{\xi}_\mu$ of length N with elements

$$\xi_a^\mu = a^{th} \text{ element of } \mu\text{th memory.} \tag{6.35}$$

The synaptic weights are calculated by the outer product

$$w_a^b = \frac{1}{N} \sum_{\mu=1}^{M} \xi_\mu^b \xi_a^\mu \text{ and } w_a^a = 0. \tag{6.36}$$

To prevent run-away in the iterated calculations, it is usual to remove the diagonal as

$$\overleftrightarrow{w} = \frac{1}{N} \sum_{\mu=1}^{M} \vec{\xi}_\mu \vec{\xi}_\mu^T - \frac{M}{N} \overleftrightarrow{I}, \tag{6.37}$$

where the synaptic weight matrix $\overleftrightarrow{w}$ is symmetric and $\overleftrightarrow{I}$ is the identity matrix. The outer product is a projection operator that projects onto a fundamental memory, and the synaptic weights perform the function of finding the closest fundamental memory to a partial or corrupted input. Once the synaptic weight matrix is defined, the storage phase is complete, and no further training is required.

6.6.2   Retrieval phase

The initial network state is set to $\overleftrightarrow{\xi}_{probe}$ which is a partial or noisy version of a fundamental memory. The network is iterated asynchronously to prevent oscillation, which means that the neurons are picked at random to update. When the jth neuron is picked, the $n+1$-st state of the neuron depends on the nth state

$$x_b^{(n+1)} = \text{sgn}\left[\sum_{a=1}^{N} x_a^{(n)} w_b^a - b_b \right].$$

(6.38)

If the new state is not equal to the old state, then it is updated. After many iterations, the states of the neurons no longer change. In this situation, the recalled state (the fixed point) is now steady. For the stable memory end state, all the output values of each neuron are consistent with all the input values. The self-consistency equation is

$$y_a = \text{sgn}\left[\sum_{b=1}^{N} y_b w_a^b + b_a \right] \quad \text{for } a = 1, 2, ..., N.$$

(6.39)

Example 6.1 Three neurons, $N = 3$, two fundamental memories, $M = 2$

Assume the fundamental memories are

$$\xi_1 = \begin{pmatrix} 1 \\ -1 \\ 1 \end{pmatrix}, \quad \xi_2 = \begin{pmatrix} -1 \\ 1 \\ -1 \end{pmatrix}.$$

The synaptic weight matrix from Eq. (6.37) is

$$\overset{\leftrightarrow}{W} = \frac{1}{3}\begin{pmatrix} +1 \\ -1 \\ +1 \end{pmatrix}\begin{pmatrix} +1 & -1 & +1 \end{pmatrix} + \frac{1}{3}\begin{pmatrix} -1 \\ +1 \\ -1 \end{pmatrix}\begin{pmatrix} -1 & +1 & -1 \end{pmatrix} - \frac{2}{3}\begin{pmatrix} 1 & 0 & 0 \\ 0 & 1 & 0 \\ 0 & 0 & 1 \end{pmatrix}$$

$$= \frac{1}{3}\begin{pmatrix} 0 & -2 & +2 \\ -2 & 0 & -2 \\ +2 & -2 & 0 \end{pmatrix}$$

As a test, try one of the fundamental memories, such as $y = (1, -1, 1)$ as an input

$$\overset{\leftrightarrow}{W}\vec{y} = \frac{1}{3}\begin{pmatrix} 0 & -2 & +2 \\ -2 & 0 & -2 \\ +2 & -2 & 0 \end{pmatrix}\begin{pmatrix} 1 \\ -1 \\ 1 \end{pmatrix} = \frac{1}{3}\begin{pmatrix} 4 \\ -4 \\ 4 \end{pmatrix}.$$

The output is

$$\text{sgn}\left[\overset{\leftrightarrow}{W}\vec{y} \right] = \begin{pmatrix} 1 \\ -1 \\ 1 \end{pmatrix} = \vec{y}$$

and the memory is correctly recalled.

Now try an input that is not a fundamental memory. For instance, use $y = (1, 1, 1)$ as an initial input. Then

$$\overleftrightarrow{W}\vec{y} = \frac{1}{3}\begin{pmatrix} 0 & -2 & +2 \\ -2 & 0 & -2 \\ +2 & -2 & 0 \end{pmatrix}\begin{pmatrix} 1 \\ 1 \\ 1 \end{pmatrix} = \frac{1}{3}\begin{pmatrix} 0 \\ -4 \\ 0 \end{pmatrix}.$$

In this evaluation, there are values $y_a = 0$ that result (a non-allowed state). When this happens during updating, the rule is to *not* change the state to zero, but to keep the previous value. The resultant vector after the first iteration is therefore

$$\text{sgn}\left[\overleftrightarrow{W}\vec{y}\right] = \begin{pmatrix} 1 \\ -1 \\ 1 \end{pmatrix} = \xi_1,$$

which is one of the fundamental memories, and the recalled memory is stable.

As another example, use $y = (1, 1, -1)$ as an initial input

$$\overleftrightarrow{W}\vec{y}_1 = \frac{1}{3}\begin{pmatrix} 0 & -2 & +2 \\ -2 & 0 & -2 \\ +2 & -2 & 0 \end{pmatrix}\begin{pmatrix} 1 \\ 1 \\ -1 \end{pmatrix} = \frac{1}{3}\begin{pmatrix} -4 \\ 0 \\ 0 \end{pmatrix}$$

$$\text{sgn}\left[\overleftrightarrow{W}\vec{y}_1\right] = \begin{pmatrix} -1 \\ 1 \\ -1 \end{pmatrix}$$

which, again, is a fundamental memory and is stable.

The Hopfield network can have spurious states that are stable recalls, but are not one of the fundamental memories. This is because when $M < N$, then W has degenerate eigenvalues of zero. The subspace spanned by the null eigenvector associated with the zero eigenvalues constitutes the null space, and the Hopfield network includes vector projectors that project onto the null space, which produces false memories.

An essential question about associative memories is how many memories can be stored and recalled effectively. To provide an estimate for the capacity of the memory, chose a probe $\vec{\xi}_{probe} = \vec{\xi}^v$ that is one of the fundamental memories. Then

$$v_a = \sum_{b=1}^{N} w_a^b \xi_b^v$$

$$= \frac{1}{N}\sum_{\mu=1}^{M} \xi_a^\mu \sum_{b=1}^{N} \xi_\mu^b \xi_b^v \qquad (6.40)$$

$$= \xi_a^v + \frac{1}{N}\sum_{\mu\neq v}^{M} \xi_a^\mu \sum_{b=1}^{N} \xi_\mu^b \xi_b^v.$$

The first term is the correct recall, while the second term is an error term. Because the state values are ± 1 (saturated outputs of the response function), the second term has zero mean. It has the further properties

$$\sigma^2 = \frac{M-1}{N} \tag{6.41}$$

for the variance, and thus

$$S/N = \frac{1}{\dfrac{M-1}{N}} \approx \frac{N}{M} \tag{6.42}$$

for the signal-to-noise ratio for large M. It is beyond the scope of the current chapter to define the desired signal-to-noise for effective operation of the memory, but numerical studies have found a rough rule-of-thumb to be

$$S/N|_c \geq 7 \tag{6.43}$$

for the critical threshold for stability.[8] Therefore, it is necessary for the size of the vector N (the number of neurons) to be about an order of magnitude larger than the number of memories M.

[8] See Haykin (1999), p. 695.

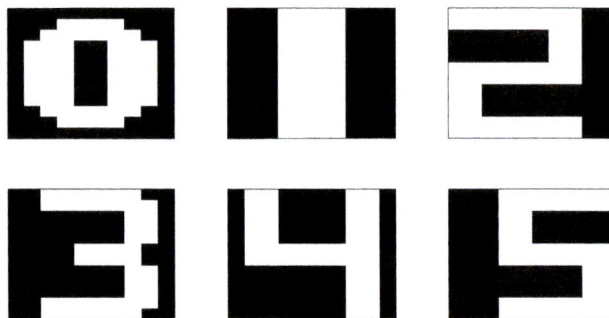

Figure 6.19 *Training set for a Hopfield network. $N = 120$ neurons and $M = 6$ memories.*

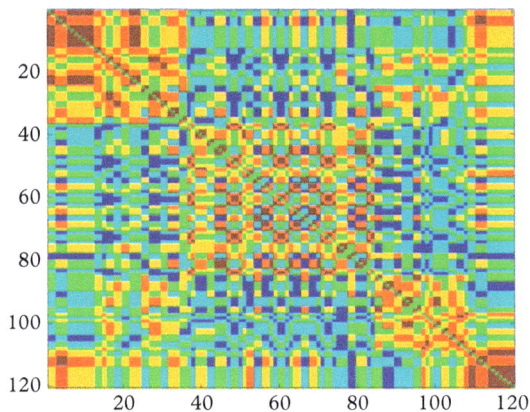

Figure 6.20 *The 120×120 synaptic weight matrix for the training set.*

6.6.3   Example of a Hopfield pattern recognizer

An example of a Hopfield network that performs number recognition is implemented with $N = 120$ neurons and $M = 6$ memories. The fundamental memories are shown in Fig. 6.19 as 10×12 arrays of ones and zeroes. These two-dimensional patterns are converted to one-dimensional vectors of 120 elements that serve as the ξ_i. The bias of each neuron is set to zero for this example. The synaptic weight matrix is calculated using Eq. (6.37) and is displayed in Fig. 6.20 for the 120×120 matrix values.

Examples of two runs are shown in Fig. 6.21 where one of the fundamental memories is scrambled with 40% probability. The first example chooses the number "4," and scrambles the pixels (shown in the middle). After 41 iterations, the correct memory is recalled. The second example does the same for the number "3."

One of the interesting aspects of neural networks is that they don't always work. An example of incorrect recall is shown in Fig. 6.22. In this case, the final steady state is not a pattern that exists in the training set. Because of the large redundancy of the network ($N/M = 20$), there are many stable solutions that are not one of the fundamental memories. These are "false" memories that were never trained, but are valid steady-state solutions.

This chapter has introduced several simple aspects of neurodynamics, beginning with single neurons that are nonlinear dynamical systems exhibiting a broad range of nonlinear behavior. They can spike or cycle or converge on stable fixed points, or participate in bifurcations. An even wider variety of behavior is possible when many neurons are coupled into networks. For instance, Hopfield networks are dissipative nonlinear dynamical systems that have multiple basins of attraction in which initial conditions may converge upon stable fixed points. Alternatively, state trajectories may be periodic, executing limit cycles, or trajectories may be chaotic, and attractors may be fractal. The wealth of behavior is the reason why artificial neural network models have attracted, and continue to attract, intense interest that is both fundamental and applied. Fundamental interest stems from the importance of natural neural systems, especially in our own consciousness and so-called intelligence. Applied interest stems from the potential for neural systems to solve complex problems that are difficult to solve algorithmically for which a holistic approach to fuzzy or ill-defined problems is often the most fruitful.

6.7   Summary

Neural dynamics provide several direct applications of the topics and tools of nonlinear dynamics that were introduced in Chapter 3. Individual neurons are modeled as nonlinear oscillators that rely on *bistability* and *homoclinic orbits* to produce spiking potentials. Simplified mathematical models, like the Fitzhugh–Nagumo and NaK models, capture successively more sophisticated behavior of

$p = 0.4$, 52 iterations

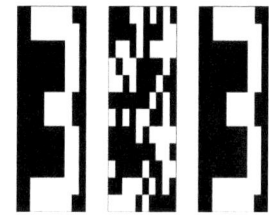

$p = 0.4$, 53 iterations

Figure 6.21 *Examples of successful pattern recall for a bit-flip probability of 40%.*

$p = 0.45$, 66 iterations

Figure 6.22 *Example of unsuccessful pattern recall. With $p = 45\%$, the 4 is translated into a non-fundamental 3. The final state is a spurious stable state.*

the individual neurons such as thresholds and spiking. Artificial neurons are composed of three simple features: summation of inputs, referencing to a threshold, and saturating output. Artificial networks of neurons were defined through specific network architectures that included the *perceptron*, *feed-forward networks* with *hidden layers* that are trained using the *Delta rule*, and *recurrent networks* with feedback. A prevalent example of a recurrent network is the *Hopfield network* that performs operations such as *associative recall*. The dynamic trajectories of the Hopfield network have *basins of attraction* in state space that correspond to stored memories.

Hodgkin and Huxley model

A mathematical model of neuron potential V based on the multiple types of ion channels in the neuron membrane. The time derivative of membrane potential is

$$C\dot{V} = I - g_K n^4 (V - E_K) - g_{Na} m^3 h (V - E_{Na}) - g_L (V - E_L), \qquad (6.1)$$

where C is the effective capacitance, V is the action potential, I is a bias current, g_K, g_{Na}, and g_L are the conductances, E_K, E_{Na}, and E_L are the equilibrium potentials, $n(V)$ is the voltage-dependent potassium activation variable, $m(V)$ is the voltage-dependent sodium activation variable (there are three sodium activation channels to four potassium channels, which determines the exponents), and $h(V)$ is the voltage-dependent sodium inactivation variable.

Fitzhugh–Nagumo model

A two-dimensional simplification of neurodynamics in the space defined by V, the membrane potential, and n, the number of activated membrane channels. The equations are similar to a van der Pol oscillator, and they tend to show limit cycles or stable fixed points.

$$\dot{V} = (V + 60) (a - V/75 - 0.8) (V/75 - 0.2) - 25n + I$$
$$\dot{n} = 0.03 (V/75 + 0.8) - 0.02n, \qquad (6.2)$$

where I is a bias current and a is a control parameter that determines the size of the limit cycle.

NaK model

The NaK model lies between the Fitzhugh–Nagumo and Hodgkin–Huxley models in complexity, adding voltage-dependent variables to Fitzhugh–Nagumo, but still simplifying the types of channels that are active in the neuron membrane with a two-dimensional dynamical space. The flow equations are

$$C\dot{V} = I - g_L (V - E_L) - g_{Na} m_\infty (V) (V - E_{Na}) - g_K n (V) (V - E_K)$$
$$\dot{n} = \frac{n_\infty (V) - n (V)}{\tau (V)}. \qquad (6.5)$$

Perceptron

A perceptron is a single-layer feed forward network. The output values are a function of the input values through a synaptic weight matrix w_k^j as

$$z_k = S\left(\sum_{j=1}^{m} w_k^j y_j - b_k\right), \qquad (6.11)$$

where the y_j are the inputs, w_k^j is the weight for the jth input to the kth output neuron, and b_k is the threshold of the kth output neuron. The transfer function $S(v_k)$ is one of the sigmoidal functions of Table 6.1.

Delta rule

The delta rule is used to determine the synaptic weights of a feed-forward network. The value of delta is

$$\Delta_k^\mu = \left[Z_k^\mu - \phi\left(v_k^\mu\right)\right]\phi'\left(v_k^\mu\right), \qquad (6.16)$$

where the prime denotes the derivative of the transfer function. The adjustments made to the weights during training are

$$\delta w_j^k = \varepsilon \sum_\mu \Delta_k^\mu Y_j^\mu$$
$$\delta b_k = -\varepsilon \sum_\mu \Delta_k^\mu, \qquad (6.17)$$

where ε is a small value that prevents the adjustments from overshooting as the process is iterated. The delta rule is extended to multilayer feed-forward networks and is called error back propagation.

Hopfield network synaptic weight

The synaptic weight matrix is constructed as the outer product of the fundamental memory vectors

$$\overleftrightarrow{w} = \frac{1}{N}\sum_{\mu=1}^{M}\overrightarrow{\xi}_\mu\overrightarrow{\xi}_\mu^{\,T} - \frac{M}{N}\overleftrightarrow{I}, \qquad (6.37)$$

where the synaptic weight matrix is symmetric. The outer product is a projection operator that projects onto a fundamental memory, and the synaptic weights perform the function of finding the closest fundamental memory to a partial or corrupted input.

240   *Introduction to Modern Dynamics*

6.8 Bibliography

S. Haykin, *Neural Networks: A Comprehensive Foundation* (Prentice Hall, 1999).
This textbook brings the topic of artificial neural networks to an easy level for students. There is a very useful set of associated program codes.

E. M. Izhikevich, *Dynamical Systems in Neuroscience: The Geometry of Excitability and Bursting* (MIT University Press, 2007).
The wide variety of mathematical models for biological neurons are reviewed in this text, along with many examples of behavior from nonlinear dynamics.

B. Müller and J. Reinhardt, *Neural Networks: An Introduction* (Springer, 1990).
This is a more advanced text that provides a deeper view into rigorous properties of neural networks, and makes useful analogies to thermodynamics and solid state physics.

6.9 Homework exercises

Analytic problems

1. **Fitzhugh–Nagumo model:** Perform a stability analysis of the fixed points and limit cycle for the Fitzhugh–Nagumo model. What are the bifurcation thresholds?
2. **Perceptron:** Design a sine-wave sensor as a single-layer perceptron. What are the weights and thresholds?
3. **Center-surround:** Starting from random weights, use a training set to train the center-surround single-layer perceptron using the delta rule.
4. **Multilayer perceptron:** Derive the delta rule for a three-layer feed-forward network with two layers of hidden neurons.
5. **Hopfield network:** Consider a simple Hopfield network made up of two neurons. There are four possible states for the network. The synaptic weight matrix of the network is

$$W = \begin{bmatrix} 0 & -1 \\ -1 & 0 \end{bmatrix}$$

   (a) Find the two stable states using the stability condition.
   (b) What is the steady-state behavior for the other two input states?
   (c) Define the energy function for infinite gain, and evaluate for all four states.
6. **Hopfield network:** Construct a Hopfield synaptic weight matrix for the three fundamental memories

$$\xi_1 = [1, 1, 1, 1, 1]^T$$
$$\xi_2 = [1, -1, -1, 1, -1]^T$$
$$\xi_3 = [-1, 1, -1, 1, 1]^T$$

(a) Use asynchronous updating to demonstrate that these three fundamental memories are stable.
(b) What happens when you flip the second element of ξ_1?
(c) What happens when you mask (set to zero) the first element of ξ_3?
(d) Find at least three spurious memories of the network.

7. **Energy function:** Show that the energy function of a Hopfield network may be expressed as

$$E = -\frac{N}{2} \sum_{\nu=1}^{M} m_\nu^2,$$

where m_ν denotes an overlap defined by

$$m_\nu = \frac{1}{N} \sum_{j=1}^{N} x_j \xi_{\nu j},$$

where x_j is the jth element of the state vector $\mathbf{x}$, $\xi_{\nu j}$ is the jth element of the fundamental memory ξ_ν, and M is the number of fundamental memories.

Computational projects[9]

8. **Bistability:** For the NaK model in the bistable regime, start the dynamics at the fixed point, and give the system an isolated membrane potential spike that takes it to the limit cycle. Once on the limit cycle, give the membrane potential a spike and track the dynamic trajectory as it relaxes back to the fixed point.

9. **Bifurcation and hysteresis:** Find appropriate parameters for the NaK model that demonstrate hysteresis. Start the dynamics of the NaK model and track an orbit. Slowly change a parameter until there is a bifurcation (sudden change in behavior). Then reverse the changes in that parameter until a bifurcation occurs back to the original dynamics. You are seeking conditions under which there is hysteresis: the two critical thresholds (one for the up transition and one for the down transition) on a single parameter are not the same.

10. **Homoclinic orbit:** In the NaK model, can you find an infinite-period homoclinic orbit? How tightly (significant figures) do you need to control the parameters to get the period to diverge?

11. **Multilayer perceptron:** Implement a 4-bit OR in a two-layer perceptron. How many hidden neurons are needed?

12. **Generalization:** Train the 4-bit AND with eight training examples. What is the probability that a network that acts correctly on the training set will generalize correctly to the remaining eight cases?

13. **Multilayer perceptron:** How well can a single hidden layer solve the "balanced bit" problem? This is when a group of bits have equal numbers of 0 and 1. What is the error probability of a well-trained network? Solve the "balanced bit" problem using two layers of hidden neurons.

[9] Matlab programming examples are found at www.works.bepress.com/ddnolte.

7 Evolutionary Dynamics

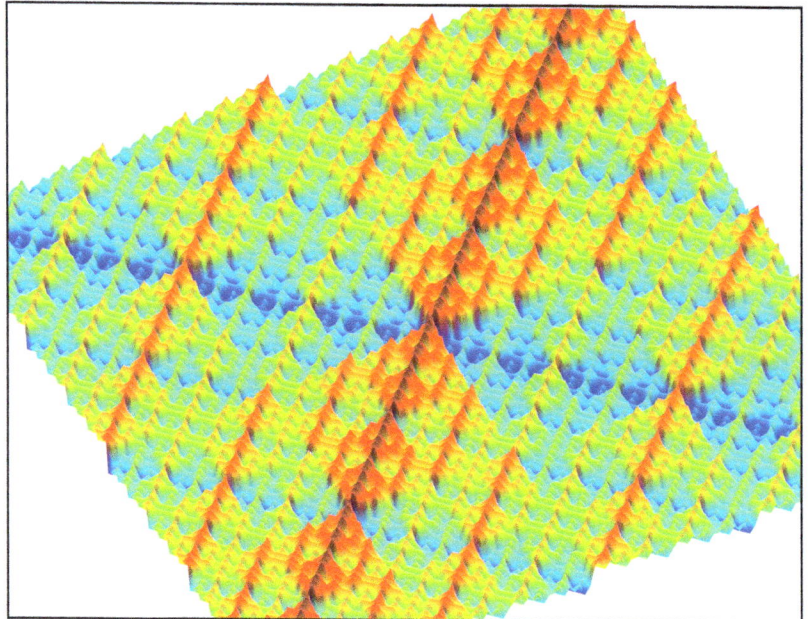

Natural evolution is a motive force of such generality that it stands as one of the great paradigms of science that transcends its original field. This chapter introduces the topic of evolutionary dynamics. In the context of dynamical systems, it is the study of dynamical equations that include growth, competition, selection, and mutations. The types of systems that have these processes go beyond living species and ecosystems, extending to such diverse topics as language evolution, crystal growth, business markets, and communication networks.

Problems in evolutionary dynamics provide some of the simplest applications of the overall themes of this book, namely a focus on nonlinear flow equations for systems with several degrees of freedom and the approach to steady-state behavior. For instance, questions of stability become questions of sustainability of ecosystems. These problems generally relate to the growth and decay of mixed populations, such as the rabbits vs. sheep model in Example 3.3 of Chapter 3 that introduced a simple one-on-one competition of two species competing for the same grazing land where the dynamics generally led to full extinction of one species or the other. In this chapter, we will see examples where species can coexist and where individual populations go through boom and bust cycles in classic predator–prey models. Zero-sum games are a pervasive example that is introduced when finite resources must be divided among multiple species. One of the important developments in modern dynamics is the idea of a quasi-species and the dynamics of quasi-species diffusing across a fitness landscape in a high-dimensional "fitness" space.

The mathematical flows in this chapter are relatively simple—capturing growth, competition, and selection. The stabilities of the fixed points are generally saddles, nodes, and centers. But the consequences of these solutions are profound, affecting not only the evolution of species, but also addressing issues of the continuing existence of ecosystems under pressure by global environmental change, our immune systems under assault by viruses, and the rise of cancer—all topics that will be of central importance for future physicists living and working in the complexities of an overcrowded world.

7.1 Population dynamics

The study of population dynamics has a long history. One of the famous early examples was the study of rabbit populations by Leonardo Pisano (1170–1250), more commonly known as Fibonacci. In his *Liber Abaci* (1202), in which he introduced Arabic numerals to Western Civilization, he considered a rabbit population with the conditions

- In the "zeroth" month, there is one pair of rabbits.
- In the first month, the first pair begets another pair.
- In the second month, both pairs of rabbits have another pair, and the first pair dies.
- In the third month, the second pair and the new two pairs have a total of three new pairs, and the older second pair dies.
- Each pair of rabbits has two pairs in its lifetime, and dies.

If the population at month n is $F(n)$, then at this time, only rabbits who were alive back to month $n-2$ are fertile and produce offspring, so $F(n-2)$ pairs are added to the current population of $F(n-1)$. Therefore, the total is $F(n) = F(n-1) + F(n-2)$. Starting with one pair of rabbits, the rabbit pairs for each month becomes

$$1, 2, 3, 5, 8, 13, 21, 34, 55, 89, 144, 233, 377, 610, 987, 1597, \ldots$$

This is the famous Fibonacci sequence that plays a role in many areas of mathematics related to the golden mean

$$\gamma = \lim_{n \to \infty} F(n)/F(n-1) = 1.618. \tag{7.1}$$

Fibonacci spiral

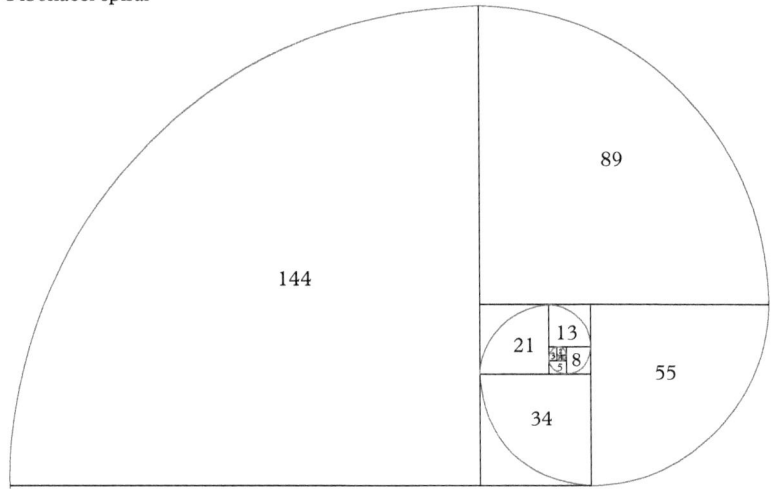

Figure 7.1 *A Fibonacci spiral enclosing areas that increase as the Fibonacci series. When θ grows by π/2, the radius grows by the golden mean.*

The golden mean also plays a role in nature, where many examples abound, such as the curve of the nautilus shell. An example of a Fibonacci spiral is shown in Fig. 7.1. The equation for this spiral (also known as the golden spiral) is

$$r = r_0 e^{b\theta} \tag{7.2}$$

for a growth rate b, where the radius of the spiral when θ is a right angle is equal to the golden mean, such that

$$e^{b\frac{\pi}{2}} = 1.618, \tag{7.3}$$

which yields the growth factor $b = 0.3063$. Although, this is not a very realistic example of population dynamics, it shows how venerable the field is, and it sets the stage for more realistic models.

7.1.1   Reproduction vs. population pressure

Consider a species that reproduces at a certain rate, and no individuals die. In this case the population grows at an exponential rate

$$\dot{x} = \alpha x$$
$$x(t) = x_0 \exp(\alpha t). \tag{7.4}$$

However, if there are limited resources, then individuals die at a rate proportional to the population, adding a limiting term to the dynamics as

$$\dot{x} = (\alpha - \beta x)\, x. \tag{7.5}$$

This has two steady-state solutions:

$$x^* = \alpha/\beta \quad \text{and} \quad x^* = 0, \tag{7.6}$$

the second of which represents extinction. The first is a stable balance between growth and population pressure. This balanced fixed point has a negative Lyapunov exponent, and the relaxation of perturbations back to steady state occurs at rate given by α. This example of population growth and stasis is at the heart of more sophisticated models of competing populations.

7.1.2   Species competition and symbiosis

The general population dynamics for two species is

$$\begin{aligned}\dot{x} &= f(x, y)\, x \\ \dot{y} &= g(x, y)\, y. \end{aligned} \tag{7.7}$$

These equations admit a wide variety of fixed points and stability conditions. The flows are always in the positive quadrant of the phase plane because the dynamics are bounded by the nullclines $x = 0$ and $y = 0$ which are also separatrixes. No flow lines can cross out of the positive quadrant of the phase plane, which guarantees that all populations are non-negative. The functional forms for $f(x, y)$ and $g(x, y)$ can be arbitrary, but are usually polynomials. For instance, the rabbit vs. sheep flow described by Eq. (3.29)

$$\begin{aligned}\dot{x} &= x\,(3 - x - 2y) \\ \dot{y} &= y\,(2 - x - y) \end{aligned} \tag{3.29}$$

in Example 3.3 of Chapter 3 used negative feedback (population pressure) on both populations because both rabbits and sheep eat the grass. These dynamics lead to a saddle point as one or the other species went extinct (see Fig. 3.5).

Intraspecies population pressure can be balanced by advantages in symbiotic relationships with other species in the ecosystem. In this case the interspecies feedback is positive, which stabilizes the populations. As an example, consider the symbiotic dynamics

$$\begin{aligned}\dot{x} &= x\,(1 - 2x + y) \\ \dot{y} &= y\left(\frac{1}{2} + \frac{1}{2}x - y\right) \end{aligned} \tag{7.8}$$

with positive interspecies feedback. There are fixed points at $(0, 0)$ and at $(1, 1)$. The first is a saddle, while the second is a stable attractor. The symbiotic relationship between the species balances the intraspecies population pressure, leading to a stable equilibrium in which both species thrive together.

7.1.3   Predator–prey models

Predator–prey dynamics describe the dynamic evolution of at least two species, at least one of which preys upon the other. One of the simplest and early equations for predator–prey systems was proposed independently by Alfred J. Lotka in 1925 and Vito Volterra in 1926. These are the Lotka–Volterra equations

$$\dot{x} = x\,(\alpha - \beta y)$$
$$\dot{y} = -y\,(\gamma - \delta x)\,, \tag{7.9}$$

where y is the number of predators and x is the number of prey. The prey reproduce at the rate α, and are eaten with a rate β times the product of the number of prey and predators. The predators reproduce at a rate δ times the product of the number of prey and predators, and die off at a rate γ. Rather than rabbits and sheep that compete for the same food stock, this is now rabbits and foxes in which one of the species is the food for the other.

The Lotka–Volterra equations have an extinction fixed point $(x^*, y^*) = (0, 0)$ and a nonzero steady-state balance (stable fixed point) at

$$x^* = \gamma/\delta$$
$$y^* = \alpha/\beta. \tag{7.10}$$

The Jacobian matrix is

$$\mathcal{J} = \begin{pmatrix} \alpha - \beta y & -\beta x \\ \delta y & -\gamma + \delta x \end{pmatrix}. \tag{7.11}$$

At the fixed point at $(x^*, y^*) = (0, 0)$ the stability is governed by

$$\mathcal{J}_{(0,0)} = \begin{pmatrix} \alpha & 0 \\ 0 & -\gamma \end{pmatrix} \tag{7.12}$$

with Lyapunov exponents $(\alpha, -\gamma)$. Therefore, this is a saddle-point (when all parameters are positive). At the steady-state fixed point, the stability is governed by

$$\mathcal{J} = \begin{pmatrix} 0 & -\dfrac{\beta\gamma}{\delta} \\ \dfrac{\delta\alpha}{\beta} & 0 \end{pmatrix} \tag{7.13}$$

and the Lyapunov exponents are $\pm i\sqrt{\alpha\gamma}$, and the fixed point is a center. Therefore, the two populations oscillate about the nonzero fixed point.

7.2   Virus infection and immune deficiency

One of the most complex ecosystems to consider is our natural immune system that guards against the invasion of our bodies by foreign bacteria and viruses. When microbes invade, they multiply. But they also trigger our immune system that mounts a defense against the invaders. The viruses become the prey and the immune response the predator. This sounds like rabbits and foxes again—but the essential difference is that new strains of viruses can arise through random mutation and can escape the original immune response. The body is resilient and can mount a new attack—unless the virus disables the immune response, which is what occurs in the case of HIV infection. Clearly, this is a complicated ecosystem with stochastic processes and measure-countermeasure escalation. The long-term health of the individual is at stake, and we can predict outcomes—remission or relapse—based on the properties of the viruses and the immune response.

To start, consider a simple population dynamics given by

$$\dot{v} = v\,(r - ax)$$
$$\dot{x} = -bx + cv, \tag{7.14}$$

where v is the viral strain and x is the specific immune response (like antibody production) to that strain. A simulation of the population response to a single strain is shown in Fig. 7.2. The fixed point is a stable spiral, and the populations oscillate as they relax to a steady state at long times as a balance is established between the virus reproduction and the immune system attack on the viruses. If there are N virus strains, and a matched number of immune responses, then the equations become

$$\dot{v}_a = v_a\,(r - ax_a) \quad a = 1:N$$
$$\dot{x}_a = -bx_a + cv_a \tag{7.15}$$

and the state space now has *2N* dimensions. However, not all the viruses will be present at the start. A single virus can attack the system, eliciting an immune response that keeps it in check, but then the virus can mutate into a different strain that is independent and escapes the initial immune response, causing a new immune response, and so on. This process adds a fundamentally new character to the flow equations. The flows studied in earlier chapters of this book were fully

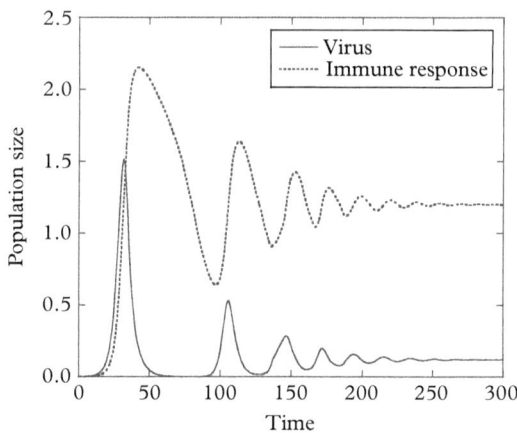

Figure 7.2 *Population size of the viral load and the immune response for a single virus strain with r = 2.4, a = 2, b = 0.1, c = 1.*

deterministic. They might exhibit complex chaotic behavior, but it was deterministic behavior without intrinsic randomness. The random emergence of virus strains introduces a stochastic process that is no longer deterministic. However, this new aspect causes no difficulty for the solution of the flow equations—the concentrations of the new strains are zero until the mutation event, and then the population evolution is governed by the deterministic equations. Therefore, adding a stochastic generation process to flow equations remains in the domain of the ODE solvers.

For the stochastic emergence of mutants, assume that new viral strains arise with a probability per time given by P. The number N of different viral strains increases in time as existing strains mutate into new strains that are unaffected by the previous immune response, but which now elicit a new immune response. The resulting population of all strains is shown in Fig. 7.3. Each subpopulation of antigen–immune pairs is independent and the total viral load grows linearly in time (on average).[1]

The immune system can be smarter and can do better at fighting antigenic variation by working globally with a response that fights all viral strains arising from a single type of virus, regardless of their mutations. This can be modeled with a cross-reactive immune response that depends on the total viral load. A new variable z is the cross-reactive response and enters into the equations as

$$\dot{v}_a = v_a\,(r - ax_a - qz)$$
$$\dot{x}_a = -bx_a + cv_a \tag{7.16}$$
$$\dot{z} = kv - bz$$

in which z is the cross-reactive response that decays at a rate given by b, and v is the total viral load $v = \sum_a v_a$. This global immune response attacks all viral strains, making the infection of each new strain that arises less and less effective, until the system establishes a steady state in which the viral load is held constant

[1] A more realistic model makes the mutation probability per time proportional to the total viral load, which would lead to exponential growth.

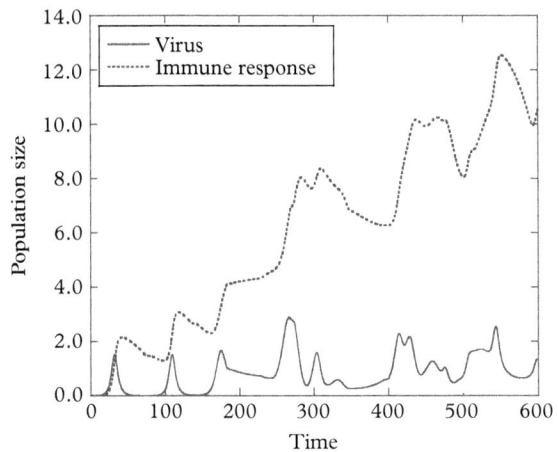

Figure 7.3 *Example of antigen diversity. The total population size of the viral load and the immune response are shown for r = 2.4, a = 2, b = 0.1, c = 1 with a random probability for the rise of new virus strains.*

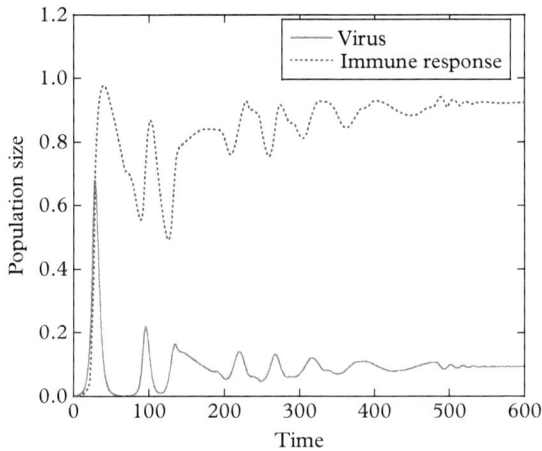

Figure 7.4 *Example of a cross-reactive immune response. The total population sizes of the viral load and the immune response are shown for r = 2.4, a = 2, b = 0.1, c = 1, q = 2.4, k = 1 with a random probability for the rise of new virus strains.*

and no new strain can escape from the global immune response, as shown in Fig. 7.4. At long times the viral load is held constant by the immune system, and the system maintains a steady state.

In this last example, the cross-reactive immune response locks the virus strains in a death struggle. Therefore, the best way for the virus to gain the upper hand would be to disable the immune defense. This is what HIV does as the precursor to AIDS. Such an attack on the immune system can be modeled by knocking down both the specific immune response, as well as the global cross-reactive immune response by adding another term to the equations that suppresses the immune response proportional to the viral load. This immune suppression is modeled as

$$\dot{v}_a = v_a\,(r - ax_a - qz)$$
$$\dot{x}_a = -bx_a + cv_a - uvx_a \qquad (7.17)$$
$$\dot{z} = kv - bz - uvz,$$

which has inhibitory terms in the second and third equations. Now, as new strains appear, they reduce the effectiveness of the immune system through the new terms. Eventually, the immune response saturates, as shown in Fig. 7.5, but the number of viral strains keeps increasing over time, and the disease eventually overwhelms the system.

This simple model of HIV infection[2] has gone through several levels, starting with a basic immune model, up through cross-reactivity, and finally to inactivation of the immune response. The phenomenon of immune escape emerges easily, despite the simple competitive models. In such open systems, there are neither constraints on population size nor any steady-state average (what might be called conserved) quantities. Open-ended growth is not sustainable in an ecosystem subject to finite resources. Real systems are bounded, which require an additional mathematical structure that goes beyond predator–prey models.

[2] See Nowak (2006), p. 167.

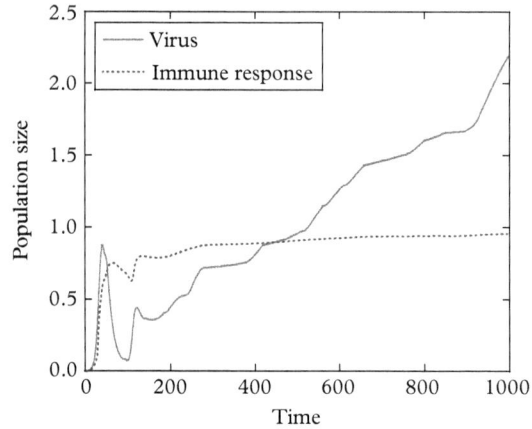

Figure 7.5 *Disabling the immune response. The total population sizes of the viral load and the immune response are shown for r = 2.4, a = 2, b = 0.1, c = 1, q = 2.4, k = 1, and u = 1 with a random probability for the rise of new virus strains.*

7.3 The replicator equation

Zero-sum games happen when there are finite resources that must be divided among several groups. The zero-sum condition is a conservation law. In the case of populations, it states that the sum of all individuals among all subpopulations is a constant. In a zero-sum game when one group wins, another group must lose. This means that the interaction matrix between these groups, also known as the payoff matrix, is asymmetric and ensures that the population growth dynamics remain bounded. There can be oscillations, and multi-group cycles as well as stable fixed points, but these all live in a bounded space that keeps the total number of individuals constant. This space is called a simplex.

An example of a simplex for four subpopulations is shown in Fig. 7.6. This is a three-simplex, a tetrahedron displayed in three-dimensional space. The tetrahedron has 4 vertexes, 6 edges, and 4 faces. The vertexes represent single subpopulations. An edge represents a continuous variation between two subpopulations. For a face, one subpopulation is zero and the total number is divided up among the three other subpopulations. Inside the simplex, any combination of the four subpopulations is possible. In all cases, the sum over all subpopulations is a constant.

The bounded dynamics of a zero-sum game can be captured by a simple population growth model called the replicator model. The replicator equation has a species growth rate that is proportional to the fitness of the species, but unlike the simpler unbounded growth models, the fitness is a function of the concentrations of all the other species. The fitness of the ath species is given by[3]

$$f^a(\vec{x}) = \sum_{b=1}^{N} x_b p_b^a, \tag{7.18}$$

[3] In this chapter, many formulas contain repeated indices, but these do *not* imply the Einstein summation convention. All summations in this chapter are written out explicitly. The state vectors are row vectors that multiply transition matrices from the left.

3-Simplex

[, ,] Face
{ , , } Edge
(, ,) Vertex

$$x + y + z + w = 1$$

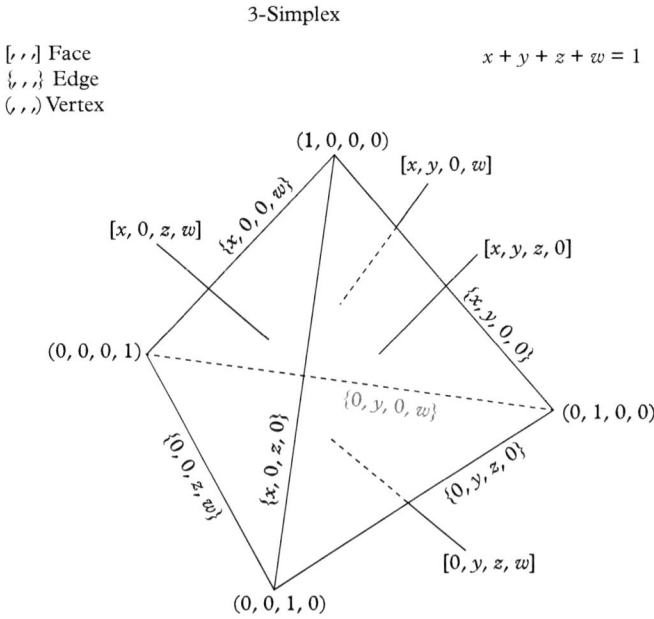

Figure *7.6 A three-simplex (tetrahedron) showing 4 vertices, 6 edges, and 4 faces. The constraint $x + y + z + w = 1$ allows all combinations to be expressed uniquely in or on the simplex.*

where x_a is a species population and p_a^b is the payoff matrix. If species a benefits over species b, then the payoff matrix element is positive. The payoff matrix is an antisymmetric matrix $p_b^a = -p_a^b$ with zero trace $Tr\{p_a^b\} = 0$. The average fitness is given by

$$\phi(\vec{x}) = \sum_{a=1}^{N} f^a x_a. \tag{7.19}$$

The replicator equation has a growth rate proportional to the difference between the species fitness and the average fitness. The replicator equation has the simple form

Replicator equation: $\dot{x}_a = x_a (f^a - \phi)$ (7.20)

(no implicit summation) with the conservation condition

$$\sum_{a=1}^{N} x_a = 1, \tag{7.21}$$

which constrains all trajectories to lie in or on the $N - 1$ simplex.

For very small N, the cases are simple. For $N = 2$, with only two competing species, often a single species takes the full share while the other goes extinct.

This is the classic "winner-take-all" competition scenario. Whenever only two businesses are competing for a fixed market, one can win everything and the other goes out of business, although a stable equilibrium between the two (usually with unequal market shares) also can occur. If $N = 3$, then one may win all, or two share the market and drive the third out of business, or they can all share in the market with a fixed point that has a stable spiral. It may seem that once the third player is driven out, then $N = 2$ dynamics would take over and one player could win it all. But the possibility of a third player re-entering the playing field can keep the dynamics from converting into winner-take-all. This may provide a motivation for some businesses to keep the competitive field open to start-ups. The start-ups may ultimately fail, but they can shield moderate-sized companies from being driven out by the biggest player.

Several examples of replicator dynamics with asymmetric payoff matrices are shown in Fig. 7.7 for $N = 8$ subpopulations. For many random choices of the payoff matrix elements, the solutions tend to have only a few characteristic behaviors. These include: (1) stable fixed points of 3 or 5 species with 5 or 3 species, respectively, condemned to extinction; (2) period-3 or period-5 cycles in which 3 or 5 species oscillate against each other while the other species go extinct; and (3) combinations of cycles and stable fixed points, usually with period-3 cycles and two stable cycles and the other species go extinct. Indeed, for N competing species, there are generally $N/2$ that go extinct, while the others are either stable or participate in cycles with other species.

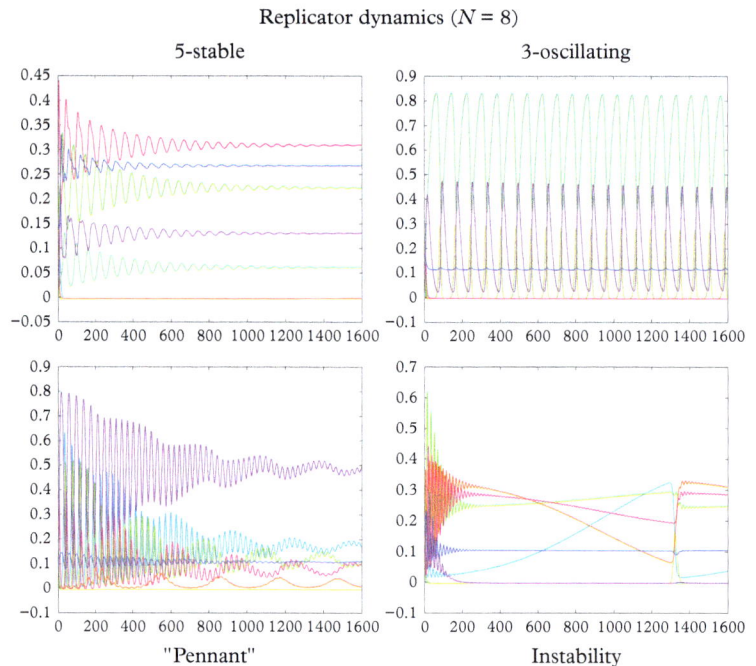

Figure 7.7 *Examples of replicator dynamics for $N = 8$. Three and five species cycles are common, as well as 3 or 5 stable species. In all cases of $N = 8$, 3 to 5 species tend to go extinct.*

7.4 The quasi-species equation

Mutation is an essential part of evolutionary dynamics by providing the adaptability needed to survive in the face of changing environmental pressures. A model in population dynamics that includes mutation among subspecies is the *quasi-species* equation, developed by Manfred Eigen and Peter Schuster in the 1980s. The term "quasi-species" is synonymous with "subspecies." This model goes beyond predator and prey and allows species and their subspecies to evolve in response to external evolutionary forces, with survival governed by subspecies fitness, often in response to predation. The mutations treated in this model are different than in the randomly arising mutations of viral infections. In that case, a new viral strain arises by mutation from an established strain, but there is no back mutation. This is appropriate for complex species with many genes. However, in the case of molecular evolution, as for DNA, point mutations in the digital code can mutate back and forth with some probability that is modeled with a transition rate in a matrix of all possible mutations.

The quasi-species equation contains a transformation term that mutates individuals into and out of a quasi-species. The transition matrix elements (mutation rates) from one species to a mutated species is governed by a stochastic mutation matrix Q_a^b. A *stochastic matrix* has the property that its rows and columns sum to unity. The transition rate at which subspecies b mutates into subspecies a is the mutation matrix multiplied by the fitness of the original species

$$W_a^{\prime b} = f^b Q_a^b. \tag{7.22}$$

The transition matrix is a symmetric matrix for which the back mutation is just as probable as the forward mutation. In the quasi-species model, individual fitness is *not* a function of the subspecies population size (known as a frequency-independent fitness which is opposite to the case in the replicator equation). The average fitness of the entire population is

$$\phi = \sum_{a=1}^{N} f^a x_a, \tag{7.23}$$

which is the weighted average, by population size (frequency x_a), of the individual fitness values f^a of each species. The quasi-species equations include a limiting growth term related to the average fitness φ of the entire population that keeps the relative populations bounded.

The quasi-species equation is

Quasi-species equation:
$$\dot{x}_a = \sum_{b=1}^{N} x_b (f^b Q_a^b) - x_a \sum_{b=1}^{N} f^b x_b$$
$$= \sum_{b=1}^{N} x_b W_a^b - \phi x_a \tag{7.24}$$

and the quasi-species fixed-point equation is

$$\dot{x}_a = \sum_{b=1}^{N} x_b^* W_a^{'b} - \phi^* x_a^* = 0. \tag{7.25}$$

In operator notation this is the eigenvalue problem

$$x^* W = \phi^* x^*. \tag{7.26}$$

Therefore, the left eigenvectors of the transition matrix W are the fixed points of the quasi-species equation.[4] These eigenvectors are used to determine the average fitness of the population distribution in $\phi^* = \phi(x^*)$.

In the time evolution of the subpopulations within the quasi-species equation, the total number of individuals grows exponentially. The number of individuals in the ath subspecies is X_a and the fraction of the total population X is given by $x_a(t)$. These are connected by

$$X_a(t) = x_a(t) e^{\psi(t)}, \tag{7.27}$$

where the function $\psi(t)$ is related to the growth of the total population. For instance, taking the derivative of Eq. (7.27) yields

$$\dot{X}_a(t) = \dot{x}_a(t) e^{\psi(t)} + x_a(t) e^{\psi(t)} \dot{\psi}(t). \tag{7.28}$$

If we define

$$\dot{\psi} = \phi \tag{7.29}$$

with the solution

$$\psi(t) = \int_0^t \phi(s) \, ds \tag{7.30}$$

then Eq. (7.28) becomes

$$\dot{X}_a = \dot{x}_a e^{\psi} + \phi x_a e^{\psi} \tag{7.31}$$

and multiplying both sides by $e^{-\psi}$ gives

$$\dot{X}_a e^{-\psi} = \dot{x}_a + \phi x_a$$
$$= \sum_{b=1}^{N} x_b \left(f^b Q_a^b \right) \tag{7.32}$$

using the second line of the quasi-species equation Eq. (7.24). The total number of individuals is

$$X = \sum_{a=1}^{N} X_a = \left(\sum_{a=1}^{N} x_a \right) e^{\psi(t)} = e^{\psi(t)} \tag{7.33}$$

---

[4] Left eigenvectors of a matrix A are the transpose of the right eigenvectors of the transpose matrix A^T.

and the time-derivative of the total population is

$$\dot{X} = \phi X, \qquad (7.34)$$

which grows exponentially with a rate equal to the average fitness. This is an important aspect of the quasi-species dynamics—the unlimited growth of the total population. There is a zero-sum game within the dynamics of the relative fractions of individuals, $x_a(t)$, but this is a consequence of the symmetry of the mutation matrix.

The two key elements of the quasi-species equation are the fitness f^a and the mutation matrix Q_a^b. The fitness parameter f^a determines the survival of the fittest, and takes on the properties of a "fitness landscape." It often happens that groups of mutations that are "close" to each other can be more advantageous to survival than mutations that are farther away. This means that there are local maxima as well as local minima as the subspecies index a moves through the populations of subspecies. This landscape plays the role of a potential surface in a random walk, as evolution climbs peaks of fitness within the fitness landscape.[5]

The mutation matrix can best be understood at the molecular level as base-pair permutations within a gene during replication. In a gene of fixed length, there can be only a finite number of possible combinations, which sets the number N of subspecies.

7.4.1   Molecular evolution

Evolutionary dynamics occurs at all levels. Top-level evolution occurs in global ecosystems, as in the slow change of species in response to global warming. Human societies participate in evolution as languages and dialects change slowly over time. More locally, species evolve in response to predator–prey selection. The evolution of species takes place within the genome where mutations of the genome may lead to improved fitness in the face of a changing environment and make it more likely to pass on its genetic code. The genome is composed of molecules— DNA—that have discrete base pairs, and the genetic code is contained within the base pair sequence. If the sequence is interrupted or changed, this is the origin of genetic mutations. At this molecular level, the mutations are discrete, and sometimes they are as simple as a single nucleotide permutation (SNP).

There are four nucleotides in DNA. These are thymine (T), cytosine (C), adenine (A), and guanine (G), as shown in Fig. 7.8. Thymine pairs only with adenine, and cytosine pairs only with guanine. A double helix has two strands, and each strand is the complement of the other, as shown in Fig. 7.9. If one strand starts CCGATTA, then the complementary terminal of the other strand is the complement (reading backwards) GGCTAAT. When DNA is read off as messenger RNA, it is translated as a digital code into proteins by ribosomes. The digital code consists of groups of three nucleotides, called codons, that translate to specific amino acids. Proteins are made up of amino acids—21 in all. A unique sequence of amino acids, built up into a polypeptide, imparts the unique properties

[5] The idea of a fitness landscape was first coined by Sewall Wright (1889– 1988) in the context of theoretical population genetics. Wright was one of the founders of theoretical population genetics, together with J. B. S. Haldane and R. A. Fisher, who combined ideas of evolutionary theory with genetics in the 1930s.

Figure 7.8 *Nucleotides that occur in DNA. Each is attached to a phosphate deoxyribose backbone. Guanine pairs only with cytosine, and adenine pairs only with thymine.*

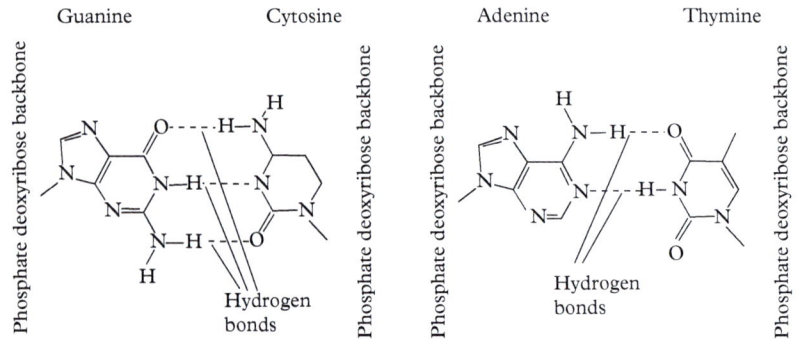

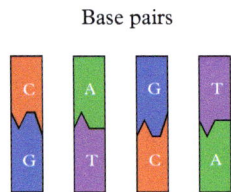

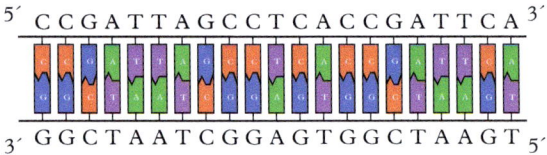

Figure 7.9 *Complementary DNA strrands. The two strands, when bound, form the double helix. The sequence of one strand is the complement of the other strand, as C only binds to G and A to T.*

of all the different proteins. The digital code has high redundancy, with several different codons corresponding to the same amino acid. But this is easily tolerated, because the total number of distinct codons is $4^3 = 64$, and the discreteness of the genetic code ensures (almost) error-free translation.

7.4.2 Hamming distance and binary genomes

To study the role of genetic mutations, calculations are simplified by using binary codes instead of the four-level DNA code. In a binary code, a mutation is a bit flip and the permutation space is a multidimensional 4D hypercube, with one dimension per bit. For instance, the permutation space for 4 bits is the hypercube shown in Fig. 7.10. Each axis (x, y, z, w) has only two values, 0 or 1. The inner cube represents $w = 0$, and the outer cube represents $w = 1$. Each edge represents the flip of a single bit. The "distance" between two vertices is the smallest number of edges that connects those two points. This four-cube is sometimes called a

Binary 4-bit hypercube

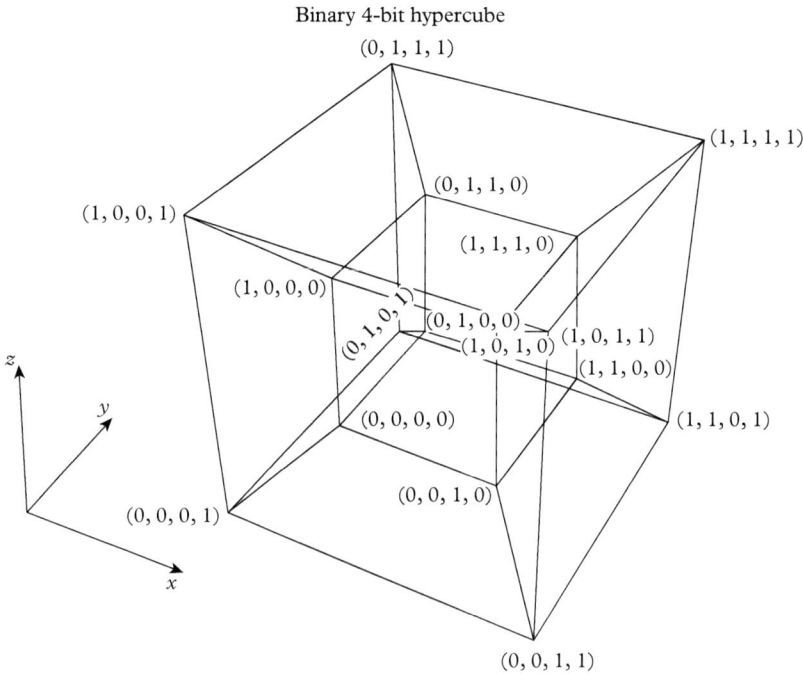

Figure 7.10 *A 4-bit hypercube repre-sents all possible bit combinations among 4 bits. The outer cube represents exten-sion into the fourth dimension.*

simplex, but is not to be confused with the three-simplex of four population values shown in Fig. 7.6. In that simplex, the four values are continuous between 0 and 1, and sum to unity—a constraint that reduces the four dimensions to three and allows the three-simplex to be drawn in ordinary three-dimensional space.

The distance between two binary numbers can be defined simply as the min-imum number of bit flips that take one number into another. On the hypercube, this is the so-called Manhattan distance, or grid distance, and is the smallest num-ber of edges that connects two points. In the case of binary numbers, this is called the Hamming distance. A gray scale map of the Hamming distance between all 7-bit numbers from 1 to 127 is shown in Fig. 7.11. The Hamming distance is important for binary mutations because it defines how many bit errors, or flips, are needed to mutate from one binary number (quasi-species) to another.

With the definition of a Hamming distance between two bit strings, there are many possible choices for the mutation matrix as a function of the Hamming distance. One simple assignment of the mutation matrix is

$$Q_a^b = \frac{1}{1 + \varepsilon^{-1} H_a^b} \left(\sum_{b=0}^{N} \frac{1}{1 + \varepsilon^{-1} H_a^b} \right)^{-1}, \qquad (7.35)$$

where the mutation probability decreases with increasing Hamming distance H_a^b between a and b, and the parameter ϵ determines how fast it decreases with distance.

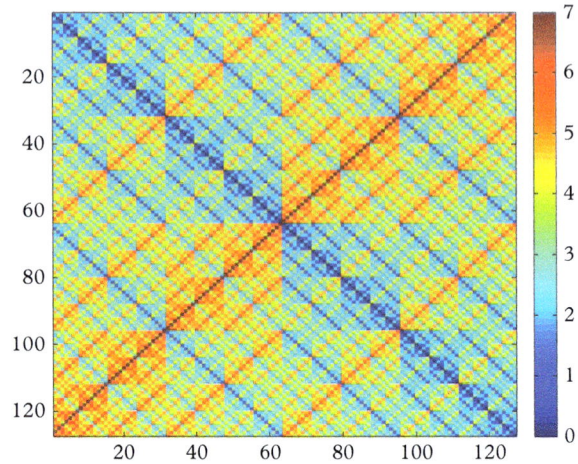

Figure 7.11 *Hamming distances H_a^b between 7-bit binary numbers from 0 to 127.*

The Hamming distance also can help define the fitness landscape, if there is a most-optimal subspecies α. A convenient way to define the fitness can be

$$f^b = \exp\left(-\lambda H_\alpha^b\right) \qquad (7.36)$$

with the parameter λ controlling the "contrast" of the landscape. Species that are farther away from the α species have lower fitness. Equations (7.35) and (7.36) establish a molecular evolution on binary strings. The bit flip probability is constant per bit, meaning that it is most likely that "close" strings will mutate into each other, while more distant strings will be less likely to mutate into each other. The most fit subspecies α will act as an attractor. As bit strings get closer to the most-fit string, they will survive and out-multiply less-fit subspecies. In this way, the population walks up the peaks in the fitness landscape.

As a concrete example, consider 7-bit strings with $n = 2^7 = 128$, using the Q_a^b and f^b equations defined by the Hamming distance in Eqs. (7.35) and (7.36). The key control parameter is the strength of the mutation rate ϵ. If the mutation rate is large, and the landscape contrast is small, then the final population of species will tend to match the fitness landscape: high-fitness populations will be the largest, and low-fitness populations will be the smallest. For instance, simulated quasi-species are shown in Fig. 7.12 for a random initial population for a high mutation rate $\epsilon = 0.5$. At long times, the most fit species dominates, but other near-by species also survive. No species die off, because it is always possible to generate distant species by a few improbable bit flips. This makes the quasi-species equation noticeably different than the replicator equation that does not include cross-mutations and hence many of the populations become extinct. The fitness landscape is shown in Fig. 7.13 along with the final population numbers for the large mutation rate. The final population matches the fitness landscape almost perfectly for this high mutation rate.

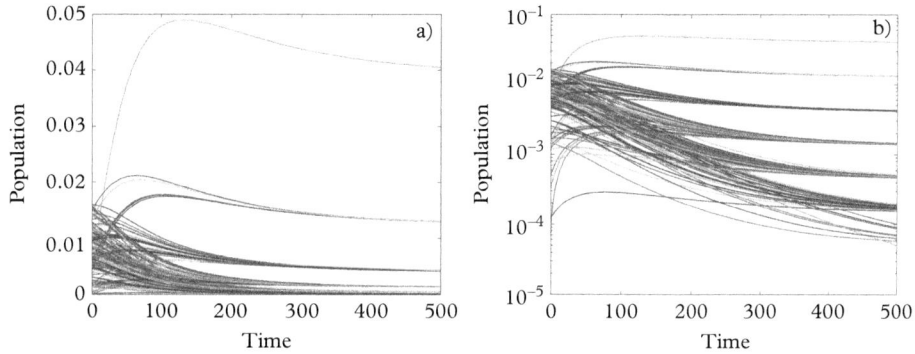

Figure 7.12 *Simulation of quasi-species frequency as a function of time for a random starting population. (a) Linear scale. (b) Log scale.*

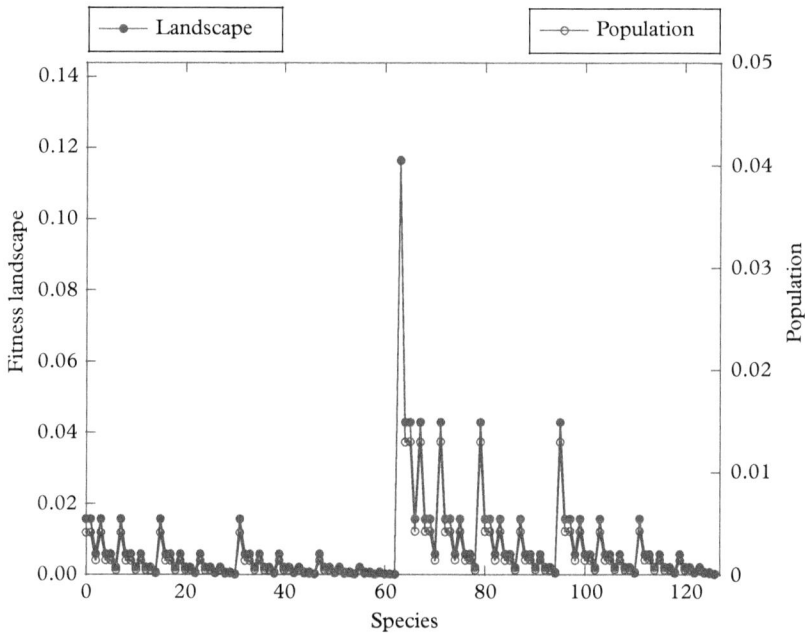

Figure 7.13 *Fitness landscape compared to final population. Solid data denote the fitness landscape, while open data are the final population. The most-fit species in this example is the 64th species.*

One of the important aspects of quasi-species behavior is the adaptation of mutating species to the fitness landscape. Not only do species evolve to climb fitness peaks in the fitness landscape, but they also abandon local peaks to climb a distant taller peak. The fitness landscape of a two-peak fitness landscape is shown in Fig. 7.14. The narrow peak is taller than the wide peak, but the wide peak contains three times as many species as the narrow peak. The mutation rate plays an important role in determining the final population distribution. For large mutation rates, the final population matches the fitness landscape. For

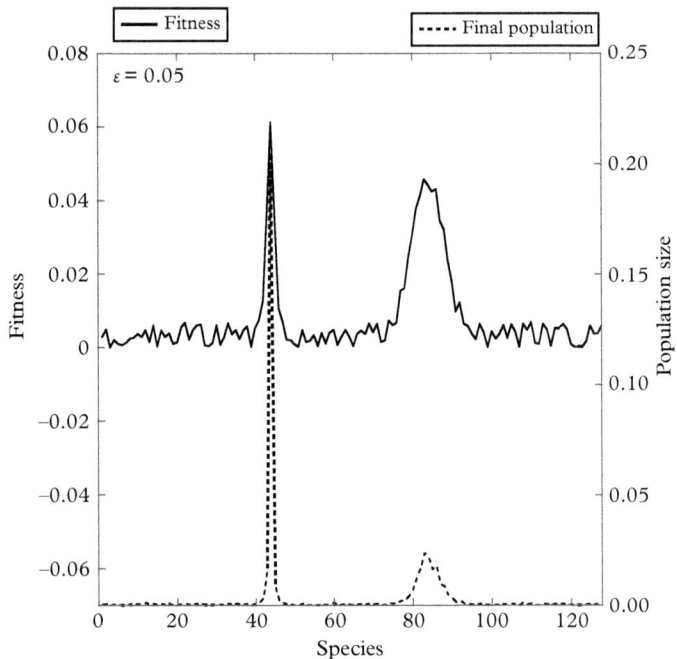

Figure 7.14 *Double-peak fitness landscape (upper trace). The narrow peak is taller, but the area under the broad peak is three times larger. At this intermediate value ε = 0.05, the most probable species is the most fit, but the number of species associated with the broad peak outnumbers those associated with the narrow peak by almost three to one.*

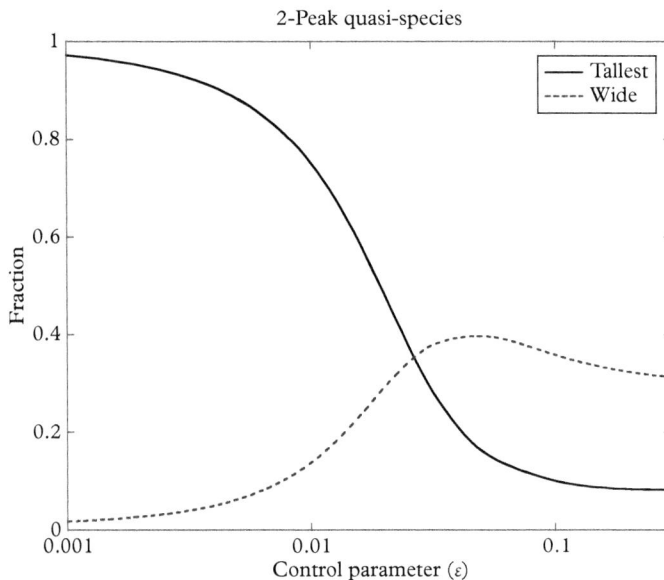

Figure 7.15 *Fraction of species associated with the tall-and-narrow relative to the wide peak of Fig. 7.14 as functions of the mutation parameter ε.*

intermediate mutation rates, the peak in the fitness landscape that has the largest height-times-width (integral) wins. For small mutation rates, the highest peak wins. For an intermediate control parameter $\epsilon = 0.05$, the populations evolve to the final state shown in the figure where the most probable fraction of the population occupies the tallest peak. However, the mutation rate is large enough in this case that 40% of the population occupies the broad peak while only 15% occupies the tallest peak. The population fractions of the two respective peaks are plotted in Fig. 7.15 as a function of the mutation parameter ϵ. For very low mutation rate, the tallest peak is favored. But as the mutation rate increases, the number of species associated with the broad fitness peak outnumbers the fittest species.

The quasi-species equation can show a wide variety of behavior depending on the choice of "distance" between two quasi-species, and depending on the topology of the fitness landscape. For instance, correlations in the fitness landscape can lead to a type of evolutionary "persistence" as a quasi-species that has a random mutation (that puts it on the edge of a fitness peak) climbs (mutates) consistently "uphill," up the fitness gradient to reach the peak. However, the quasi-species equation is still relatively simple, and from a biological point of view may be best suited to study chemical processes, such as evolution of DNA or RNA or rapidly-mutating viruses, rather than the evolution of complex organisms. To study more complex systems requires an extension of the quasi-species equation to include groups or cliques that are self-supportive and enhance survival by security in numbers. This extension is accomplished with the replicator–mutator equation.

7.5   The replicator–mutator equation

In the quasi-species equations, the fitness for each subspecies is independent of the population size. A straightforward extension of the quasi-species model introduces a frequency-dependent fitness. This can occur when groups of subspecies form alliances that work together. Belonging to a group increases the chances of survival. Frequency-dependent fitness also can model bandwagon effects, such as a "popular" group or a fad that induces others to change allegiance and draws them to it.

The quasi-species equation with frequency-dependent fitness is called the replicator–mutator equation

Replicator–mutator equation:
$$\dot{x}_a = \sum_{b=1}^{N} x_b \left(f^b Q_a^b \right) - \phi x_a$$
$$f^b(\vec{x}) = \sum_{c=1}^{N} x_c p_c^b \qquad (7.37)$$
$$\phi(\vec{x}) = \sum_{b=1}^{N} f^b x_b$$

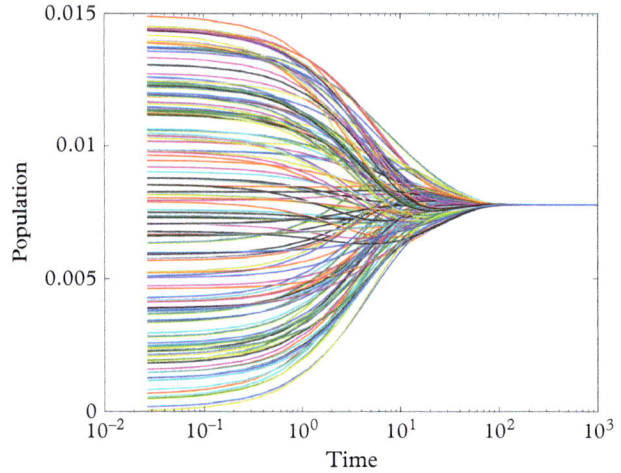

Figure 7.16 *The evolution of 50 quasi-species with time until all species are equally probable.*

This equation is a combination of the replicator equation and the quasi-species equation. The pay-off matrix is p_a^b, and the mutation matrix is Q_a^b for an individual in species b mutating into species a. The fitness of the ath species is f^a, which now depends on the size of the group. The coefficients of the pay-off matrix p_a^b are all positive numbers and do not need to be antisymmetric. One convenient payoff matrix uses the Hamming distance through

$$p_a^b = \exp\left(-\beta H_a^b\right) \tag{7.38}$$

which produces a "clique" effect in which nearby populations have higher payoff and bolster the fitness of the local group of species. Equation (7.38) for payoff should be contrasted to Eq. (7.36) which defined fitness for the quasi-species model.

The replicator–mutator equation, with a symmetric mutation matrix Q_a^b, has a striking property at high mutation rates (large ϵ): there is a global fixed point in which every population is equally probable. The time-evolution of a distribution of initial population probabilities is shown in Fig. 7.16 for high mutation rate. The original spread of population probabilities constricts over time until every population is equally probable. In this case, the frequency-dependent fitness combined with frequent mutation produces a collective effect that equalizes selection for all species.

As the mutation rate decreases, the populations are more difficult to balance, as shown in Fig. 7.17. For a mutation rate below a critical value (in this case $\epsilon = 0.029$), the balance is destroyed, and a single species dominates and contains most of the individuals. Therefore, there is a bifurcation from an equitable balance to a winner-take-nearly-all end state shown in Fig. 7.18. What makes this transition interesting is that higher mutations cause stabilization of the populations. While at lower mutation rates, the still nonzero mutation rate eventually allows most species to evolve into the dominant one.

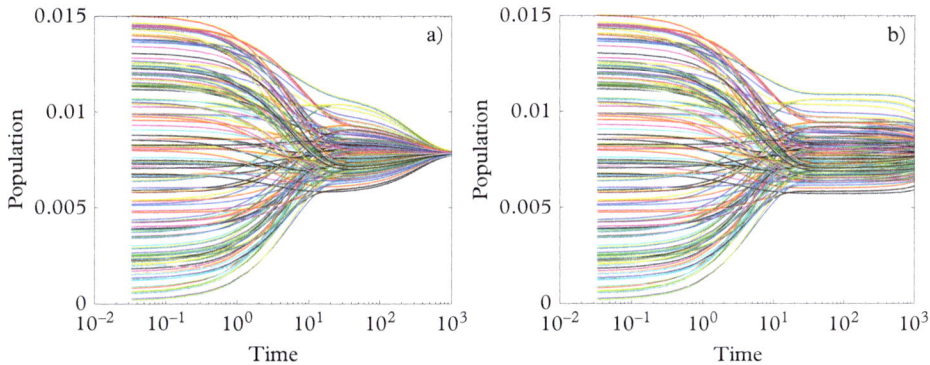

Figure 7.17 *The evolution of 50 quasi-species vs. time for (a) $\epsilon = 0.03$, (b) $\epsilon = 0.0291$.*

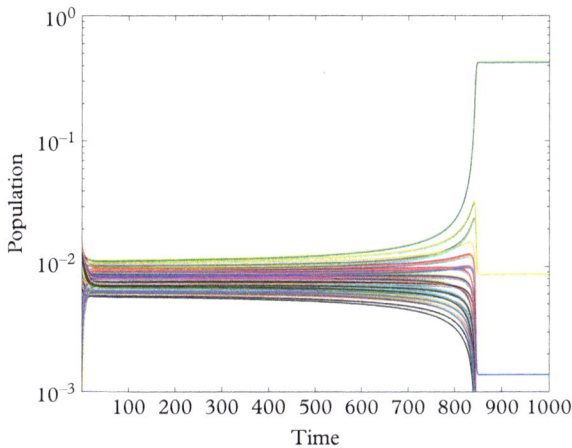

Figure 7.18 *Evolution of 50 quasi-species for strong mutation with $\epsilon = 0.0290$. One species wins after exponential growth.*

7.6 Dynamics of finite numbers (optional)

The examples of evolutionary dynamics that have been presented in this chapter so far have taken a continuous approach. The replicator, mutator, quasi-species, and replicator–mutator equations are all deterministic flows on continuous variables. However, there is another side to evolutionary dynamics based on the dynamics of finite numbers and stochastic processes. This is also the arena of game theory. It is beyond the scope of this book to go in this direction, but this facet of evolutionary theory is too important to omit entirely in a book on modern dynamics. The simplest consequence of having a finite population is the certainty of extinction after sufficient time. This certainty is a simple result of the finite probability for a neutral mutation to take over an entire population.

Consider a death-reproduction model in a mixed population of fixed size N that is completely neutral. Neutrality means that who dies and who reproduces is selected at random. For simplicity, assume that the mixed population consists of only two species: type A and type B individuals. There are N_A individuals of species A and $N_B = N - N_A$ individuals of species B. At every time step an individual is selected at random to reproduce, and an individual is selected at random to die. The two can be the same, when the individual first reproduces and then dies. As the model proceeds, the number of A and B individuals fluctuates randomly, but not indefinitely.

There are two fixed points in this stochastic dynamical model. They are when $N_A = N$ or $N_B = N$, that is, when every individual of one of the species has died. Once this state is reached, there is no longer any chance for the other species to reemerge (in the absence of mutation). The important point is that for a finite population size N, there is a nonzero probability that random drift eventually will bring the system to one of the fixed points.

In the field of evolutionary biology, this process of neutral drift has provided a valuable tool for measuring the rate of genetic change. Most genetic mutations are neutral. This is because the probability is very low that a mutation will improve on functions that have been optimized over millions of years of natural selection. Conversely, if a mutation is deleterious, then those individuals will not survive. Therefore, most mutations that do not kill an individual have no overt influence on the survivability of an individual—they are neutral. From the law of finite probabilities, once a mutation occurs, there is a finite chance that it eventually will spread over the entire species. The rate at which neutral genetic mutations in DNA occur has remained fairly constant throughout evolutionary history. This is the so-called theory of the "molecular clock." Therefore, by measuring the "distance" between two genomes, let's say of humans and chimpanzees, the time of divergence of the two species can be predicted with some accuracy.

The probability that a single mutation will spread over a species and become fixed in its genome is called the fixation probability. For neutral drift, the probability that a single A individual in a population of N individuals will eventually take over the species is simply

$$P = 1/N. \tag{7.39}$$

When there is a reproductive advantage of A over B given by the factor $r > 1$, the fixation probability for species A becomes

$$P_A = \frac{1 - \dfrac{1}{r}}{1 - \dfrac{1}{r^N}} \tag{7.40}$$

and for species B becomes

$$P_B = 1 - P_A = \frac{1 - r^{N-1}}{1 - r^N}.$$ (7.41)

For N_A and N_B starting individuals, the fixation probabilities are

$$P_{A,N_A} = \frac{1 - \dfrac{1}{r^{N_A}}}{1 - \dfrac{1}{r^N}}, \qquad P_{B,N_B} = \frac{1 - r^{N_B}}{1 - r^N}.$$ (7.42)

Even if A has a reproductive advantage over B, there is still a chance that B will take over the species. This is a consequence of the discrete and finite numbers, and would not be possible in a continuous system. Therefore, the dynamics of finite systems have a stochastic character with improbable outcomes still being possible. While the results presented here are based on the assumption that the total number N is constant, many of the arguments based on finite probabilities on finite numbers continue to hold even when N is not a constant, but changes slowly.

This model of neutral drift and fixation probabilities between two species is called a Moran process. As simple as this model is, the Moran process shows up in many aspects of game theory and evolutionary dynamics of finite populations. It is a winner-take-all dynamic that is common in fields as diverse as economics and language evolution. More complex finite processes continue to share many of the qualitative results of the Moran process, including natural selection that is not neutral. One of the outstanding problems in molecular evolution is the emergence of new functions with selective advantages of increased fitness. For instance, it is difficult to envision how a molecule as complicated as ATP synthase, that has multiple complex functioning parts, some of which move like mechanical gears, could have emerged from random mutations of DNA. It would seem that deep valleys in the fitness landscape would separate existing function from new advantageous functions. However, the theory of nearly neutral networks in high-dimensional spaces of DNA base-pair permutations suggests that evolution is not required to bridge deep valleys, nor take direct paths, especially when the fitness landscape has correlations. These are topics at the forefront of evolutionary biology, and they draw from the rich phenomena of evolutionary dynamics.

7.7   Summary

Evolutionary dynamics treats the growth and decay of populations under conditions that influence reproduction, competition, predation, and selection. The *Lotka–Volterra equations* were early population equations that describe population cycles between predators and prey. *Symbiotic* dynamics can lead to stable ecosystems among a wide variety of species. *Natural selection* operates through the principle of relative fitness among species. Fitness can be frequency dependent (if it depends on the size of the population) or frequency independent (intrinsic fitness of the species). A simple frequency-dependent fitness model is the *replicator equation* that follows a *zero-sum game* among multiple species whose total populations reside within a *simplex*. Many closely related species, connected by mutation, are called quasi-species. The *quasi-species equation* includes the mutation among quasi-species through a *mutation matrix* which is a *stochastic matrix*. The quasi-species have an intrinsic fitness that can be viewed as a multidimensional *fitness landscape*, and the quasi-species diffuse and climb up fitness peaks within this landscape. Adding frequency-dependent fitness to the quasi-species equation converts it to the *replicator–mutator equation* that captures collective group dynamics.

Lotka–Volterra equations

The simple predator–prey population dynamics equations are

$$\dot{x} = x\,(\alpha - \beta y)$$
$$\dot{y} = -y\,(\gamma - \delta x)\,, \tag{7.9}$$

where y is the number of predators and x is the number of prey. The prey reproduce at the rate α, and are eaten with a rate β times the product of the number of prey and predators. The predator reproduces at a rate δ times the product of the number of prey and predators, and dies off at a rate γ.

Replicator equation

The replicator equation is the growth equation for the population x_i of the ith species based on its relative fitness (*with* frequency-dependent fitness through the pay-off matrix p_a^b) relative to the average fitness in the ecosystem

$$\dot{x}_a = x_a \cdot (f^a - \phi) \quad \text{(no implicit summation)} \tag{7.20}$$

$$\text{Species fitness: } f^a\,(\vec{x}) = \sum_{b=1}^{n} x_b p_b^a$$

$$\text{Average fitness: } \phi\,(\vec{x}) = \sum_{a=1}^{n} f^a x_a$$

Quasi-species equation

This equation describes the growth of numerous sub-species (quasi-species) under the action of a mutation matrix Q_a^b *without* frequency-dependent fitness

$$\dot{x}_a = \sum_{b=0}^{N} x_b\,(f^b Q_a^b) - x_a \sum_{b=0}^{N} f^b x_b$$
$$= \sum_{b=1}^{N} x_b W_a^{\prime b} - \phi x_a \tag{7.24}$$

Transition matrix: $W_a^b = f^b Q_a^b$ (no implicit summation)

Replicator–mutator equation

The growth equation with *both* mutation and frequency-dependent fitness is a replicator–mutator equation

$$\dot{x}_a = \sum_{b=1}^{N} x_b f^b Q_a^b - \phi x_a$$

$$f^b(\vec{x}) = \sum_{c=1}^{N} x_c p_c^b \qquad (7.37)$$

$$\phi(\vec{x}) = \sum_{a=1}^{n} f^a x_a$$

Moran process

A Moran process is a stochastic process in a finite population that exhibits neutral drift and fixation of a single subspecies, and extinction of all others, at long times.

Stochastic matrix

A stochastic matrix is a random matrix whose columns and rows each sum to unity.

7.8 Bibliography

M. Eigen and P. Schuster, *The Hypercycle* (Springer, New York, 1979).

M. Eigen, *Steps towards Life: A Perspective on Evolution* (Oxford University Press, Oxford, 1992).

M. Eigen, J. McCaskill and P. Schuster, The molecular quasi-species. *Advances in Chemical Physics*, vol. 75, pp. 149–263, 1989.

S. Gravilet, *Fitness Landscapes and the Origins of Species* (Princeton University Press, 2004).

S. Kauffman, *The Origins of Order: Self-Organization and Selection in Evolution* (Oxford University Press, Oxford, 1993).

Motoo Kimura, *The Neutral Theory of Molecular Evolution* (Cambridge University Press, 1968).

M. Nowak, *Evolutionary Dynamics: Exploring the Equations of Life* (Harvard University Press, Cambridge, MA, 2006).

P. Schuster and P. F. Stadler, *Networks in molecular evolution. Complexity*, vol. 8(1), pp. 34–42, 2002.

Sewall Wright, *Evolution: Selected Papers* (University of Chicago Press, 1986).

7.9 Homework exercises

Analytic problems

1. **Species competition and symbiosis for two species:** How many different types of stability are possible for Eq. (7.7) when the function $f(x, y)$ and $g(x, y)$ are each first-order (linear) in the variables? What are the competition/symbiosis conditions for each?

2. **Species competition and symbiosis for three species:** Add a third species to Eq. (7.7) and assume only linear functions. Classify all the possible types of fixed point that occur, and give the competition/symbiosis conditions for each.

3. **N-Species:** Assume a large but finite number N of species. Under what competition/symbiosis conditions can all species coexist in stable equilibrium at finite population size?

4. **Quasi-species:** In the quasi-species (QS) equation, show that small mutation parameter ϵ favors the highest peak rather than the "strongest" peak.

5. **Quasi-species:** Construct an analytical three-species model for the quasi-species equation in terms of f^a and Q_a^b. Find the fixed points from the quasi-species eigenvalue equation. Evaluate their stability. Enumerate the types of behavior that can occur in a three-species model.

6. **Moran process:** For a Moran process, derive the fixation probability for a reproductive advantage $r > 1$ of A over B.

7. **Virus:** For the virus problem, what needs to be added to the immune–response equations (7.16) to completely eradicate the virus?

8. **Quasi-species:** In the QS equation, if the initial population is all in a single subspecies, what is the charastric time for the emergence of other subspecies? How is this affected by the choice for ϵ and λ?

9. **Quasi-species:** Under what conditions can the quasi-species equation have oscillatory solutions?

Computational projects[6]

10. **Predator–prey:** Play with the parameters of the predator-prey model. What different types of behavior happen?

11. **Replicator:** For replicator dynamics, classify the types of behavior for $N = 4$ and for $N = 5$. Can you estimate the probabilities for each type of behavior? What is the most common type of behavior? What type is behavior is very rare?

12. **Quasi-species:** Compare the final populations for a Hamming distance mutation matrix vs. a random mutation matrix. Does the character of the possible solutions depend on the choice of mutation matrix?

[6] Matlab programming examples are found at www.works.bepress.com/ddnolte.

13. **Quasi-species:** For the Hamming distance mutation matrix, at fixed ϵ, does the final population depend on the landscape contrast λ? Do the results scale as a function of the ratio ϵ/λ?

14. **Quasi-species:** For the quasi-species equation, assume there are 128 species connected by a network graph with adjacency matrix A_{ij}. Assume the mutation distance is given by distance on the graph. Explore the differences among small-world, scale-free, and ER graphs as functions of the mutation control parameter ϵ.

15. **Quasi-species:** Explore what happens if the mutation matrix in the quasi-species model is not a symmetric matrix (but it is still a stochastic matrix). This model would have asymmetric mutation rates among species.

16. **Quasi-species:** Set up a dynamic competition between mutation and payoff. For instance, mutation can be larger at small distances, but payoff can be larger at long distances.

8 Economic Dynamics

Dow Jones average and trading volume February 2007 to February 2012.

Economies are highly dynamic complex systems. Look at the price of stocks. What drives the wild fluctuations? We hear about supply and demand, and about corporate competition, but more is going on than just this. There are inflation, and unemployment, and crops being impacted by weather. There are governmental fiscal policies that sometimes are at odds with governmental monetary policies. And there are trade imbalances, and unfair subsidies along with trade wars. On top of all of this there are human expectations, both rational and irrational, and those who game the system, gambling on futures and on derivatives.

The surprise is that some of this can be captured in dynamical flows or as iterated maps, just like the ones we have been studying throughout this book. This is not to say that economic models can make you rich. The hard part is finding what the dynamical variables are, with what coefficients, and what trends to include and which to leave out. In the end, predicting stock prices is like predicting the weather. There is some rational basis for short-term predictions, but sensitivity to initial conditions (SIC) ruins most long-term predictive power.

Nonetheless, much can be learned and understood by modeling the complexities of economies, and it is a good application of the methods of nonlinear dynamics. There are stable and unstable fixed points for stable and unstable aspects of economies. Limit cycles capture business cycles, and saddle points can govern inflation. Evolutionary dynamics plays a role in corporate competition and the rise and fall of companies and their markets, as well as distributed dynamical networks of businesses with interwoven supply chains and interdependent goods and money markets. This chapter takes us far from physical systems of particles subject to forces—the usual topic of physics texts. But *econophysics* is an emerging field that benefits from the insights we have gained from the physical systems that were studied in the preceding chapters.

8.1   Micro- and macroeconomics

Before beginning the study of economic dynamics, it is necessary to establish the basic terminology and facts of economics. Economics is a mature field with an overload of jargon (at least from an outsider's point of view), and there are many subfields within economic studies. The major division is between microeconomics and macroeconomics.

Microeconomics is concerned with economics in the small—households and businesses. The variables are household or corporate income, and the prices of products that vary as a function of the quantities supplied and quantities demanded. The demand for goods, and the ability to supply these goods in competitive markets with more than one supplier and many buyers often leads to stable equilibria. Perturbations from equilibrium relax with characteristic times (Lyapunov exponents). The price and quantity relationships can be nonlinear, but linearized approximations often yield the basic behavior.

Microeconomics becomes more complex when behavior is driven by expectations rather than by actual factors, and when there are delays. Expectations and delays are often related to human responses, and their origins may lie more in the fields of sociology or psychology than in quantitative sciences like physics. However, once we admit that expectations differ from real values, and that there are real delays in the way people and markets respond, then these processes can be incorporated into the mathematical modeling, sometimes leading to chaotic behavior.

Macroeconomics deals with economics in the large—the sum of all the microeconomic processes that ultimately determine unemployment and inflation and the availability of investment money. It also deals with regional and global economies in which nations are linked by international trade networks. Here, too, there can be linear models with stable equilibria, but also fundamental causes of inflation, such as saddle points with unstable manifolds. This is the scale of global competition that affects consumer prices and trade imbalances.

The macroeconomy is primarily driven by the private sector, but with important input from the federal government to help support research and development and provide a fertile environment to allow new technologies and their associated businesses to emerge. An essential part of macroeconomics are fiscal policies set by the central government and monetary policies set by the central bank. Fiscal policies relate to government taxes and spending, while monetary policies relate to federal interest rates. These policies are set in place to stabilize the economy in the face of inefficiencies in markets and business cycles. However, these policies also directly influence factors such as inflation, wages and unemployment. This chapter first studies simple microeconomic models of supply and demand, business cycles and competition. Then it turns to the macroeconomics of markets.

8.2 Supply and demand

The simplest supply and demand model assumes that supply S is a linearly increasing function of price p (if there are higher profits, suppliers will make more goods), and demand D is a linearly decreasing function of price (if the price is too high, consumers stop buying). In other words

$$D = a - bp$$
$$S = c + dp. \tag{8.1}$$

These simple relations are obvious, and do not yet constitute a dynamical system. Dynamics arise when the price adjusts in response the current conditions between supply and demand. In economics, the time rate of change of a variable is called an *adjustment*. A simple price adjustment depends on excess demand $E = D - S$ as

$$\dot{p} = kE$$
$$= k\,(D - S). \tag{8.2}$$

Prices increase with demand, but fall with supply for $k > 0$. This simple dynamical system has a fixed point when $D = S$ at the price

$$p^* = \frac{a - c}{d + b}, \tag{8.3}$$

shown in Fig. 8.1, and it has a Lyapunov exponent

$$\lambda = -k\,(b + d). \tag{8.4}$$

The time dependence for the price is

$$p\,(t) = \frac{a - c}{b + d} + \left[p_0 - \frac{a - c}{b + d} \right] \exp\,(-k\,(b + d)\,t) \tag{8.5}$$

and for the supply and demand are

$$D\,(t) = \left[\frac{ad + bc}{b + d} \right] - b \left[p_0 - \frac{a - c}{b + d} \right] \exp\,(-k\,(b + d)\,t)$$

$$S\,(t) = \left[\frac{ad + bc}{b + d} \right] + d \left[p_0 - \frac{a - c}{b + d} \right] \exp\,(-k\,(b + d)\,t), \tag{8.6}$$

Linear supply and demand

$$D = a - bp$$
$$S = c + dp$$

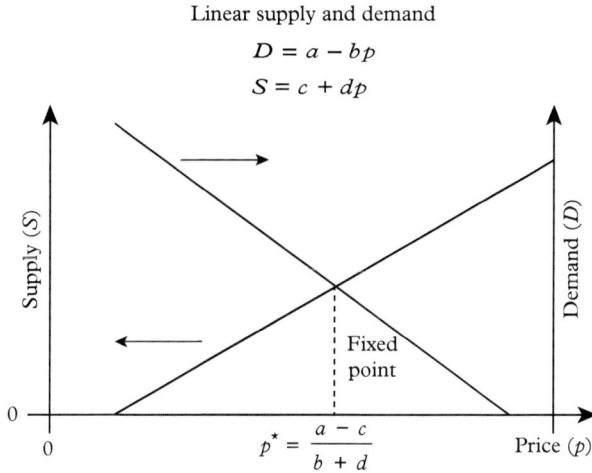

Figure 8.1 *A simple linear supply and demand model. Prices rise if there is excess demand, but fall if there is excess supply. The fixed point is a stable equilibrium.*

where p_0 is the initial price. The fixed point is stable, and fluctuations in price relax towards the fixed point with a time constant given by

$$\frac{1}{\tau} = k\,(b + d) \tag{8.7}$$

that is the same for price, demand, and supply.

Now consider a competitive market, still governed by linearized equations, but with multiple products. These products have an excess demand given by $E_a = D_a - S_a$ with a dynamic price adjustment

$$\dot{p}_a = k_a E_a \tag{8.8}$$

with positive k_a. For two products, the excess demand can be written

$$\begin{aligned} E_1 &= a + bp_1 + cp_2 \\ E_2 &= d + ep_1 + fp_2 \end{aligned} \tag{8.9}$$

with the two-dimensional flow

$$\begin{aligned} \dot{p}_1 &= k_1\,(a + bp_1 + cp_2) \\ \dot{p}_2 &= k_2\,(d + ep_1 + fp_2). \end{aligned} \tag{8.10}$$

The Jacobian matrix is

$$\mathcal{J} = \begin{pmatrix} k_1 b & k_1 c \\ k_2 e & k_2 f \end{pmatrix} \tag{8.11}$$

with trace and determinant

$$\begin{aligned} \tau &= k_1 b + k_2 f \\ \Delta &= k_1 k_2\,(bf - ec). \end{aligned} \tag{8.12}$$

If the trace is negative and the determinant is positive, then there is a stable equilibrium. This requires

$$k_1b + k_2f < 0$$
$$bf > ec. \tag{8.13}$$

The first condition is automatically satisfied if both b and f are negative (stable feedback). The second condition is satisfied if the slope of the $\dot{p}_1$ nullcline is larger than the slope of the $\dot{p}_2$ nullcline. If this second condition is not satisfied, then the fixed point is a saddle point with an unstable manifold that either will cause an uncontrolled price increase, or cause one product to vanish from the marketplace.

Example 8.1 Two-product competition

Consider a market with two products. If the excess demand is

$$E_1 = 3 - 6p_1 + 3p_2$$
$$E_2 = 16 + 4p_1 - 8p_2 \tag{8.14}$$

with price adjustments

$$\dot{p}_1 = 2E_1$$
$$\dot{p}_2 = 3E_2 \tag{8.15}$$

then this model leads to a stable fixed point for both products in a competitive equilibrium. The phase plane is shown in Fig. 8.2 for prices p_1 and p_2. The stable equilibrium is at $p^* = [2,3]$.

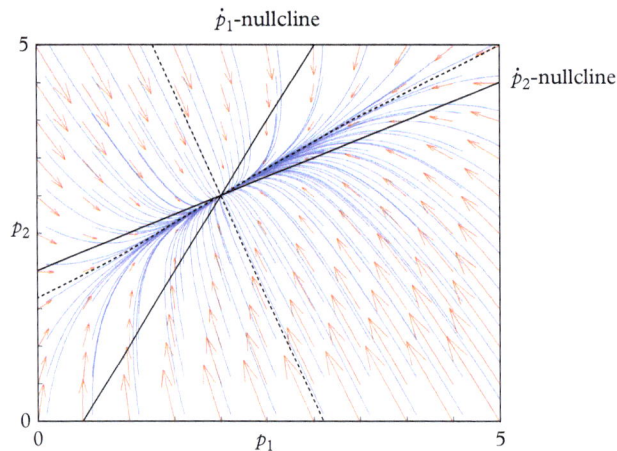

Figure 8.2 *Phase plane for prices p_1 and p_2 in a competitive model. The solid lines are nullclines. The dashed curves are the separatrixes.*

8.3 Business cycles

Business cycles are a common phenomenon in microeconomics, when prices and quantities oscillate in a limit cycle. Some business cycles are driven literally by the changing seasons, which is a phase-locked process. For instance, winter creates a demand for winter coats, and summer creates a demand for swim suits. Other types of business cycles are driven by nonlinear cost of materials or labor and by time lags in manufacturing or distribution networks.

8.3.1 Price and quantity adjustment

A business cycle can occur between price and quantity when there is a limiting feedback through a finite labor force. For instance, demand, in addition to depending on price, must also depend on the number of consumers. As more people get jobs making more product, they have the money to buy the product—the labor force becomes the consumer force. On the other hand, if people are spending too much time working, they have less time to buy and enjoy the product. This feedback can lead to a business cycle between price and quantity, mediated by nonlinear properties of the labor force.

Consider price p and quantity q (another word for supply S) adjustments

$$\dot{p} = \alpha \left[D\left(p, L\right) - q \right]$$
$$\dot{q} = \beta \left[p - L'\left(q\right) \right] \tag{8.16}$$

(α and β are both positive), where $D(p, L)$ is an aggregate demand function that depends on both price p and labor L. Labor cost is a function of quantity, and the cost of labor per quantity produces $L'\left(q\right) = \frac{\partial L}{\partial q}$, which is called the marginal wage cost and is a positive-valued function. The fixed point occurs when demand equals output, and when price equals the marginal wage cost

$$D\left(p^*\right) = q^*$$
$$p^* = L'\left(q^*\right). \tag{8.17}$$

The quantity produced is assumed to depend linearly on price through

$$q = q_0 + q_1 p \tag{8.18}$$

and the labor function is assumed to be quadratic

$$L\left(q\right) = l_0 + l_1 q + l_2 q^2$$
$$L'\left(q\right) = l_1 + 2 l_2 q \tag{8.19}$$

$L(q)$ is a nonlinear function in which more product requires a disproportionally larger workforce.[1] The aggregate demand is a function of both the price and labor force

$$D\left(p, L\right) = D_0 - h\left(L\right) p$$
$$= D_0 - \left(h_0 + h_1 L\right) p, \tag{8.20}$$

[1] This situation can occur in manufacturing if the manufacturing costs are not scalable. An *economy of scale* is preferred, for which increasing quantity can be produced with only marginal additional cost, as is often the case for software companies.

where all coefficients are positive. The decrease in demand with increasing price is augmented by having more people working with less time to buy. By combining Eqs. (8.16)–(8.20) the nonlinear flow is

$$\dot{p} = \alpha \left[a - bp + cp^2 - dp^3 - q \right]$$
$$\dot{q} = \beta \left[p - e - fq \right]$$

(8.21)

(collecting the previous terms into quantities a through f) with the nullclines

$$q = a - bp + cp^2 - dp^3 \quad \dot{p}\text{-nullcline}$$
$$p = e + fq \qquad\qquad\quad \dot{q}\text{-nullcline}$$

(8.22)

The $\dot{q}$-nullcline is a linear function, and the $\dot{p}$-nullcline is a third-order curve. The phase plane is similar to the Fitzhugh–Nagumo model of single neurons in Chapter 6 (see Fig. 6.4) with stable fixed points that convert to stable limit cycles by changing the control parameter a.

Example 8.2 The price of labor

An example, using the parameters $a = 50$, $b = 9$, $c = 0.8$, $d = 0.02$, $e = 0$, and $f = 0.5$, is shown in Fig. 8.3 for $\alpha = 1$ and $\beta = 2$. The dynamics are very similar to the Fitzhugh–Nagumo neuron model of Chapter 6. When the $\dot{q}$-nullcline passes either the lower or the upper extrema of the $\dot{p}$-nullcline, the limit cycle converts to a stable fixed point. In this model, changing the rates α or β does not change the nullclines or the fixed point, but does change the relaxation rates and the shape of the limit cycle.

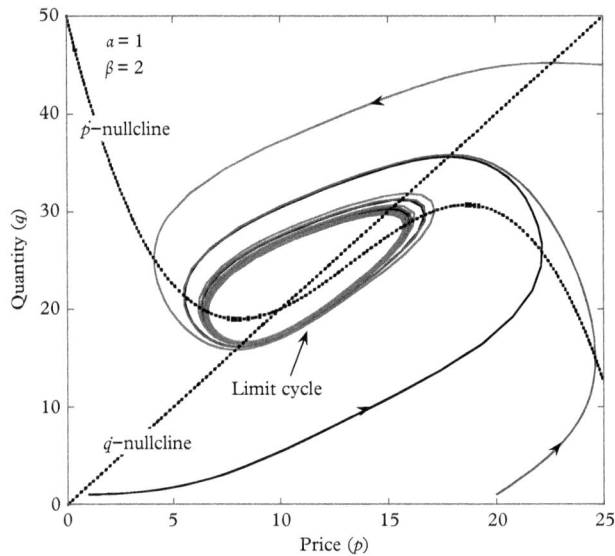

Figure 8.3 *Business cycle with price and quantity adjustment mediated through a labor force.*

8.3.2   Inventory business cycles

In inventory businesses, there can be cycles in inventory stocks in response to production rates. The cycles are driven by time lags and nonlinearity. There is often a time lag between desired inventory stock and the actual stock. A simple lag equation assumes that there is a lag time τ between a desired inventory stock and the actual stock B

$$\dot{y} = \frac{1}{\tau}\left(by_e - B\right), \tag{8.23}$$

where y is the sales, and the desired stock is given by the product by_e, where y_e is the expected sales. The expected sales depends on the level of sales y, but also can depend on the time rate of change of sales as well as the acceleration in sales. This is captured through the equation

$$y_e = a_2\ddot{y} + a_1\dot{y} + y. \tag{8.24}$$

The nonlinearity in the problem arises from an optimal production rate for a nominal investment. If sales and production are low, then the rate of change of the stock increases with production. However, if the production is too high (high sales), then there is insufficient investment to support increased stocks. This leads to an adjustment equation for the inventory stocks that depend on sales as

$$\dot{B} = my\left(1 - y\right). \tag{8.25}$$

These equations define a three-dimensional flow

$$\begin{aligned}
\dot{V} &= c_3 B - c_2 V - c_1 y \\
\dot{B} &= \frac{r}{c_3} y\left(1 - y\right) \\
\dot{y} &= V,
\end{aligned} \tag{8.26}$$

where the coefficients are

$$c_3 = \frac{r}{m} = \frac{1}{ba_2} > 0, \qquad c_2 = \frac{(ba_1 - \tau)}{ba_2}, \qquad c_1 = \frac{1}{a_2} > 0.$$

There are two equilibria, one at $y = 0$ and one at $y = 1$. The Lyapunov exponents at $y = 1$ are either all negative (stable), or else have one real negative value and two complex pairs, leading to cyclic behavior and the potential for chaos. The condition for the transition from stable behavior to periodic solutions (through a Hopf bifurcation) is

$$c_1 c_2 < mc_3.$$

Note that c_2 decreases as the lag increases. If the lag is very short (rapid adjustment to sales) then there is a stable equilibrium. However, in real distribution systems there is always a time lag between production and sales, which can lead to a limit cycle oscillation in the dynamics between sales and stock inventory.

Example 8.3 Inventory cycles

The dynamics of sales and stocks in Eq. (8.26), as the control parameter r increases, is shown in Fig. 8.4. When $r = 0.3$ (small gain on the nonlinearity for a fixed c_3 and c_2), there is a stable fixed point. As the nonlinear gain r increases to $r \approx 0.4$, a Hopf bifurcation occurs as the fixed point becomes unstable, but the trajectories are entrained by a limit cycle. This limit cycle expands with increasing gain until a period doubling event occurs at $r = 0.715$. With increasing r, a bifurcation cascade brings the system to chaotic behavior above $r = 0.8$. This example shows how a stable equilibrium converts to a business cycle and then to chaotic cycles with the increase in the nonlinear gain of the stock inventory adjustments. Business cycles and chaos are possible in this nonlinear problem because of the acceleration term in Eq. (8.24) that increases the dimensionality of the dynamics to three dimensions (required for chaos) and because of the lag between production and sales in Eq. (8.23).

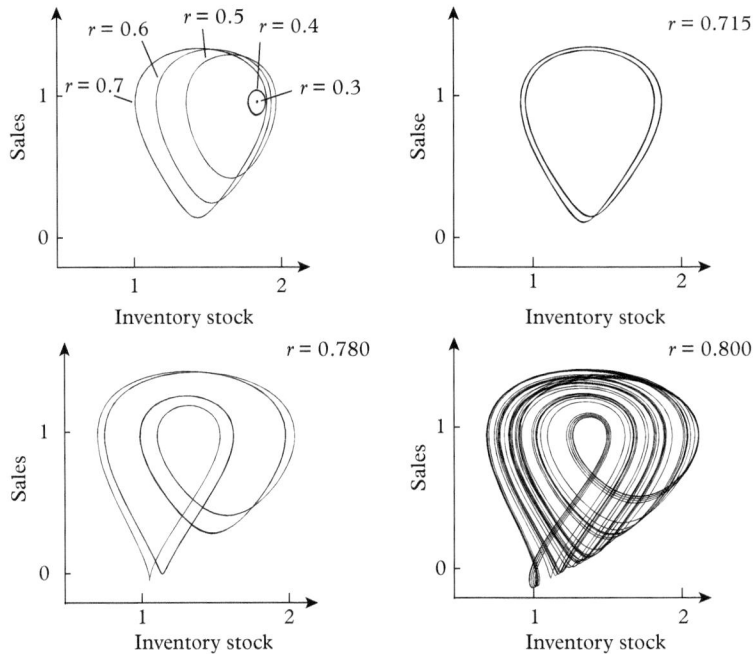

Figure 8.4 *Production and stock cycles for increasing values of r for $c_1 = 1$, $c_2 = 0.4$, and $c_3 = 0.5$. There is a Hopf bifurcation at $r = 0.4$. Period doubling begins at $r = 0.715$ and leads through a bifurcation cascade to chaos.*

8.3.3   Cobweb model for delayed adjustments

Many economic time series (price trajectories) have a natural discrete-time nature because of day-to-day cycles. For instance, the opening price today depends on the closing price yesterday, and current demand may depend on the expected price tomorrow. Discrete maps naturally have delays, and when combined with nonlinearities, can lead to chaotic behavior. In the language of economics, discrete models are called cobweb models, so-named because of the "cobweb" look of bifurcation diagrams.

Supplies are often nonlinear functions of expected price. For instance, when prices are low, supplies increase slowly with increasing price because of fixed costs for manufacturing. At the other extreme, when prices are high, supplies again can only increase slowly because of limited resources. Therefore, the supply vs. price curve is typically sigmoidal. As an example, supplies can be modeled as

$$q^s = \operatorname{atan}(rp^e),\qquad(8.27)$$

for which supplied quantities depend on the current expected price. The atan function is antisymmetric, and hence prices and supplies can be negative relative to a reference level.

The demand can be assumed to be a linear function of current actual price

$$q^d = a - bp\qquad(8.28)$$

and it can further be assumed that supply instantaneously satisfies demand

$$q^d = q^s.\qquad(8.29)$$

The delay, in this model, occurs between the expected price and the previous actual price. A typical adaptive expectations relation is given by

$$p^e_{n+1} = \lambda p_n + (1-\lambda)\, p^e_n,\qquad(8.30)$$

where the mixing parameter λ trades off between reality versus expectations. Combining these equations gives

$$p^e_{n+1} = \lambda \left(\frac{a - \operatorname{atan}(rp^e_n)}{b} \right) + (1-\lambda)\, p^e_n\qquad(8.31)$$

as the discrete cobweb map. The nonlinear gain for the map is r, and the control parameter for demand is given by a.

Example 8.4 A cobweb model

A numerical example for this supply and demand cobweb model is shown in Fig. 8.5 for $\lambda = 0.305$, $b = 0.25$, and $r = 4.5$. The control parameter in the model is the parameter a that controls the demand as a function of the current actual price.

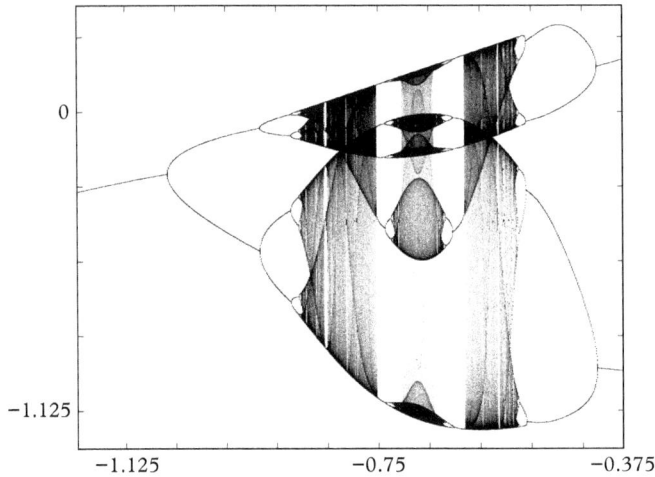

Figure 8.5 *Nonlinear supply-and-demand cobweb for $\lambda = 0.305$, $b = 0.25$, and $r = 4.5$. The expected price is plotted against the demand control parameter a.*

It is important to keep in mind that models that display chaos are often sensitive to the parameters, and even minor shifts in the value of the parameters can move the model outside the chaotic regime. Therefore, while chaotic behavior *can* occur in a given model, it is not necessarily *likely* to occur, except under special conditions. For instance, the value of λ in the adaptive expectations model plays a crucial role. If λ is too small, then there is insufficient nonlinearity for the model to display chaotic behavior. On the other hand, there are certain universal behaviors to look for, such as any model that can be approximated with a quadratic map, such as the logistic map of Chapter 3. These are likely to produce bifurcations and cascades to chaos.

8.4 Consumer market competition

The consumer product market is fickle, with trends that seem to blow with the prevailing wind. What makes a customer buy an iPhone over an Android? Why did Blackberry virtually disappear after being the early leader in the marketplace? What happened to Myspace that caused it to be overtaken by Facebook? What will overtake Facebook? Many of the answers to these questions depend on what is in fashion, what is considered "cool" or what your friends and family are using. Some of the central questions of microeconomics are whether a market is stable against competition, whether a winner-takes-all dynamics dominates, or whether a market may die out entirely.

8.4.1   Competition in oligopolies

The competition among N businesses for a single market is a situation that can be modeled with simple price and profit functions, yet can lead to complicated equilibria, especially for $N > 2$. To start, consider two competing companies that are competing in the same market with different versions of the same product. The total quantity manufactured is

$$Q(t) = q_1(t) + q_2(t), \tag{8.32}$$

where q_1 and q_2 are the quantities manufactured by each company, respectively. A simple price model assumes a maximum allowable price p_{max} and a linear dependence on total quantity supplied, such as

$$p(t) = p_{max} - Q(t). \tag{8.33}$$

The revenues for each company are

$$\begin{aligned} R_1(t) &= p(t)q_1(t) = [p_{max} - q_1(t) - q_2(t)]q_1(t) \\ R_2(t) &= p(t)q_2(t) = [p_{max} - q_1(t) - q_2(t)]q_2(t). \end{aligned} \tag{8.34}$$

The costs for production (known as marginal cost) are

$$\begin{aligned} C_1(t) &= m_1 q_1(t) \\ C_2(t) &= m_2 q_2(t), \end{aligned} \tag{8.35}$$

where m_1 and m_2 are nearly equal (for viable competition). Therefore, the profits are

$$\begin{aligned} \pi_1(t) &= [p_{max} - q_1(t) - q_2(t)]q_1(t) - m_1 q_1(t) \\ \pi_2(t) &= [p_{max} - q_1(t) - q_2(t)]q_2(t) - m_2 q_2(t). \end{aligned} \tag{8.36}$$

To find the desired quantities x_1 and x_2 that maximize profits, the partial derivatives of the profits are taken with respect to the respective manufactured quantity

$$\begin{aligned} \frac{\partial \pi_1}{\partial q_1} &= (p_{max} - m_1) - 2q_1 - q_2 = 0 \\ \frac{\partial \pi_2}{\partial q_2} &= (p_{max} - m_2) - 2q_2 - q_1 = 0 \end{aligned} \tag{8.37}$$

which are solved for

$$x_1 = \frac{1}{2}\left(p_{\max} - m_1\right) - \frac{1}{2}q_2$$
$$x_2 = \frac{1}{2}\left(p_{\max} - m_2\right) - \frac{1}{2}q_1.$$

(8.38)

These are the optimal production quantities. The price adjustments are proportional to the difference between the desired price that maximizes profits and the actual price. Assuming that there is a simple proportionality between the cost and the amount manufactured, the quantity adjustment equation is

$$\dot{q}_1 = k_1\left(x_1 - q_1\right)$$
$$\dot{q}_2 = k_2\left(x_2 - q_2\right).$$

(8.39)

The dynamic equations for quantity manufactured and sold (no inventory stock) are finally

$$\dot{q}_1 = \frac{1}{2}\left(p_{\max} - m_1\right)k_1 - k_1 q_1\left(t\right) - \frac{k_1}{2}q_2\left(t\right)$$
$$\dot{q}_2 = \frac{1}{2}\left(p_{\max} - m_2\right)k_2 - k_2 q_2\left(t\right) - \frac{k_2}{2}q_1\left(t\right).$$

(8.40)

In this model, each company sets their optimal production assuming their competitor is also operating at their optimal production. Therefore, this model precludes predatory price wars. Also, this is a model of constant marginal cost, which means there is a simple proportionality between the cost and the amount manufactured. This model has a competitive equilibrium, as shown in Fig. 8.6. The equilibrium is stable for the selected parameters, allowing the two companies to coexist.

8.4.2   Consumer product coherence

Many consumer product markets are disposable markets that are not essential. Choosing an online social network is not like buying an appliance or food. Choices of consumer products like VHS over Beta, Macintosh over PCs, iPhones over BlackBerrys are matters of taste. In such a case, the likelihood that a consumer will switch from one product to another depends on the trends which, in terms of the models we studied in Chapter 7 on evolutionary dynamics, means that there is a frequency-dependent fitness that contributes to the survival of a product in a competitive market.

As an example, consider the smartphone consumer market. Smartphones have many features in common, but the details of operating them can be quite different. These details have to do with which buttons to push or where to find certain menu items. When two products have similar procedures, then a user can switch

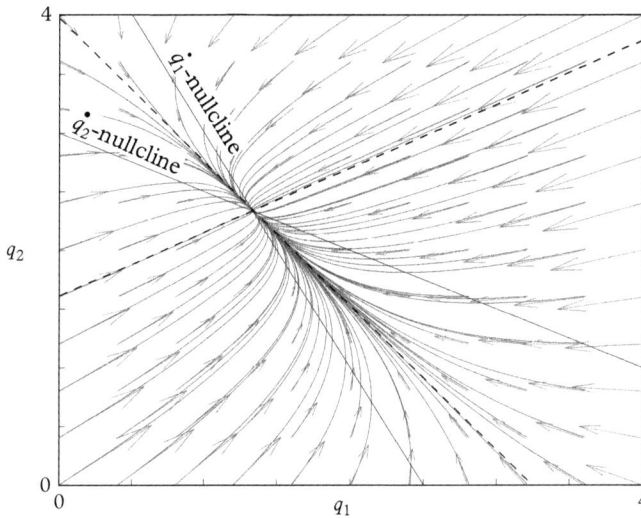

Figure 8.6 *Competitive phase plane for quantities q_1 and q_2 manufactured by two competing companies. The equilibrium is stable for the parameters: $m_1 = 4$, $m_2 = 3$, $p_{max} = 9$, $k_1 = 0.75$, and $k_2 = 0.5$.*

fairly easily from one to the other. But when two products have very different procedures, then there is a barrier for a user to switch.

Consider two specific products G_a and G_b. The probability that a user who uses product G_a uses a procedure that is compatible with the procedure of product G_b is given by the probability p_a^b. This probability defines a matrix whose values lie in the range between zero and unity with $p_a^a = 1$ (users using the same product have perfect mutual understanding). The advantage (or payoff) of a user using a procedure on G_a being able to use the same procedure on product G_b is given by the symmetric matrix

$$F_a^b = \frac{1}{2}\left(p_a^b + p_b^a\right),\tag{8.41}$$

where the matrix p_a^b is not necessarily symmetric. Among a population of users, the frequency of users who use product G_a is denoted by the variable x_a. The average payoff (loyalty) for this subpopulation of users is

$$f^a = \sum_b x_b F_b^a,\tag{8.42}$$

where the payoff leads to differing levels of advantage for different subpopulations. The average fitness of the entire population is

$$\phi = \sum_a x_a f^a,\tag{8.43}$$

which is called the product coherence of the population. When $\phi = 1$, then all products have converged to the same universal procedures. This is like Microsoft DOS evolving into Microsoft Windows in response to the Apple Macintosh graphical user interface.

The key element in the model is the ability of populations of users to move, meaning the ability for one set of users to change randomly (on a whim) to the competing product. This is given by the stochastic matrix Q_a^b that defines the probability that a user of product G_b converts into a user of G_a.

The dynamical equation, the flow, describing the evolution of the subpopulations of users is now

$$\dot{x}_a = \sum_b x_b \left(f^b Q_a^b \right) - \phi x_a, \tag{8.44}$$

which is the replicator–mutator equation, Eq. (7.37), of Chapter 7. It allows instances of perfect replication (groups of friends tend to adopt the same products) as well as instances of mutation (switching). The second term, that has the minus sign, ensures that the total population size remains constant during the dynamical evolution. Depending on the procedural overlap, the fixed points can range from each individual using a different product (zero product coherence) to the situation where all users use the same product (perfect product coherence). The question is: what overlap allows the existence of product coherence? In other words, is there a threshold for a Universal Product (standardization)?

To explore this question, consider a symmetric case when all N products have the same probability of feature and procedural overlap,

$$p_{aa} = 1 \quad p_{ab} = p, \tag{8.45}$$

where p is between 0 and 1. The conversion rates are

$$Q_{aa} = q \quad Q_{ab} = \frac{1-q}{N-1}, \tag{8.46}$$

where q is the probability of buying the same product and

$$u = \frac{1-q}{N-1} \tag{8.47}$$

is the probability of buying a different product. The replicator–mutator equation now is

$$\dot{x}_a = x_a \left[(q - u) f_a - \phi \right] + u\phi \tag{8.48}$$

(no summation implied). The payoff is

$$f_a = p + (1-p) x_a \tag{8.49}$$

and the product coherence is

$$\phi = p + (1-p) \sum_a x_a^2. \tag{8.50}$$

One solution for this symmetric case (all $p_{ab} = p$) is when all products have an equal number of users $x_a = 1/N$, as in Fig. 7.16. Another solution is when one product is dominant, while all others share equally in the remaining population

$$x_a = X \quad x_b = \frac{1 - X}{N - 1} \tag{8.51}$$

for a value X that is a function of the overlap rate q. The interesting part of these conclusions is that, for a Universal Procedure Set to evolve, a finite overlap among the feature procedures is needed. Zero overlap tends to prevent the evolution of a Universal Procedure Set.

8.5   Macroeconomics

The productivity of a nation, and the financial well-being of its citizens, depends on many factors with hidden dependencies that are often nonlinear. For instance, fiscal policy that raises corporate taxes boosts short-term income to a government that seeks to reduce deficits, but can decrease long-term income if businesses fail to succeed. With monetary policy, the government can lower interest rates to try to boost consumer spending, but lower gains on investment can stall economic recovery. Just how strong the feedback might be is usually not known a priori, and a transition to qualitatively different behavior can occur in a national economy with even relatively small adjustments in rates.

The complexities of macroeconomics have challenged economic predictions from the beginning of civilization, and even today great debates rage over fiscal and monetary policies. Part of the challenge is the absence of firm physical laws that are true in any absolute sense. There are many trends that hold historically for many situations, but there are always exceptions that break the rule. When the rules of thumb break down, economists scramble to get the economy back on track. These complexities go far beyond what this chapter can explore. However, there are some simple relationships that tend to be robust, even in the face of changing times.

8.5.1   IS–LM models

One of the foundations of macroeconomics is the relationship between *investment-savings* (IS) in the goods markets on the one hand and *liquidity-money* (LM) in the money markets on the other. As the interest rate on borrowed money increases, companies are less likely to make investment expenditures, and consumers are more likely to save—the IS part of the economy. This is balanced by the demand for money in the economy that decreases with increasing interest rates—the LM part of the economy. These two trends lead to the so-called IS–LM models, of which there are many variants that predict a diverse range of behaviors from stable nodes (equilibrium models) to saddle-point trajectories (unbounded inflation or deflation).

Let us begin with a simple IS–LM model that is purely linear and that balances expenditures against real income, and the dynamics are adjusted according to simple linear relaxation. The dynamics take place in a two-dimensional space of income, g, and interest rate, r. The expenditure as a function of time is assumed to be

$$E(t) = G + c(1 - T)g(t) + ig(t) - ur(t) \qquad (8.52)$$

for positive G, u, $i > 0$, with $0 < c < 1$ and $0 < T < 1$, where the meanings of the variables and coefficients are given in Table 8.1.

Governmental fiscal policies affect the amount of government spending G and the tax rate T. These are called exogenous variables because they are controlled externally by the government. The other parameters, such as i the response of investment to the interest rate, are called endogenous because they cannot directly be controlled and are properties of the economy. The endogenous parameters need to be tuned in the model to fit observed economic behavior.

The demand for money is assumed to be related to g (the real income or the gross domestic product (GDP)) and negatively related to the rate of interest r as

$$D(t) = kg(t) - vr(t) \qquad (8.53)$$

for positive coefficients $k > 0$, $v > 0$. The adjustments in the real income and the interest rate are

$$\begin{aligned}
\dot{g} &= \alpha\,[E(t) - g(t)] \\
\dot{r} &= \beta\,[D(t) - m_0]
\end{aligned} \qquad (8.54)$$

Table 8.1 *Variables and coefficients in the IS–LM model*

E	Real expenditure
g	Real income (gross domestic product)
r	Interest rate
T	Tax rate
G	Autonomous (government) expenditure
c	Propensity to consume
i	Increased investment in response to income g
u	Suppressed investment in response to interest rate r
D	Demand for money
k	Enhanced demand for money in response to income g
v	Suppressed demand for money in response to interest rate r

for positive rates α, $\beta > 0$, where m_0 is the nominal available money (liquidity preference equal to M/P, the amount of money M relative to price of money P). The amount of money available to the economy is controlled by governmental monetary policies of the central bank (Federal Reserve). Plugging the expenditures and demand for money into the adjustment equations yields a two-dimensional flow in the space of income and interest rates. Defining the consumer consumption-investment index as

$$C = c(1 - T) + (i - 1),$$ (8.55)

the two-dimensional flow is

$$\dot{g} = \alpha(Cg - ur + G)$$
$$\dot{r} = \beta(kg - vr - m_0).$$ (8.56)

The two isoclines are linear functions of g. The LM curve is the $\dot{r}$-nullcline, and the IS curve is the $\dot{g}$-nullcline

$$r = \frac{1}{u}(G + Cg) \quad \dot{g}\text{-nullcline} \quad \text{IS curve}$$
$$r = \frac{1}{v}(kg - m_0) \quad \dot{r}\text{-nullcline} \quad \text{LM curve}$$ (8.57)

The LM curve is a line with positive slope as a function of income (interest rate is an increasing function of income). The IS curve can have negative or positive slopes, depending on the sign of the consumption-investment index C.

In macroeconomics, one of most important consequences of changes in rates or coefficients is the conversion of stable equilibria into unstable equilibria or saddle points. In economic terms, this is the difference between a stable national economy and run-away inflation (or deflation). The stability of the IS–LM model is determined by the determinant and trace of the Jacobian matrix, which are

$$\Delta = \alpha\beta[uk - vC]$$
$$Tr = \alpha C - \beta v.$$ (8.58)

Nodes are favored over saddle points for positive Δ, and stable behavior is favored over unstable behavior for negative trace. Therefore, stability in national economies require positive Δ and negative trace—both of which result for smaller values of the consumption-investment index C. The IS curve has negative slope (negative C) when propensity to consume c is low (for a fixed tax rate T), and investment i in response to income g is also low, which provides stability to the economy.

Example 8.5 IS–LM model

A numerical model with a stable equilibrium is shown in Fig. 8.7. The LM curve is the $\dot{r}$ isocline, and the IS curve is the $\dot{y}$ isocline. The parameters of the model are: $G = 20$; $k = 0.25$; $c = 0.75$; $m_0 = 2$; $T = 0.25$; $v = 0.2$; $i = 0.2$; $u = 1.525$; $\alpha = 0.2$; and $\beta = 0.8$. The stability is determined by $\Delta = 0.096$ and $Tr = -0.488$, which determines the fixed point to be a stable spiral.

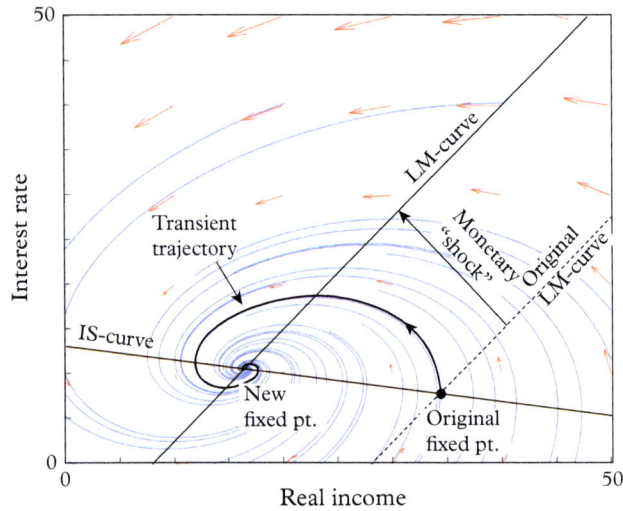

Figure 8.7 *IS–LM model with decreasing real income vs. increasing interest rates. The IS curve and the LM curve are the isoclines. The fixed point is a stable spiral. The result of a sudden decrease in money m_0 leads to a "shock" and a transient trajectory that spirals to the new fixed point at lower GDP and higher interest rate.*

The IS–LM model makes it possible to predict the effect of governmental fiscal and monetary policies on the national economy. Fiscal policy controls the government expenditure G and the tax rate T. Monetary policy controls the amount of money m_0 available to the economy. Controlling these quantities affects interest rates on money and affects the GDP.

An important feature of the IS–LM model is the ability to model "shocks" to the economy caused by sudden changes in government spending or taxes, or by sudden changes in the available money. When any of these are changed discontinuously, then a previous equilibrium point becomes the initial condition on a trajectory that either seeks a new stable equilibrium, or that becomes unstable. An example of a shock is shown in Fig. 8.7 when the money supply m_0 is suddenly decreased. The original fixed point under large supply becomes the initial

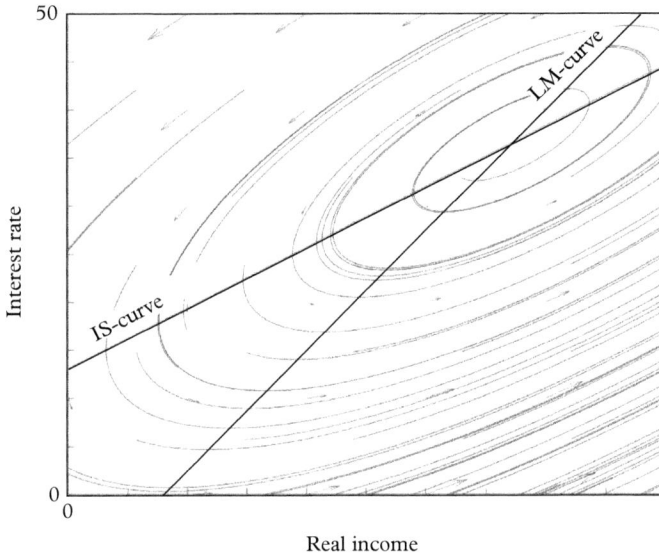

Figure 8.8 *IS–LM model with increasing real income vs. interest rate, but with an overheated economy causing an unstable spiral. The IS curve and the LM curve are the isoclines.*

value for relaxation to the new stable fixed point. The transient trajectory shown in Fig. 8.7 spirals as it converges on the new equilibrium.

Stability can be destroyed if there is too much increased expenditure in response to income, which occurs when an economy is "overheated." This effect is captured by increasing the parameter i in Eq. (8.52) and is illustrated in Fig. 8.8 with $i = 1.5$. It is also possible to have saddle nodes if the IS curve has a higher slope than the LM curve. In either case, the economy becomes unstable.

8.5.2 Inflation and unemployment

Inflation and unemployment are often (but not always) linked by an inverse relationship in which government policy, that seeks to increase employment above its current demand, leads to higher inflation. This relationship is known as the *Phillips curve* and is captured mathematically by the linearized form

Phillip's curve: $\qquad \pi = -a\,(u - u_n) + \pi_e,$ $\qquad\qquad$ (8.59)

where $\pi = \dot{p}$ is the current rate of inflation—the rate of change in the price p of money, u is the unemployment rate, π_e is the expected inflation rate, and u_n is the *non-accelerating inflation rate of unemployment* (NAIRU). This somewhat unwieldy acronym stands for the employment rate that keeps inflation constant (non-accelerating). Unemployment, in turn, has an inverse relationship to the *gross domestic product* (GDP) through

$$-a\,(u - u_n) = \alpha\,(g - g_n),\tag{8.60}$$

where g is the GDP and g_n is the GDP under the condition that $u = u_n$. Therefore, expected inflation (predicting the future) differs from actual inflation (reality) to a degree determined by the GDP. If an economy is "overheated," then national productivity is higher than the normal GDP and inflation exceeds the expected inflation. In tandem with the Phillips curve, expected inflation adjusts according to the difference of current from expected inflation:

Adaptive expectations: $\dot{\pi}_e = \beta\,(\pi - \pi_e).$ (8.61)

Finally, the GDP depends on the money markets and on the goods markets that are characterized, respectively, by the difference $(m - p)$ between money availability and money price and by the expected rate of inflation π_e. This is captured by the *aggregate demand* equation:

$$g = a_0 + a_1\,(m - p) + a_2 \pi_e\tag{8.62}$$

with all positive coefficients. Combining the Phillips curve, Eq. (8.59), with adaptive expectations, Eq. (8.61), leads to

$$\dot{\pi}_e = \beta \alpha\,(g - g_n),\tag{8.63}$$

and taking the derivative of the aggregate demand equation Eq. (8.62) leads to

$$\dot{g} = a_1\,(\dot{m} - \pi) + a_2 \dot{\pi}_e,\tag{8.64}$$

where $\dot{m}$ is the growth of the money stock that is printed by the federal government (an exogenous parameter controlled by fiscal policy) with inflation $\pi = \dot{p}$. Equations (8.63) and (8.64) combine into the two-dimensional flow

$$\begin{aligned}\dot{g} &= a_1\,(\dot{m} - \pi_e) + \alpha\,[a_2\beta - a_1]\,(g - g_n)\\ \dot{\pi}_e &= \alpha\beta\,(g - g_n)\end{aligned}\tag{8.65}$$

in the space defined by the GDP and the expected rate of inflation. The nullclines are

$$\begin{aligned}\pi_e &= \frac{\alpha\,[a_2\beta - a_1]\,(g - g_n)}{a_1} + \dot{m} \quad &\dot{g}\text{-nullcline}\\ g &= g_n \quad &\dot{\pi}_e\text{-nullcline}\end{aligned}\tag{8.66}$$

where the $\dot{g}$-nullcline is a line with negative slope for $a_1 > a_2\beta$, and the $\dot{\pi}_e$-nullcline is a vertical line.

Example 8.6 Gross domestic product and expected inflation

The dynamics of inflation and unemployment are similar in character to the linearized IS–LM models and share the same generic solutions. One difference is seen in the $\dot{\pi}_e$ isocline in Fig. 8.9 that is simply a vertical line at $y = y_n$ that maintains the NAIRU. The example uses the parameters $a_1 = 2$, $\dot{m} = 1$, $a_2 = 1$, $y_n = 1$, $\alpha = 2$, and $\beta = 0.5$. The figure also shows a governmental money shock as it increases $\dot{m}$ to increase the GDP and hence lower unemployment. However, the previous fixed point becomes the initial value to a transient trajectory that relaxes back to the original unemployment level. Therefore, the increased rate of money availability only increases employment temporarily (perhaps preceding an election), but ultimately leads to long-term inflation without long-term employment.

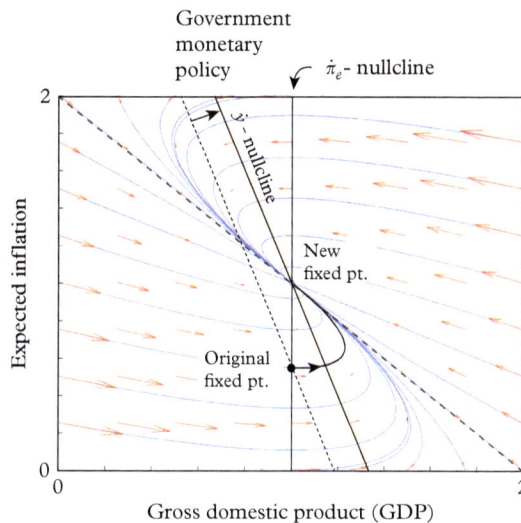

Figure 8.9 *Expected rate of inflation vs. the gross domestic product related to the Phillips curve for unemployment. The vertical line is the $\dot{\pi}_e$ isocline which is the condition that the GDP has the value needed to maintain the NAIRU. Government monetary policy (increasing $\dot{m}$) can increase GDP and decrease unemployment temporarily, but the system relaxes to the original unemployment rate, meanwhile incurring an increase in inflation.*

The historical relationship between inflation and unemployment rarely obeys the Phillips curve. The historical values are plotted in Fig. 8.10 between 1941 (just after the depression) to 2013. The Phillips curve states that increasing unemployment is accompanied by lower inflation, which is not easily observed in the figure, except for fairly short periods. Larger forces on the national economy tend

Figure 8.10 *Historical inflation and unemployment from 1941 to 2013. The current NAIRU is assumed to be around 6%, and the target inflation rate is around 2%. There is no evidence for a long-term Phillips curve in these data.*

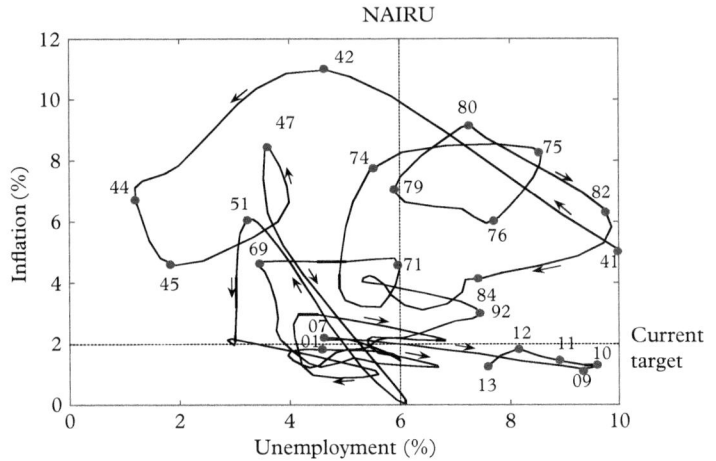

to overwhelm the Phillip's curve. On the other hand, for more than 70 years, both unemployment and inflation in the US has generally been below 10% without the wild increases that have been seen in some economies (such as southern Europe).

The simple linearized examples of macroeconomics that have been described in this chapter are just the tip of the ice-berg for economic modeling. There are many complexities, such as nonlinear dependences and delays in adjustments, that lead to complex dynamics. There are often unexpected consequences that defy intuition—and that make it difficult to set fiscal and monetary policies. The lack of precise laws (the Phillips curve is just a rule-of-thumb, and only works over short time frames), and ill-defined functional dependences opens the door to subjective economic philosophies, i.e., politics. In this sense, economics does not need to be entirely "rational," but can be driven by perceived benefits as well as real benefits. However, the astute student should note that even philosophy and political inclinations can be captured by probabilistic payoff matrices of the type that were studied in Chapter 7 on evolutionary dynamics, and hence even "irrational" behavior sometimes can be predicted quantitatively.

8.6 Stochastic dynamics and stock prices (optional)

[2] The character of this section is substantially different than the rest of the text. Stochastic calculus and random walks are not continuous functions like the flow equations that are the main themes of this book. This section is included here as an optional section because of the importance of this topic to econophysics, and because it introduces the idea of a non-continuous trajectory.

The price of a stock is the classic example of a stochastic time series. The trajectory of a stock price has many of the properties of a random walk, or Brownian motion. Although the underlying physics of fluctuating stock prices is not as easy to see as the underlying physics of Brownian motion, the description in terms of random forces still applies. The analogy of a stock price to a random walk underlies many of today's investment strategies for derivatives and hedge funds.[2]

8.6.1 Random walk of a stock price: efficient market model

Consider a stock that has a long-term average rate of return given by μ. The average value of the stock increases exponentially in time as

$$\bar{S} = S_0 e^{\mu t}. \tag{8.67}$$

However, superposed on this exponential growth (or decay if the rate of return is negative) are day-to-day price fluctuations. The differential change in relative stock price, in the presence of these fluctuations, is given by

$$\frac{dS}{S} = d(\ln S) = \mu\, dt + \sqrt{2D}\, dW, \tag{8.68}$$

where D plays the role of a diffusion coefficient, and the differential dW is a *stochastic variable* that represents the fluctuations.[3] Fluctuations can arise from many sources, and may have underlying patterns (correlations) or not. The simplest models of stock prices assume that there are no underlying patterns in the fluctuations and that the stock price executes a random walk. This condition of randomness and history independence is called a Markov process, and the differential dW is known as a Wiener process.

The trajectory of a stock price, with day-to-day fluctuations, is modeled as a discrete process by the finite difference approximation of Eq. (8.68):

$$x_{n+1} = x_n + \mu\Delta t + \sqrt{2D\Delta t}\, \xi_n, \tag{8.69}$$

where the variable $x = \ln(S/S_0)$ and ξ_n is a random value from a normal distribution with standard deviation of unity. Examples of stochastic stock-price trajectories are shown in Fig. 8.11. The dashed line is the average exponential growth. This type of trajectory is called a geometric random walk on the price S (or an arithmetic random walk on the variable x).

8.6.2 The Langevin equation

This section takes a look at fundamental aspects of stochastic processes that apply to broad areas of physics. The motivation of understanding stock price fluctuations provides a convenient excuse to explore what happens when deterministic flows, which we have studied so thoroughly up to now in this textbook, acquire a stochastic element.

One of the most important conclusions of chaos theory and nonlinear dynamics is that not all random-looking processes are actually random. In deterministic chaos, one could argue that structures such as strange attractors are not random at all. But sometimes, in nature, processes really are random, or at least have to be treated as such because of their complexity. Brownian motion is a perfect example of this. At the microscopic level, the jostling of the Brownian particle can

[3] A stochastic variable, or Wiener process, is an ordered set of random values that is nowhere differentiable. The character of stochastic processes is fundamentally different than for continuous-variable flows. However, the time evolution of stochastic systems can still be defined, and stochastic processes can be added to continuous flows.

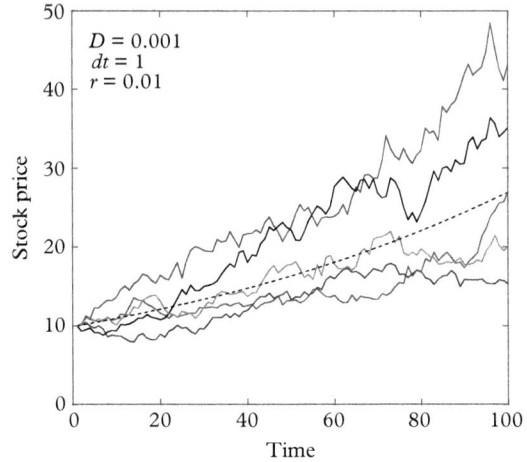

Figure 8.11 *Examples of geometric random walks on a stock price. The dashed line is the average growth of the stock.*

be understood in terms of deterministic momentum transfers from liquid atoms to the particle. But there are so many liquid particles that their individual influences cannot be directly predicted. In this situation, it is more fruitful to view the atomic collisions as a stochastic process with well-defined physical parameters, and then study the problem statistically.

Stochastic processes are understood by considering a flow that includes a random function. The resulting set of equations are called the Langevin equation, namely

$$\dot{x}_a = f_a(x_1, x_2, \ldots, x_N) + \sigma_a \xi_a(t), \qquad (8.70)$$

where $\xi_a(t)$ is a set of N stochastic functions and σ_a is the standard deviation of the ath process. The stochastic functions are in general non-differentiable, but are integrable. They have zero mean and no temporal correlations. The solution of Eq. (8.70) is an N-dimensional trajectory that has some properties of a random walk superposed on the dynamics of the underlying flow.

As an example, take the case of a particle moving in a one-dimensional potential, subject to drag and to an additional stochastic force

$$\begin{aligned} \dot{x} &= v \\ \dot{v} &= -\gamma v - \frac{1}{m}\frac{dU}{dx} + \sqrt{2B}\xi(t), \end{aligned} \qquad (8.71)$$

where γ is the drag coefficient, U is a potential function, and B is the velocity diffusion coefficient. Take a double-well potential as an example:

$$U(x) = \frac{\alpha}{2}x^2 + \frac{\beta}{4}x^4. \qquad (8.72)$$

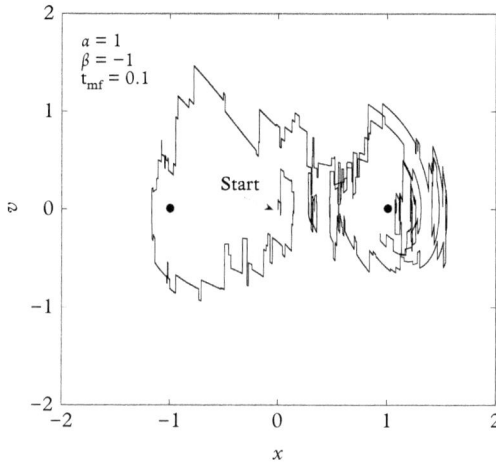

Figure 8.12 *Stochastic trajectory of a particle in a double-well potential. The start position is at the unstable fixed point between the wells, and the two stable fixed points (well centers) are the solid dots.*

One stochastic trajectory is shown in Fig. 8.12 that applies discrete velocity jumps using a normal distribution of jumps of variance $2B$ in a mean time τ. The notable character of this trajectory, besides the random-walk character, is the ability of the particle to jump the barrier between the wells. In the deterministic system, the initial condition would dictate which stable fixed point would be approached. In the stochastic system, there are random fluctuations that take the particle from one basin of attraction to the other.

The density from a simulation of $N = 4000$ particles is shown in Fig. 8.13 for the double-well potential. The probability distribution for an ensemble of particles at long times (much larger than the relaxation time) is given by

$$p(x) = \frac{e^{-2V(x)/\sigma^2}}{\int\limits_{-\infty}^{\infty} e^{-2V(x)/\sigma^2}\, dx}. \tag{8.73}$$

The density is obtained by integrating over the trajectories and is shown in Fig. 8.14 with a fit to Eq. (8.73). Larger fluctuations σ tend to smooth the probability function, while smaller σ leads to strongly localized densities near the stable fixed points.

The stochastic long-time probability distribution $p(x, v)$ in Fig. 8.13 introduces an interesting new view of trajectories in state space that have a different character than many of the state-space flows that we have studied in this book. If we think about starting a large number of systems with the same initial conditions, and then letting the stochastic dynamics take over, we can define a time-dependent probability distribution $p(x, v, t)$ that describes the likely end-positions of an ensemble of trajectories on the state plane as a function of time. This introduces the idea of the trajectory of a probability cloud in state space, which has a strong analogy to time-dependent quantum mechanics. The

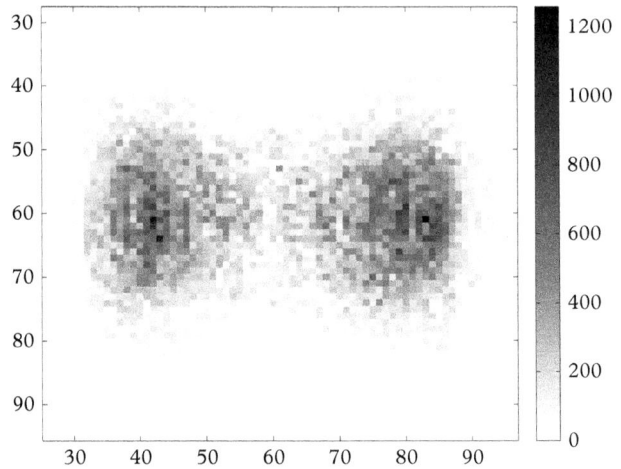

Figure 8.13 *Density of N = 4000 random walkers in the double-well potential with σ = 1.*

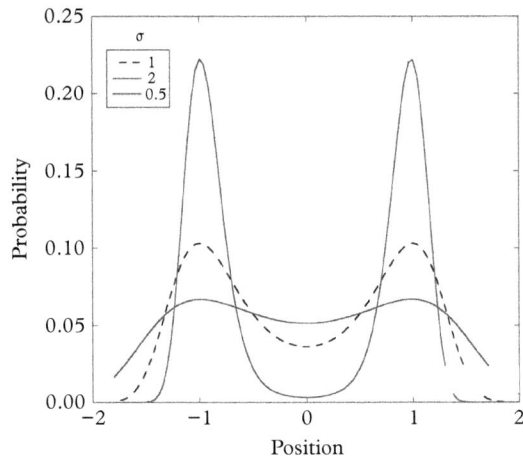

Figure 8.14 *Density of N = 4000 random walkers in the double-well potential for the same parameters as Fig. 8.13.*

Schrödinger equation can be viewed as a diffusion equation in complex time, which is the basis of a technique known as quantum Monte Carlo that solves for ground state wave functions using concepts of random walks. This goes beyond the topics of this textbook, but it shows how such diverse fields as econophysics, diffusion, and quantum mechanics can share common tools and language.

8.6.3 Hedging and the Black–Scholes equation

Hedging is a venerable tradition on Wall Street. To hedge means that a broker sells an option (to purchase a stock at a given price at a later time) assuming that the stock will fall in value (selling short), and then buys, as insurance against the price rising, a number of shares of the same asset (buying long). If the broker balances enough long shares with enough short options, then the portfolio's value is insulated from the day-to-day fluctuations of the value of the underlying asset. This type of portfolio is one example of a financial instrument called a *derivative*. The name comes from the fact that the value of the portfolio is *derived* from the value of the underlying assets. The challenge with derivatives is finding their "true" value at any time before they mature. If a broker knew the "true" value of a derivative, then there would be no risk in the buying and selling of derivatives.

To be risk free, the value of the derivative needs to be independent of the fluctuations. This appears at first to be a difficult problem, because fluctuations are random and cannot be predicted. But the solution actually relies on just this condition of randomness. If the random fluctuations in stock prices are equivalent to a random walk superposed on the average rate of return, then perfect hedges can be constructed with impunity.

The equation for a geometric random walk of a stock price is

$$\frac{dS}{S} = d\left(\ln S\right) = \mu dt + \sigma dW, \tag{8.74}$$

where S is the price of the stock, μ is the average rate of return, and σ is the size of the fluctuations. A derivative based on the stock has a "true" value $V(S, t)$ that depends on the price of the stock and on the time that the option will be called. To determine this value $V(S, t)$, it is necessary to draw from a fundamental result of stochastic calculus known as Ito's lemma.

Consider a random variable that executes a drift and random walk:

$$dx = adt + bdW. \tag{8.75}$$

If there is a function $f(x, t)$ that depends on the random variable, then the differential of that function is given by

$$\begin{aligned} df &= \frac{\partial f}{\partial t} dt + \frac{\partial f}{\partial x} dx + \frac{1}{2}\frac{\partial^2 f}{\partial x^2} dx^2 \\ &= \frac{\partial f}{\partial t} dt + \frac{\partial f}{\partial x}\left(adt + bdW\right) + \frac{1}{2}\frac{\partial^2 f}{\partial x^2}\left(adt + bdW\right)^2 \\ &= \left[\frac{\partial f}{\partial t} + a\frac{\partial f}{\partial x}\right] dt + b\frac{\partial f}{\partial x} dW + \frac{b^2}{2}\frac{\partial^2 f}{\partial x^2} dW^2 \end{aligned} \tag{8.76}$$

The Wiener process, W, is a function that is differentiable nowhere, but it has the property

$$dW^2 = dt. \tag{8.77}$$

Using this result, and keeping only the lowest orders in differentials, gives

Ito's formula: $\qquad df = \left[\frac{\partial f}{\partial t} + a\frac{\partial f}{\partial x} + \frac{b^2}{2}\frac{\partial^2 f}{\partial x^2}\right] dt + b\frac{\partial f}{\partial x} dW,$ $\qquad\qquad$ (8.78)

which is also known as *Ito's lemma*. Ito's lemma forms the foundation of many results from stochastic calculus.

To make the connection between Ito's lemma and derivative pricing, take the correspondences

$$x = S, \qquad a = \mu S, \qquad b = \sigma S, \tag{8.79}$$

and Eq. (8.75) gives the geometric random walk

$$dS = \mu S dt + \sigma S dW. \tag{8.80}$$

The differential of the value of a derivative value $dV(S, t)$ is obtained from Ito's lemma

$$dV = \left(\mu S \frac{\partial V}{\partial S} + \frac{\partial V}{\partial t} + \frac{1}{2} \sigma^2 S^2 \frac{\partial^2 V}{\partial S^2} \right) dt + \sigma S \frac{\partial V}{\partial S} dW. \tag{8.81}$$

The goal of a hedge is to try to zero-out the uncertainties associated with fluctuations in the price of assets. The fluctuations in the value of the derivative arise in the last term in Eq. (8.81). The strategy is to combine the purchase of stocks and the sale of options in such a way that the stochastic element is removed.

To make a hedge on an underlying asset, create a portfolio by selling one call option (selling short) and buying a number N shares of the asset (buying long) as insurance against the possibility that the asset value will rise. The value of this portfolio is

$$\pi(t) = -V(S, t) + NS(t). \tag{8.82}$$

If the number N is chosen correctly, then the short and long positions will balance, and the portfolio will be protected from fluctuations in the underlying asset price. To find N, consider the change in the value of the portfolio as the variables fluctuate,

$$d\pi = -dV + NdS, \tag{8.83}$$

and insert Eqs. (8.80) and (8.81) into Eq. (8.83) to yield

$$d\pi = -\left(\mu S \frac{\partial V}{\partial S} + \frac{\partial V}{\partial t} + \frac{1}{2} \sigma^2 S^2 \frac{\partial^2 V}{\partial S^2} \right) dt - \sigma S \frac{\partial V}{\partial S} dW + N (\mu S dt + \sigma S dW)$$

$$= \left(N\mu S - \mu S \frac{\partial V}{\partial S} - \frac{\partial V}{\partial t} - \frac{1}{2} \sigma^2 S^2 \frac{\partial^2 V}{\partial S^2} \right) dt + \left(N - \frac{\partial V}{\partial S} \right) \sigma S dW \tag{8.84}$$

Note that the last term contains the fluctuations. These can be zeroed-out by choosing

$$N = \frac{\partial V}{\partial S}, \tag{8.85}$$

and then

$$d\pi = -\left(\frac{\partial V}{\partial t} + \frac{1}{2}\sigma^2 S^2 \frac{\partial^2 V}{\partial S^2}\right) dt. \tag{8.86}$$

The important observation about this last equation is that the stochastic function W has disappeared. This is because the fluctuations of the N share prices balance the fluctuations of the short option.

When a broker buys an option, there is a guaranteed rate of return r at the time of maturity of the option which is set by the value of a risk-free bond. Therefore, the price of a perfect hedge must increase with the risk-free rate of return. This is

$$d\pi = r\pi\, dt \tag{8.87}$$

or

$$d\pi = r\left(-V + \frac{\partial V}{\partial S}S\right) dt. \tag{8.88}$$

Upon equating Eq. (8.86) with Eq. (8.88), this gives

$$
\begin{aligned}
d\pi &= \left(-\frac{\partial V}{\partial t} - \frac{1}{2}\sigma^2 S^2 \frac{\partial^2 V}{\partial S^2}\right) \Delta t \\
&= r\left(-V + \frac{\partial V}{\partial S}S\right) \Delta t
\end{aligned}
\tag{8.89}
$$

Simplifying, this leads to a PDE for $V(S, t)$:

Black–Scholes equation: $$\frac{\partial V}{\partial t} + \frac{1}{2}\sigma^2 S^2 \frac{\partial^2 V}{\partial S^2} + rS\frac{\partial V}{\partial S} - rV = 0, \tag{8.90}$$

which is known as *the Black–Scholes equation.*[4]

The Black–Scholes equation is a PDE whose solution, given boundary conditions and time, defines the "true" value of the derivative, and determines how many shares to buy at $t = 0$ at a specified guaranteed return rate r (or, alternatively, stating a specified stock price $S(T)$ at the time of maturity T of the option). It is a diffusion equation that incorporates the diffusion of the stock price with time. If the derivative is sold at any time t prior to maturity, when the stock has some value S, then the value of the derivative is given by $V(S, t)$ as the solution to the Black–Scholes equation. It was first derived by Fischer Black and Myron Scholes in 1973 with contributions by Robert Merton. Scholes and Merton won the Nobel Prize in economics in 1997 (Black died in 1995).

One of the interesting features of this equation is the absence of the mean rate of return μ (in Eq. (8.74)) of the underlying asset. This means that any stock of any value can be considered, even if the rate of return of the stock is negative! This

[4] Black-Scholes is a partial differential equation (PDE) and not an ordinary differential equation (ODE). This book has exclusively focused on ODEs (flows), but deviates here because of the importance of Black-Scholes to econophysics.

type of derivative looks like a truly risk-free investment. You would be guaranteed to make money even if the value of the stock falls, which may sound too good to be true...which of course it is. The success (or failure) of derivative markets depends on fundamental assumptions about the stock market. These include that it would not be subject to radical adjustments or to panic or irrational exuberance, which is clearly not the case. Just think of booms and busts. The efficient and rational market model, and ultimately the Black–Scholes equation, assumes that fluctuations in the market are governed by Gaussian random statistics. When these statistics fail, the derivatives fail, which they did in the market crash of 2008.

The main focus of this textbook has been on trajectories, which are captured in simple mathematical flow equations. The Black–Scholes equation does not describe a trajectory. It is a PDE that must be solved satisfying boundary conditions, which goes beyond the topics of this book. However, because it plays a central role in econophysics, an example of its solution is given here. One of the simplest boundary conditions is for a European call option for which there is a spot price that is paid now and a strike price that will be paid at a fixed time in the future. The value of the derivative for a given strike price K is a function of the spot price S and the time to maturity T as

$$V(S, T) = \frac{1}{2}\left[1 + erf\left(\frac{d_1}{\sqrt{2}}\right)\right]S - \frac{e^{-rT}}{2}\left[1 + erf\left(\frac{d_2}{\sqrt{2}}\right)\right]K, \qquad (8.91)$$

where r is the guaranteed rate of return. The arguments of the error functions are

$$d_1 = \frac{1}{\sigma\sqrt{T}}\left[\ln\left(\frac{S}{K}\right) + \left(r + \frac{1}{2}\sigma^2\right)T\right]$$
$$d_2 = d_1 - \sigma\sqrt{T}, \qquad (8.92)$$

where σ is the stock volatility. A solution of $V(S, T)$ for a strike price $K = 1$ as a function of the spot price and time to maturity is shown in Fig. 8.15 for a guaranteed rate of return $r = 0.15$ and a volatility $\sigma = 0.5$.

This chapter on economic dynamics applied many of the topics of nonlinear dynamics that we have explored in this textbook. Aspects of stability analysis (Chapter 3) were applied to predict whether competitions could be stable, or whether a winner-take-all dynamics would take over. Hopf bifurcations showed up in inventory stock business cycles, and cobweb models of chaotic cascades were a natural tool for discrete-time dynamics in the day-to-day changes in prices. Stochastic dynamics were introduced in the random walk of stock prices, which had strong analogs to the physics of Brownian motion and to the time-evolution of quantum probability distributions. We even invoked equations from evolutionary dynamics (Chapter 7) to describe some of the dynamics of the smart-phone consumer market. A topic that we did not have time to cover in this chapter is supply-chain networks that use the topics of Chapter 5. Econophysics is a broad topic that draws heavily from many aspects of modern dynamics. For this reason, many of the so-called "quants" on Wall Street have degrees in physics.

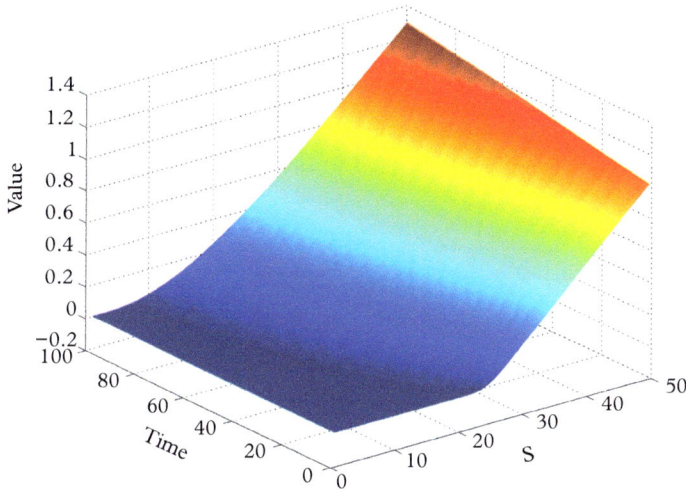

Figure 8.15 *The Black–Scholes solution for an European call option. r = 0.15, time = 0 to 2, price = 0 to 2, σ = 0.5.*

8.7 Summary

Economic theory in this chapter provided several examples of nonlinear dynamics. In *microeconomics*, forces of *supply and demand* led to stable competition as well as *business cycles*. Continuous systems with *price and quantity adjustment* as well as cost of labor exhibited *Hopf bifurcation* and a *bifurcation cascade* to chaos. Discrete *cobweb models* captured delayed adjustments that also could exhibit bifurcation cascades. In *macroeconomics* investment-savings (IS) and liquidity-money (LM) capture dynamics related to interest rates. Inflation and unemployment are also coupled through the *Phillips curve* with *adaptive expectations*. The *stochastic dynamics* of the stock market was introduced in an optional chapter. *Stochastic variables* can be added to continuous-time price models as a random contribution to ODEs, or they can drive diffusion of prices that are PDEs. An important example of a PDE in econophysics is the *Black–Scholes equation* for options pricing.

Supply and demand

Supply of quantities and their demand have an inverse relationship on price. Supply increases with increasing price (profit), while demand falls. A simple price adjustment equation depends on excess demand $E = D - S$ as

$$\begin{aligned} \dot{p} &= kE \\ &= k\,(D - S) \end{aligned} \tag{8.2}$$

for positive coefficient k.

IS–LM

In macroeconomic theory, investment-savings (IS) and liquidity-money (LM) trade off against each other as a function of interest rate. As interest rates rise, there is less investment and more spending (the IS side of the economy), while the demand for money decreases (the LM side of the economy). Expenditure E and demand D adjust as

$$\dot{g} = \alpha \left[E\left(t\right) - g\left(t\right) \right]$$
$$\dot{r} = \beta \left[D\left(t\right) - m_0 \right] \tag{8.54}$$

for gross domestic product g and money availability m_0.

Phillips curve

The Phillips curve predicts a linear relationship between inflation and unemployment. In linearized form this is

$$\pi = -a\left(u - u_n\right) + \pi_e, \tag{8.59}$$

where $\pi = \dot{p}$ is the current rate of inflation—the rate of change in the price p of money, u is the unemployment rate, π_e is the expected inflation rate, and u_n is the *non-accelerating inflation rate of unemployment* (NAIRU).

Adaptive expectations

Expected prices or inflation usually do not match actual prices or inflation, and the time-rate-of-change of these quantities often adjust linearly with the difference between expected and actual values. Adaptive expectations appear in iterative cobweb models as in Eq. (8.30) in microeconomics, and in models of inflation as in

$$\dot{\pi}_e = \beta\left(\pi - \pi_e\right) \tag{8.61}$$

in macroeconomics.

Geometric Brownian motion

The random walk in a relative property (in one dimension) is given by

$$\frac{dx}{x} = \mu dt + \sqrt{2D}dW, \tag{8.68}$$

where D is a diffusion coefficient and dW is a Wiener process with the property $dW^2 = dt$.

Ito's formula

A very useful equation from stochastic analysis of the equation

$$dx = adt + bdW \tag{8.75}$$

is

$$df = \left[\frac{\partial f}{\partial t} + a\frac{\partial f}{\partial x} + \frac{b^2}{2}\frac{\partial^2 f}{\partial x^2}\right]dt + b\frac{\partial f}{\partial x}dW \tag{8.78}$$

which is known as Ito's formula.

Black–Scholes equation

A risk-free hedge has a value V determined by the Black–Scholes equation:

$$\frac{\partial V}{\partial t} + \frac{1}{2}\sigma^2 S^2 \frac{\partial^2 V}{\partial S^2} + rS\frac{\partial V}{\partial S} - rV = 0, \tag{8.90}$$

where r is the guaranteed rate of return and σ is the volatility in the stock value.

8.8   Bibliography

E. Allen, *Modeling with Ito Stochastic Differential Equations* (Springer, 2007).

N. F. Johnson, P. Jefferies and P. M. Hui, *Financial Market Complexity: What Physicists Can Tell Us about Market Behavior* (Oxford University Press, 2003).

R. Mahnke, J. Kaupuzs and I. Lubashevsky, *Physics of Stochastic Processes: How Randomness Acts in Time* (Wiley-VCH, 2009).

A. Medio, *Chaotic Dynamics: Theory and Applications to Economics* (Cambridge University Press, New York, 1992).

M. A. Nowak, *Evolutionary Dynamics* (Harvard University Press, 2006).

A. J. Roberts, *Elementary Calculus of Financial Mathematics* (SIAM, 2009).

J. Rudnick and G. Gaspari, *Elements of the Random Walk: An Introduction for Advanced Students and Researchers* (Cambridge University Press, 2004).

R. Shone, *Economic Dynamics: Phase Diagrams and their Economic Application* (Cambridge University Press, 2002).

8.9   Homework exercises

Analytic problems

1. **Supply and demand:** Nonlinearities in supply and demand can be captured qualitatively by

$$D = \frac{a}{1+p}$$
$$S = \frac{cp}{1+p}$$

for a and c positive. What is the fixed point and Luapunov exponent?

2. **Supply and demand for three companies:** Extend Eqs. (8.9) and (8.10) to three companies. What are all possible stabilities of the fixed point? Characterize the market conditions that lead to each type of stability. (For instance, negative feedback on price, as higher price leads to less excess demand, with cooperative feedback from competitors, as higher competitor prices produces more demand for your product, leads to an attractor, etc.).

3. **Walras' law:** In a closed market, a zero-sum rule applies, called *Walras' law.* If there is an excess value in one market, then there must be a deficit in another. Mathematically this is expressed as

$$\sum_{i=0}^{N} P_i E_i = 0,$$

where P_i is the price of the ith product. The price P_0 can be used as a numeraire to divide through to give

$$E_0 + \sum_{i=1}^{N} p_i E_i = 0,$$

where $p_i = P_i/P_0$. Use Walras' law as a constraint to convert the dynamics of three competing companies to an equivalent two-dimensional (2D) dynamics.

4. **IS–LM model:** What are all possible stability conditions for the 2D IS–LM model? Characterize the conditions for each type of stability.

5. **Universal Procedure Set:** Derive Eqs. (8.51).

Computational projects[5]

6. **Price adjustment:** Explore the behavior of the quantity and price adjustment model (S182.m) by changing the rate β and the parameter q_0. What happens when β is large? Small? What happens when q_0 is positive? Negative?

7. **Logistic supply and demand:** Explore the robustness of the chaotic behavior of the logistic supply and demand model. How easy is it to display chaotic behavior as the parameters are altered? Is this a robust chaotic system that shows chaotic behavior for a broad range of model parameters?

8. **Replicator–mutator:** In the replicator–mutator equation (see Chapter 7), perform a simulation to explore what happens when the payoff of one product is larger than the others? Can the others evolve to emulate this first product and raise their fitness?

9. **Replicator–mutator:** Include an adjacency matrix (Chapter 5) into the replicator–mutator equation. This models the likelihood of mutating depending on the "social network" of links to the ith consumer.

[5] Matlab programming examples are found at www.works.bepress.com/ddnolte.

Part 4

Relativity and Space–Time

The special and general theories of relativity are quintessential examples of the geometric view of mechanics. The key elements of relativity theory are the geometric invariants (in particular the *invariant interval*) as relativity is best understood through those quantities that remain the same regardless from which frame they are being observed. The structure of *metric spaces*, defined through the *metric tensor*, provides the geometric structure in which trajectories are embedded, and the geodesic equation, constructed on the metric tensor, defines force-free motions through a metric space. In general relativity, mass and energy act to modify the structure of space time, and new geodesics emerge that are the force-free motions through the warped space. In this way, the "force" of gravity is replaced by force-free trajectories through warped space–time.

Metric Spaces and Geodesic Motion

9

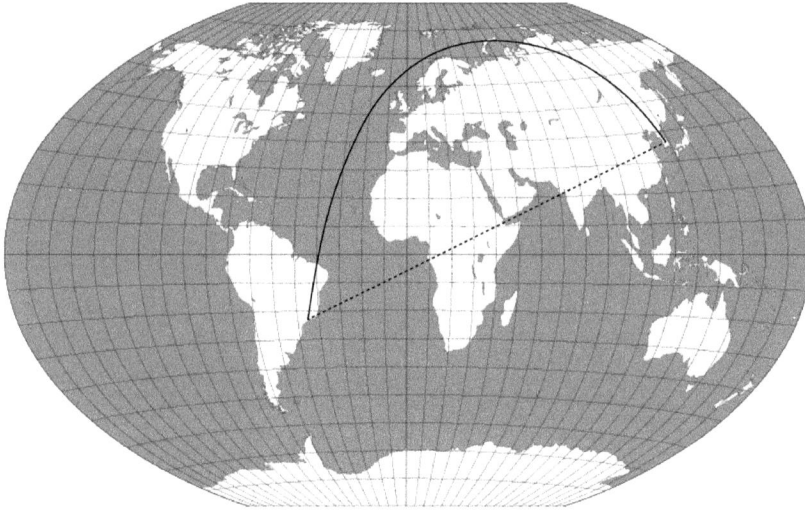

Great circle route from Rio de Janeiro, Brazil, to Seoul, Korea, on a Winkel tripel projection.
<http://commons.wikimedia.org/wiki/File:World_map_Winkel_Tripel_proj-0deg_centered.svg>

Everyone is familiar with geodesics as great circle routes on the Earth. Take a string and stretch it between two points on a globe, let's say between Rio de Janeiro and Seoul. On a Winkel tripel projection map, this path looks very strange and very long, but on the globe it is the shortest distance between those two points. Locally, the Earth is flat, which means that orthogonal Cartesian grids work fine as maps, as long as distances between two points don't get too big. The whole surface of the Earth can be covered by a set of Cartesian maps, although slight twists are necessary between every two maps. In the language of multidimensional geometry, the surface of the Earth is a manifold.[1] Within a manifold, geodesic curves are the shortest paths between two points.

Manifolds occur everywhere in dynamics. For instance, the configuration space in Lagrangian mechanics of any system with spatial constraints is a

[1] Manifold, which comes from German Mannigfältigkeit, is a term coined by Carl Friedrich Gauss (1851).

manifold, as for a particle restricted to move on a surface like an ellipsoid or a torus. For *n*-particle systems, the configuration space is $3n$-dimensional, but imposed constraints may cause the trajectories to occur on a lower-dimensional submanifold.

We saw in Chapter 2 that the equations of Lagrangian dynamics emerge from a principle of stationarity. The path along which the mechanical action has a stationary value is the physical path taken by a particle. In many common examples, the stationary value generates the path of least action—like the least distance of a geodesic. This analogy between least action and least distance in a manifold has more than a superficial similarity. By choosing the appropriate metric distance in a dynamical manifold, one indeed finds that the actual mechanical trajectories are geodesic trajectories.

This chapter defines the properties of manifolds and the metric distances within them. The metric tensor of a manifold takes center stage to help define the quantifiable properties of trajectories in configuration space. The metric tensor is used to convert vectors into a dual type of entity called covectors that span reciprocal spaces and that combine with vectors to form invariant quantities that are invariant under coordinate transformations. In addition, by defining appropriate metrics within the configuration space of conservative dynamical systems, trajectories can be understood as geodesic paths.

The concepts and mathematical tools developed in this chapter occur in many of the chapters in this book, especially the following ones. In the theory of relativity (both special and general, in Chapters 10 and 11) the metric tensor is used to define invariants as well as to find the properties of curved spaces around massive bodies.

9.1   Manifolds and metric tensors

Manifold is roughly synonymous with the concept of an "*n*-dimensional coordinate space." The simplest manifolds are Cartesian coordinate frames, but many interesting manifolds are not Cartesian, nor are they Euclidean, although a manifold is locally Euclidean. By being locally Euclidean, it can be described by Cartesian coordinate frames that are applicable over restricted regions. Any map of a city or a topographic map of a county uses a Cartesian grid. In the language of manifolds, such a map is called a *chart*. However, because the surface of the Earth is not flat, to tile such charts over a continent requires slight adjustments and misalignments at the borders of each chart at which the Cartesian grid of one has a slight twist relative to the next. Ideally, the charts overlap a little, and in this overlap region, there are two different coordinate systems that define the same landmarks. A coordinate transformation exists between the two coordinate frames that translates one into the other.

If you start in New York City with a stack of topographic maps and end up in Fairbanks, Alaska, you will move through many such coordinate transformations,

Charts and Atlas:
The circle

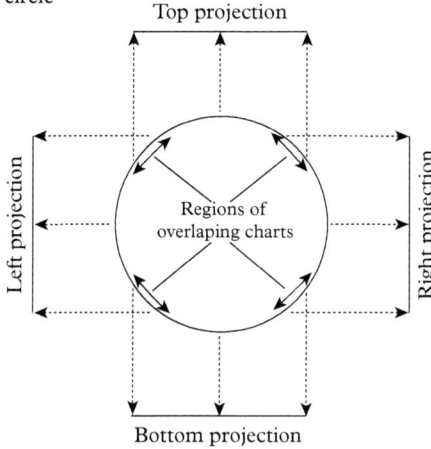

Figure 9.1 *A circle can be captured by several flat charts. Each is a projection of an open interval of the circle onto a line. The collection of charts is an atlas. The overlap between charts provides local coordinate transformations that take one to the next.*

but there will be no gaps—there is a smooth progression from one coordinate frame to the next. General manifolds require more than one chart to cover it fully, and the set of all charts is called the *atlas*. The mappings of a manifold onto multiple charts can be nonlinear, as for the projection of a circle onto an atlas of flat charts, shown in Fig. 9.1.[2]

9.1.1 Metric tensor

The geometry of manifolds is captured by tensors. For instance, distance in a manifold depends on the *metric tensor*. For a Cartesian coordinate frame, the distance element is the well-known expression

$$ds^2 = dx^2 + dy^2 + dz^2 \tag{9.1}$$

that is more generally expressed as (see Eq. 1.55)

$$ds^2 = g_{11}dx^2 + g_{22}dy^2 + g_{33}dz^2, \tag{9.2}$$

where $g_{11} = g_{22} = g_{33} = 1$.

The line element is a scalar quantity, which means that it does not have any indices associated with it. Furthermore, it depends on the product of differentials, which means it is a quadratic form. The most general quadratic form that yields a scalar quantity is

$$g_{ab}dx^a dx^b, \tag{9.3}$$

which is summed over the repeated indices.[3] Therefore, the square of the line element is defined to be

[2] The local Euclidean nature of a manifold is important for the application of tensor analysis. This is because the coordinate systems, between which coordinate transformations operate, usually include the conventional orthogonal coordinate system used to define flat spaces, such as Cartesian coordinates.

[3] In this and the following chapters, the implicit Einstein summation is assumed.

$$\textit{Line element:} \quad ds^2 = g_{ab}dx^a dx^b. \tag{9.4}$$

This form of the line element is important for special and general relativity.

A general coordinate transformation connects local Cartesian coordinates (x, y, z) of a chart with the general curvilinear coordinates (q^1, q^2, q^3). The coordinate functions can be expressed as

$$\begin{aligned} x &= x\left(q^1, q^2, q^3\right) \\ y &= y\left(q^1, q^2, q^3\right) \\ z &= z\left(q^1, q^2, q^3\right) \end{aligned} \tag{9.5}$$

The metric tensor g_{ab} is defined to be

$$\textit{Metric tensor:} \quad g_{ab} = \frac{\partial x}{\partial q^a}\frac{\partial x}{\partial q^b} + \frac{\partial y}{\partial q^a}\frac{\partial y}{\partial q^b} + \frac{\partial z}{\partial q^a}\frac{\partial z}{\partial q^b}. \tag{9.6}$$

It is a rank-2 tensor, which means that it has two indices. However, it is neither a matrix nor a coordinate transformation, because both indices are together (both below). The most general expression for the line element is

$$\begin{aligned} ds^2 = &g_{11}dx^2 + g_{12}dxdy + g_{13}dxdz + g_{21}dydx + g_{22}dy^2 + g_{23}dydz \\ &g_{31}dzdx + g_{32}dzdy + g_{33}dz^2 \end{aligned} \tag{9.7}$$

although symmetry usually causes many of the terms to be zero. The metric tensor is a fundamental aspect of the metric space it defines. The underlying properties of a space are contained uniquely in the metric tensor. It specifies the geometry of the space, whether it is flat or curved, and defines geodesic curves.

Example 9.1 Cartesian coordinates

This is the simplest case. The coordinate functions are

$$\begin{aligned} x &= q^1 \\ y &= q^2 \\ z &= q^3 \end{aligned} \tag{9.8}$$

with the line element and metric tensor given by

$$ds^2 = dx^2 + dy^2 + dz^2 \quad \text{and} \quad g_{ab} = \begin{pmatrix} 1 & 0 & 0 \\ 0 & 1 & 0 \\ 0 & 0 & 1 \end{pmatrix}. \tag{9.9}$$

Example 9.2 Polar-cylindrical coordinates

Polar-cylindrical coordinates (Fig. 9.2) are just a different way to describe a Euclidean three-dimensional space. The coordinate functions are

$$x = q^1 \cos q^2 = r \cos \theta$$
$$y = q^1 \sin q^2 = r \sin \theta \tag{9.10}$$
$$z = q^3$$

and the line element and metric tensor are

$$ds^2 = dr^2 + r^2 d\theta^2 + dz^2 \quad \text{and} \quad g_{ab} = \begin{pmatrix} 1 & 0 & 0 \\ 0 & r^2 & 0 \\ 0 & 0 & 1 \end{pmatrix}, \tag{9.11}$$

which is diagonal, but now with a position dependence.

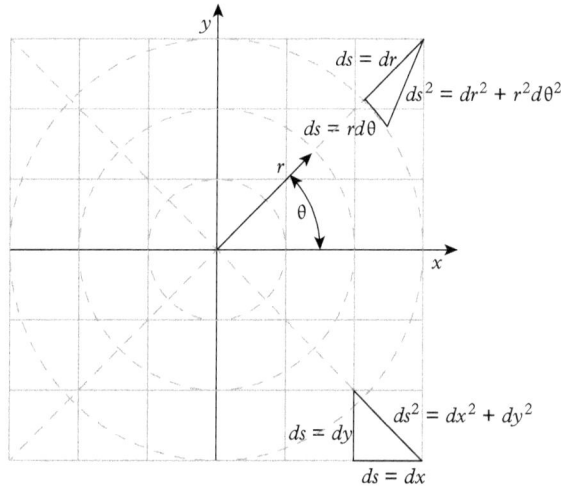

Figure 9.2 *Cartesian and polar coordinates in two dimensions (2D).*

Example 9.3 Spherical coordinates

Spherical coordinates also span three-dimensional (3D) Euclidean space, which is a flat manifold (no curvature). The coordinate functions are

continued

Example 9.3 *continued*

$$x = q^1 \sin q^2 \cos q^3 = r \sin\theta \cos\phi$$
$$y = q^1 \sin q^2 \sin q^3 = r \sin\theta \sin\phi \qquad (9.12)$$
$$z = q^1 \cos q^2 = r \cos\theta$$

and the line element and metric tensor are

$$ds^2 = dr^2 + r^2 d\theta^2 + r^2\sin^2\theta d\phi^2 \quad \text{and} \quad g_{ab} = \begin{pmatrix} 1 & 0 & 0 \\ 0 & r^2 & 0 \\ 0 & 0 & r^2\sin^2\theta \end{pmatrix}. \qquad (9.13)$$

Example 9.4 Spherical surface

It is important to keep in mind that subspaces need not be flat. The surface of a sphere, which has the constraint $r =$ const., is not a flat space. The coordinate functions are

$$x = q^1 \sin q^2 \cos q^3 = r \sin\theta \cos\phi$$
$$y = q^1 \sin q^2 \sin q^3 = r \sin\theta \sin\phi \qquad (9.14)$$
$$z = q^1 \cos q^2 = r \cos\theta$$

and the line element and metric tensor of the spherical surface are

$$ds^2 = r^2 d\theta^2 + r^2\sin^2\theta d\phi^2 \quad \text{and} \quad g_{ab} = r^2 \begin{pmatrix} 1 & 0 \\ 0 & \sin^2\theta \end{pmatrix}. \qquad (9.15)$$

As simple as this difference is from the three-sphere, we will see in the chapter on gravitation that the spherical metric in Eq. (9.13) has no curvature and describes a flat space, while the metric in Eq. (9.15) does indeed have curvature, and hence violates Euclid's parallel line theorem and all the consequences that derive from it.

Example 9.5 2D and 3D Tori

There are two common representations of a torus. One is flat (2D), and the other is curved (embedded in 3D).

Flat torus: This is the same as the 2D plane spanned by two angles that range from zero to 2π. The metric tensor components are

$$g_{11} = 1, \quad g_{12} = 0, \quad g_{22} = 1, \quad g_{ab} = \begin{pmatrix} 1 & 0 \\ 0 & 1 \end{pmatrix}. \qquad (9.16)$$

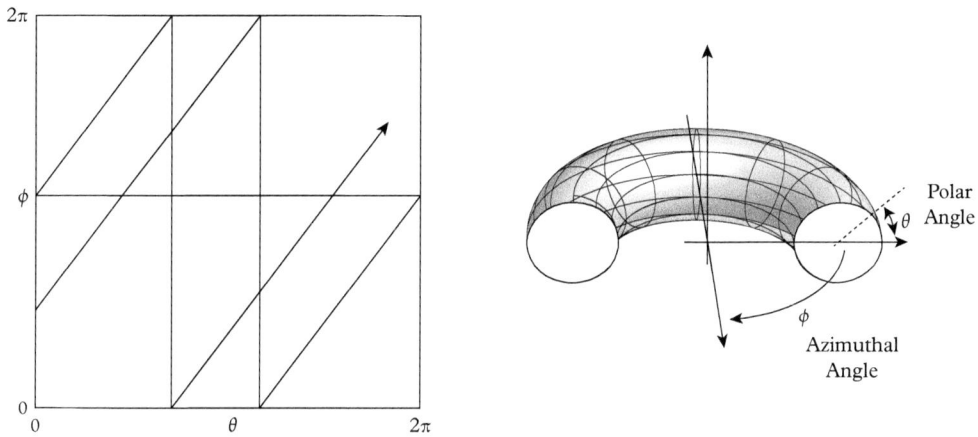

Figure 9.3 *A flat torus is defined by polar angle θ and azimuthal angle φ. A "straight line" is a helix around the torus.*

A straight trajectory on a flat torus is shown in Fig. 9.3 that is defined by the equation

$$\theta_1 = \mathrm{mod}\left(\theta_{1,0} + \omega_1 t, 2\pi\right)$$
$$\theta_2 = \mathrm{mod}\left(\theta_{2,0} + \omega_2 t, 2\pi\right)$$

(9.17)

with a slope = ω_1/ω_2. The trajectory is defined modulo 2π, and "winds" at each boundary. Trajectories on a 2D torus have been seen to play an important role in Hamiltonian chaos in Chapter 3 and in the synchronization of nonlinear systems in Chapter 5.

Curved torus: Embedded in three dimensions

The surface of this torus is defined by the parameterized set of points in 3D:

$$[(R + r\cos\theta)\cos\phi, (R + r\cos\theta)\sin\phi, r\sin\theta],$$

(9.18)

with R and r both constants (Fig. 9.4). The metric tensor is

$$g_{ab} = \begin{pmatrix} r^2 & 0 \\ 0 & (R + r\cos\theta)^2 \end{pmatrix}.$$

(9.19)

Although the flat torus and the curved torus are represented visually in the same way, they have different metrics and different ways to measure distance. The calculation of distance on the flat torus is trivial, but on the embedded torus is an integral.

continued

Example 9.5 *continued*

3-Torus

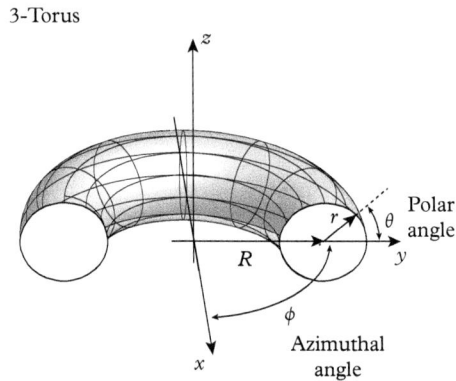

Figure 9.4 *A 3D torus embedded in three-space. The major and minor radii R and r are constants. This surface is non-Euclidean.*

Example 9.6 Affine coordinates

Affine coordinates (Fig. 9.5) are interesting for several reasons. First, the coordinate axes are not orthogonal, causing the metric tensor to have off-diagonal components. Such non-orthogonal coordinate systems are common in skewed lattices in solid state physics. Second, the Lorentz transformation of special relativity is an affine transformation that couples space and time. The affine coordinate functions are

$$
\begin{aligned}
x &= \alpha q^1 + \beta q^2 \\
y &= \gamma q^1 + \delta q^2 \\
z &= q^3
\end{aligned}
\tag{9.20}
$$

and the line element is

$$
ds^2 = \left(\alpha^2 + \gamma^2\right)\left(dq^1\right)^2 + 2\left(\alpha\beta + \gamma\delta\right) dq^1 dq^2 + \left(\beta^2 + \delta^2\right)\left(dq^2\right)^2 + \left(dq^3\right)^2
\tag{9.21}
$$

with the metric tensor

$$
g_{ab} = \begin{pmatrix} \alpha^2 + \gamma^2 & \alpha\beta + \gamma\delta & 0 \\ \alpha\beta + \gamma\delta & \beta^2 + \delta^2 & 0 \\ 0 & 0 & 1 \end{pmatrix}.
\tag{9.22}
$$

This has off-diagonal components as long as $\alpha\beta + \gamma\delta \neq 0$.

Affine coordinates

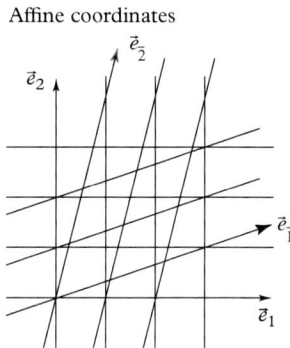

Figure 9.5 *Affine coordinates are not orthogonal. However, every point in space still has a unique decomposition into the basis vectors.*

9.1.2 Basis vectors

Although physics (phenomena) must be independent of the choice of coordinate system, the way we describe physical phenomena does depend on the coordinate frame. Furthermore, switching (transforming) from one coordinate system to another can be useful because the physical description may be simplified in one particular frame or coordinate system, in particular when such a frame is the "rest" frame of an object (or observer).

Any vector $\vec{A}$ can be decomposed into basis vectors, denoted as $\vec{e}_a$. The vector is described as a linear combination of these basis vectors as

$$\begin{aligned} \vec{A} &= A^1\vec{e}_1 + A^2\vec{e}_2 + A^3\vec{e}_3 \\ &= A^a\vec{e}_a, \end{aligned} \tag{9.23}$$

where the implicit Einstein summation over repeated indices is employed on the second line. The basis vectors of a general coordinate system Eq. (9.5) expressed in terms of a Cartesian system are

$$\textit{Basis vector:} \quad \vec{e}_a = \left(\frac{\partial x}{\partial q^a}, \frac{\partial y}{\partial q^a}, \frac{\partial z}{\partial q^a} \right). \tag{9.24}$$

The "basis vectors" are not "unit vectors." They have magnitude and direction.

The metric tensor of a transformed set of coordinates can be expressed in terms of the basis vectors. It is straightforward to see from the definition of the basis vector Eq. (9.24) and the definition of the squared line element

$$ds^2 = (dq^a \vec{e}_a) \cdot (dq^b \vec{e}_b) \tag{9.25}$$

that the metric tensor is the inner product of the basis vectors:

Metric tensor: $g_{ab} = \vec{e}_a \cdot \vec{e}_b,$ (9.26)

which is consistent with the definition of the metric tensor in Eq. (9.6).

Although the vector quantity $\vec{A}$ is an object of reality and cannot depend on the choice of coordinate frame, the components of $\vec{A}$, that are denoted as A^a (with a superscript), *do* depend on the coordinate frame, and these components transform into one another as different coordinate choices are made. As we slip from one coordinate description to another, the A^a change and the $\vec{e}_a$ (denoted with the subscript) change too, but the physical vector $\vec{A}$ itself remains invariant. This immediately tells us that A^a and the basis vectors $\vec{e}_a$ must transform inversely to each other to keep $\vec{A}$ invariant. Consider a vector space spanned by basis vectors $\vec{e}_a$. Any choice of coordinate frame is equivalent

$$\vec{A} = A^a \vec{e}_a = A^{\bar{a}} \vec{e}_{\bar{a}}, \tag{9.27}$$

where the bar refers to a different frame. To see how basis vectors transform, rewrite the right-side term as

$$A^{\bar{a}} \vec{e}_{\bar{a}} = \left(R^{\bar{a}}_b A^b \right) \vec{e}_{\bar{a}}, \tag{9.28}$$

where $R^{\bar{a}}_b$ is the coordinate transformation. Rearranging the implicit sums gives

$$A^b \left(R^{\bar{a}}_b \vec{e}_{\bar{a}} \right) = A^b \vec{e}_b. \tag{9.29}$$

Change the indices (they are called "dummy" indices because they merely refer to the set of dimensions of the space and can be called anything) by changing b to a and $\bar{a}$ to $\bar{b}$

$$A^a R^{\bar{b}}_a \vec{e}_{\bar{b}} = A^a \vec{e}_a$$
$$A^a \left(R^{\bar{b}}_a \vec{e}_{\bar{b}} - \vec{e}_a \right) = 0 \tag{9.30}$$

with the result

$$\vec{e}_a = R^{\bar{b}}_a \vec{e}_{\bar{b}}, \tag{9.31}$$

which inverts to (see Eq. (1.61))

$$\vec{e}_{\bar{b}} = R^a_{\bar{b}} \vec{e}_a, \tag{9.32}$$

where

$$R^{\bar{b}}_a = \left(R^a_{\bar{b}}\right)^{-1} \tag{9.33}$$

provides the simple relation for inverse. Therefore, the components of a basis vector transform inversely to the components of a vector:

Transformation properties:
$$A^{\bar{b}} = R^{\bar{b}}_a A^a$$
$$\vec{e}_{\bar{b}} = R^a_{\bar{b}} \vec{e}_a \tag{9.34}$$

In terms of rotation transformations, the new basis vectors are rotated clockwise from the original basis, while a vector would be seen within this new frame to have been rotated counterclockwise. The vector is invariant to the coordinate transformation because the components of the vector transform oppositely to the basis vectors (see Fig. 9.6). The result that vectors are independent of coordinate transformations goes back to the view that vectors are "real" and cannot depend on the way that they are described in different coordinate systems.

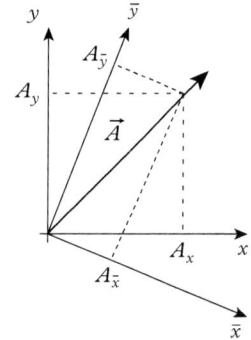

Figure 9.6 *Rotated coordinate frames. The vector $\vec{A}$ is invariant to coordinate transformation.*

Example 9.7 Basis vectors

Cylindrical coordinates:

The transformation functions are

$$q^1 = r = \left(x^2 + y^2\right)^{1/2} \qquad x = q^1 \cos q^2$$
$$q^2 = \theta = \tan^{-1}(y/x) \quad \text{and} \quad y = q^1 \sin q^2 \tag{9.35}$$
$$q^3 = z \qquad z = q^3$$

The Jacobian matrix and Jacobian determinant are

$$\mathcal{J} = \begin{pmatrix} \cos\theta & \sin\theta & 0 \\ -r\sin\theta & r\cos\theta & 0 \\ 0 & 0 & 1 \end{pmatrix} \quad \text{and} \quad |\mathcal{J}| = r. \tag{9.36}$$

The basis vectors of the cylindrical coordinate frame are

$$\vec{e}_r = (\cos\theta, \sin\theta, 0)$$
$$\vec{e}_\theta = (-r\sin\theta, r\cos\theta, 0) \tag{9.37}$$
$$\vec{e}_r = (0, 0, 1)$$

The basis vectors are not all unit vectors, nor does $\vec{e}_\theta$ have the same units as $\vec{e}_r$.

continued

Example 9.7 *continued*

Spherical:

The transformation functions are

$$q^1 = r = \left(x^2 + y^2 + z^2\right)^{1/2} \qquad x = q^1 \sin q^2 \cos q^3$$
$$q^2 = \theta = \cos^{-1}\left(z/(x^2 + y^2 + z^2)^{1/2}\right) \quad y = q^1 \sin q^2 \sin q^3 \qquad (9.38)$$
$$q^3 = \phi = \tan^{-1}(y/x) \qquad z = q^1 \cos q^2$$

The Jacobian matrix and the Jacobian are

$$\mathcal{J} = \begin{pmatrix} \sin\theta \cos\phi & \sin\theta \sin\phi & \cos\theta \\ r\cos\theta \cos\phi & r\cos\theta \sin\phi & -r\sin\theta \\ -r\sin\theta \sin\phi & r\sin\theta \cos\phi & 0 \end{pmatrix} \text{ and } |\mathcal{J}| = r^2 \sin\theta, \qquad (9.39)$$

and the basis vectors are

$$\vec{e}_r = (\sin\theta \cos\phi, \sin\theta \sin\phi, \cos\theta)$$
$$\vec{e}_\theta = (r\cos\theta \cos\phi, r\cos\theta \sin\phi, -r\sin\theta) \qquad (9.40)$$
$$\vec{e}_\phi = (-r\sin\theta \sin\phi, r\sin\theta \cos\phi, 0)$$

9.1.3 Covectors

In the language of vectors as rank-one tensors, the entity that operates on a vector to yield a real number is known as a dual-vector. These dual-vectors also have other names, such as covariant vectors or covectors, and also as one-forms. The name *covector* comes from the fact that components of covectors transform like the basis vectors—they co-transform. In contrast, the components of vectors transform contrary to the basis vectors, and hence vectors are sometimes called contravariant vectors. This book will usually use the names *vectors* and *covectors*.

Covectors have components described with a subscript, such as B_a. A mnemonic to remember this is "co goes below." Because covectors map vectors onto real numbers, they are also described as functions of vectors and are sometimes denoted with a "tilde" $\tilde{B}$ to distinguish them from vectors $\vec{A}$, so that

$$\tilde{B}(\vec{A}) = \tilde{B}(A^a \vec{e}_a)$$
$$= A^a \tilde{B}(\vec{e}_a) \qquad (9.41)$$
$$= A^a B_a,$$

where the components of a covector transform as

$$B_{\tilde{b}} = \tilde{B}(\vec{e}_{\tilde{b}}) = \tilde{B}(\Lambda^a_{\tilde{b}} \vec{e}_a)$$
$$= \Lambda^a_{\tilde{b}} \tilde{B}(\vec{e}_a) \qquad (9.42)$$
$$= \Lambda^a_{\tilde{b}} B_a.$$

The components of covectors transform inversely to the components of a vector

$$A^{\bar{b}} = \Lambda^{\bar{b}}_a A^a$$
$$B_{\bar{b}} = \Lambda^a_{\bar{b}} B_a.$$

(9.43)

While vectors are visualized as arrows with direction and length, covectors are often visualized as parallel planes with direction and spacing (Fig. 9.7). This visualization is particularly relevant for plane waves $e^{ik_a x^a}$ that have plane wave fronts and associated k-vectors.

A covector is obtained from a vector through the metric tensor that lowers the index

$$x_a = g_{ab} x^b,$$

(9.44)

where the Einstein summation is implied.[4] An inner product is then expressed as

$$x_a x^a = g_{ac} x^c x^a,$$

(9.45)

which is invariant under coordinate transformation.

The metric tensor has an inverse given by

$$(g_{ab})^{-1} = g^{ab},$$

(9.46)

which transforms a covector back to a vector

$$x^b = g^{ab} x_a.$$

(9.47)

The metric tensor is also used to find the inverse of a transformation. Starting from

$$x^{\bar{a}} = \Lambda^{\bar{a}}_b x^b$$

(9.48)

and lowering the index with the metric tensor gives

$$x_{\bar{b}} = g_{\bar{a}\bar{b}} x^{\bar{a}} = g_{\bar{a}\bar{b}} \Lambda^{\bar{a}}_b x^b$$
$$x_{\bar{b}} = \left(g_{\bar{a}\bar{b}} \Lambda^{\bar{a}}_b g^{cb}\right) g_{cb} x^b$$
$$= \Lambda^c_{\bar{b}} x_c.$$

(9.49)

Therefore the transformation for a covector is

$$x_{\bar{b}} = \Lambda^c_{\bar{b}} x_c,$$

(9.50)

where the inverse transformation is obtained through the metric tensor as

$$\Lambda^c_{\bar{b}} = g_{\bar{a}\bar{b}} \Lambda^{\bar{a}}_b g^{cb} = \left(\Lambda^{\bar{b}}_c\right)^{-1}.$$

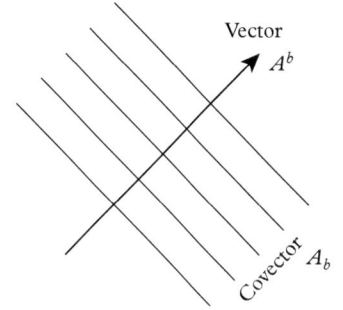

(9.51)

Figure 9.7 *Vectors are represented as arrows with direction and magnitude. Covectors are represented as parallel planes with direction and spacing.*

[4] The metric tensor is symmetric, so that $g_{ab} = g_{ba}$. This symmetry is used implicitly in many of the following derivations.

Example 9.8 Minkowski metric

Consider the invariant interval

$$s^2 = -c^2 t^2 + x^2 + y^2 + z^2, \tag{9.52}$$

which can be expressed as an inner product between a position four-vector and its corresponding four-covector

$$
\begin{aligned}
s^2 = {} & g_{00}\left(x^0\right)^2 + g_{11}\left(x^1\right)^2 + g_{22}\left(x^2\right)^2 + g_{33}\left(x^3\right)^2 \\
& + g_{01}x^0 x^1 + g_{02}x^0 x^2 + \cdots \\
& + g_{10}x^1 x^0 + g_{12}x^1 x^2 + \cdots
\end{aligned}
\tag{9.53}
$$

Comparing these expressions, the metric tensor for the space–time four-vectors is

$$g_{ab} = \begin{pmatrix} -1 & 0 & 0 & 0 \\ 0 & 1 & 0 & 0 \\ 0 & 0 & 1 & 0 \\ 0 & 0 & 0 & 1 \end{pmatrix}, \tag{9.54}$$

which is diagonal in the coordinate representation of Minkowski space. Because the eigenvalues are not all positive, this metric is called "pseudo-Riemannian." A metric with only positive eigenvalues is "Riemannian." In pseudo-Riemannian spaces, squared distances, such as Δs^2, can be negative (without invoking products of imaginary numbers).

The dual position four-covector is obtained as

$$x_a = g_{ab}x^b = \left[\begin{pmatrix} -1 & 0 & 0 & 0 \\ 0 & 1 & 0 & 0 \\ 0 & 0 & 1 & 0 \\ 0 & 0 & 0 & 1 \end{pmatrix} \begin{pmatrix} ct \\ x \\ y \\ z \end{pmatrix} \right]^T = \begin{pmatrix} -ct & x & y & z \end{pmatrix}, \tag{9.55}$$

which is represented as a row vector. The inner product is then

$$s^2 = x_a x^a = \begin{pmatrix} -ct & x & y & z \end{pmatrix} \begin{pmatrix} ct \\ x \\ y \\ z \end{pmatrix} = -c^2 t^2 + x^2 + y^2 + z^2. \tag{9.56}$$

Note that the use of contravariant and covariant indices for vectors is required for general spaces, but that Cartesian coordinates are a special case where $x^a = x_a$ because

$$g_{ab} = \begin{pmatrix} 1 & 0 & 0 \\ 0 & 1 & 0 \\ 0 & 0 & 1 \end{pmatrix}. \tag{9.57}$$

9.2 Reciprocal spaces in physics

Reciprocal spaces abound in math and physics with many well-known examples. Fourier space is perhaps the best known example. Any function of spatial variables can be transformed into its Fourier transform. The Fourier transform has spatial frequency, or k-vector, or wavenumber, as its coordinates. These coordinates span a space called Fourier space. Fourier space is a dual-space of the space of spatial functions and is expressed in terms of spatial frequencies k_a. The spatial frequency that is dual to a position vector $\vec{r}$ is given by the relation

$$\tilde{k} \cdot \vec{r} = k_a x^a = const. \tag{9.58}$$

Legendre transforms in Chapter 2 were dual transforms that would take a scalar function of position and velocity and transform it to a scalar function of position and momentum and then back again

$$\mathcal{L}L(x, \dot{x}) = H(x, p)$$
$$\mathcal{L}H(x, p) = L(x, \dot{x}). \tag{9.59}$$

Therefore, state space and phase space are a type of reciprocal spaces.

As another example, in quantum mechanics wave functions come in two types, as functions and as Hermitian conjugates. The inner product of a wave function with its Hermitian conjugate produces a real number that is related to a probability. In the "bra–ket" notation, eigenvectors look like

$$\psi = |\psi\rangle = \psi^a, \qquad \phi^\dagger = \langle\phi| = \phi_a, \qquad P = \langle\phi \mid \psi\rangle = \phi_a \psi^a.$$

In matrix notation, the eigenvectors are column matrices and the Hermitian conjugates are row vectors. The Hilbert space of Hermitian conjugates is the reciprocal space to the eigenvectors.

An important example arises in solid state physics with crystal lattices. Crystal lattices have basis vectors. Diffraction from these lattices is described as a new lattice with new basis vectors in reciprocal space, which is the dual space to the original spatial lattice (see Fig. 9.8). The important entities known as Brillouin zones, which control many of the electronic properties of crystals, live entirely in reciprocal space. If the spatial basis vectors are $\vec{a}_1, \vec{a}_2, \vec{a}_3$ (known as primitive vectors) then the reciprocal basis vectors are

$$\vec{b}_1 = 2\pi \frac{\vec{a}_2 \times \vec{a}_3}{\vec{a}_1 \cdot (\vec{a}_2 \times \vec{a}_3)}$$
$$\vec{b}_2 = 2\pi \frac{\vec{a}_3 \times \vec{a}_1}{\vec{a}_1 \cdot (\vec{a}_2 \times \vec{a}_3)} \tag{9.60}$$
$$\vec{b}_3 = 2\pi \frac{\vec{a}_1 \times \vec{a}_2}{\vec{a}_1 \cdot (\vec{a}_2 \times \vec{a}_3)}.$$

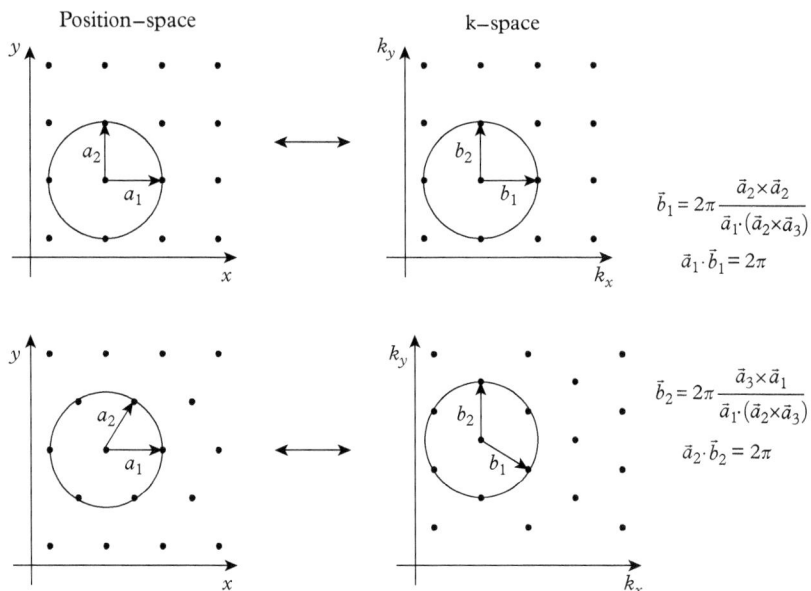

Figure 9.8 *Two-dimensional lattices and their reciprocal space. A square lattice (top) transforms to a square lattice, and a hexagonal lattice (bottom) transforms to a different hexagonal reciprocal lattice.*

The primitive vectors and their reciprocals have the important property

$$e^{i\vec{K}\cdot\vec{R}} = 1, \tag{9.61}$$

where

$$\vec{R} = n_1\vec{a}_1 + n_2\vec{a}_2 + n_3\vec{a}_3$$
$$\vec{K} = m_1\vec{b}_1 + m_2\vec{b}_2 + m_3\vec{b}_3 \tag{9.62}$$

are linear combinations with integer coefficients of the primitive vectors.

The feature common to all of these examples is the definition of a vector space, and the identification of an operation that converts vectors into real numbers (inner product, exponentiation of an inner product, etc.) using an entity (reciprocal vector, Hermitian conjugate, etc.) that forms its own vector space. This inverse vector space is the dual space of the original vector space.

9.3 Derivative of a tensor

Differential geometry is all about derivatives, and so derivatives of tensors play a central role in the description of metric spaces.

9.3.1   Derivative of a vector

To define the general derivative of a vector, it is instructive to begin with an explicit example in polar coordinates.

$$
\begin{aligned}
x &= r\cos\theta & r &= \sqrt{x^2 + y^2} \\
y &= r\sin\theta & \theta &= \operatorname{atan}\left(\frac{y}{x}\right)
\end{aligned}
\tag{9.63}
$$

The polar coordinate basis vectors may be defined in terms of the basis vectors in Cartesian coordinates as

$$
\begin{aligned}
\vec{e}_r &= \frac{\partial x}{\partial r}\vec{e}_x + \frac{\partial y}{\partial r}\vec{e}_y = \cos\theta\,\vec{e}_x + \sin\theta\,\vec{e}_y \\
\vec{e}_\theta &= \frac{\partial x}{\partial\theta}\vec{e}_x + \frac{\partial y}{\partial\theta}\vec{e}_y = -r\sin\theta\,\vec{e}_x + r\cos\theta\,\vec{e}_y,
\end{aligned}
\tag{9.64}
$$

where these are *not* unit vectors

$$
\vec{e}_r \cdot \vec{e}_r = 1, \qquad \vec{e}_\theta \cdot \vec{e}_\theta = r^2.
\tag{9.65}
$$

Now define a general vector in polar coordinates

$$
\vec{V} = \left(V^r, V^\theta\right) = V^r\vec{e}_r + V^\theta\vec{e}_\theta
\tag{9.66}
$$

and take its derivative

$$
\begin{aligned}
\frac{\partial\vec{V}}{\partial r} &= \frac{\partial}{\partial r}\left(V^r\vec{e}_r + V^\theta\vec{e}_\theta\right) \\
&= \frac{\partial V^r}{\partial r}\vec{e}_r + V^r\frac{\partial\vec{e}_r}{\partial r} + \frac{\partial V^\theta}{\partial r}\vec{e}_\theta + V^\theta\frac{\partial\vec{e}_\theta}{\partial r}
\end{aligned}
\tag{9.67}
$$

with a similar expression for θ,

$$
\begin{aligned}
\frac{\partial\vec{V}}{\partial\theta} &= \frac{\partial}{\partial\theta}\left(V^r\vec{e}_r + V^\theta\vec{e}_\theta\right) \\
&= \frac{\partial V^r}{\partial\theta}\vec{e}_r + V^r\frac{\partial\vec{e}_r}{\partial\theta} + \frac{\partial V^\theta}{\partial\theta}\vec{e}_\theta + V^\theta\frac{\partial\vec{e}_\theta}{\partial\theta}.
\end{aligned}
\tag{9.68}
$$

In index notation

$$
\frac{\partial\vec{V}}{\partial x^\beta} = \frac{\partial V^\alpha}{\partial x^\beta}\vec{e}_\alpha + V^\alpha\frac{\partial\vec{e}_\alpha}{\partial x^\beta}.
\tag{9.69}
$$

From this expression, it is clear that the derivative of a vector in curvilinear coordinates is not simply the derivative of its components. This is because in general coordinates, the changes in the basis vectors must be considered as well.

9.3.2   Christoffel symbols

The second term in the last equation is so important it is expressed with a new symbol

$$\text{Christoffel symbol of the second kind:} \quad \frac{\partial \vec{e}_\alpha}{\partial x^\beta} = \Gamma^\mu_{\alpha\beta}\vec{e}_\mu. \tag{9.70}$$

The symbol $\Gamma^\mu_{\alpha\beta}$ is interpreted as the μth component of the derivative of the αth basis vector with respect to the βth variable. It is called a Christoffel symbol of the second kind.

Using the Christoffel symbol the covariant derivative is defined as

$$\frac{\partial \vec{V}}{\partial x^\beta} = \frac{\partial V^\alpha}{\partial x^\beta}\vec{e}_\alpha + V^\alpha \Gamma^\mu_{\alpha\beta}\vec{e}_\mu$$

$$= \frac{\partial V^\alpha}{\partial x^\beta}\vec{e}_\alpha + V^\mu \Gamma^\alpha_{\mu\beta}\vec{e}_\alpha \tag{9.71}$$

$$= \left(\frac{\partial V^\alpha}{\partial x^\beta} + V^\mu \Gamma^\alpha_{\mu\beta}\right)\vec{e}_\alpha.$$

The derivative form in the parentheses is given a symbol ∇_β:

$$\text{Covariant derivative:} \quad \begin{aligned} \nabla_\beta V^\alpha &= \frac{\partial V^\alpha}{\partial x^\beta} + V^\mu \Gamma^\alpha_{\mu\beta} \\ \nabla_\beta V_\alpha &= \frac{\partial V_\alpha}{\partial x^\beta} - V_\mu \Gamma^\mu_{\alpha\beta}, \end{aligned} \tag{9.72}$$

where the derivative of a covariant vector is also included. The first term arises from the variation of V^α. The second term arises from the variation of the basis vector.

Example 9.9 Polar coordinates

Four derivatives are sufficient to define eight Christoffel symbols for polar coordinates

$$\frac{\partial \vec{e}_r}{\partial \theta} = \Gamma^r_{r\theta}\vec{e}_r + \Gamma^\theta_{r\theta}\vec{e}_\theta \qquad \frac{\partial \vec{e}_r}{\partial r} = \Gamma^r_{rr}\vec{e}_r + \Gamma^\theta_{rr}\vec{e}_\theta$$

$$\frac{\partial \vec{e}_\theta}{\partial \theta} = \Gamma^r_{\theta\theta}\vec{e}_r + \Gamma^\theta_{\theta\theta}\vec{e}_\theta \qquad \frac{\partial \vec{e}_\theta}{\partial r} = \Gamma^r_{\theta r}\vec{e}_r + \Gamma^\theta_{\theta r}\vec{e}_\theta \tag{9.73}$$

These are compared to the derivatives of Eq. (9.64),

$$\frac{\partial \vec{e}_r}{\partial \theta} = -\sin\theta\vec{e}_x + \cos\theta\vec{e}_y = \frac{1}{r}\vec{e}_\theta \qquad \frac{\partial \vec{e}_r}{\partial r} = 0$$

$$\frac{\partial \vec{e}_\theta}{\partial \theta} = -r\cos\theta\vec{e}_x - r\sin\theta\vec{e}_y = -r\vec{e}_r \qquad \frac{\partial \vec{e}_\theta}{\partial r} = -\sin\theta\vec{e}_x + \cos\theta\vec{e}_y = \frac{1}{r}\vec{e}_\theta \tag{9.74}$$

which define eight Christoffel symbols

$$\begin{aligned}
\Gamma^r_{rr} &= 0 & \Gamma^\theta_{rr} &= 0 \\
\Gamma^r_{r\theta} &= 0 & \Gamma^\theta_{r\theta} &= 1/r \\
\Gamma^r_{\theta r} &= 0 & \Gamma^\theta_{\theta r} &= 1/r \\
\Gamma^r_{\theta\theta} &= -r & \Gamma^\theta_{\theta\theta} &= 0
\end{aligned} \tag{9.75}$$

These derivatives are visualized in Fig. 9.9. For instance, in (a) the radius basis vector tilts as varies, while in (b) the tangential basis vector tilts as varies. In (c) the radius vector simply translates with increasing radius and hence $d\vec{r} = 0$. In (d) the tangent basis vector increases in magnitude with increasing radius, but maintains its orientation.

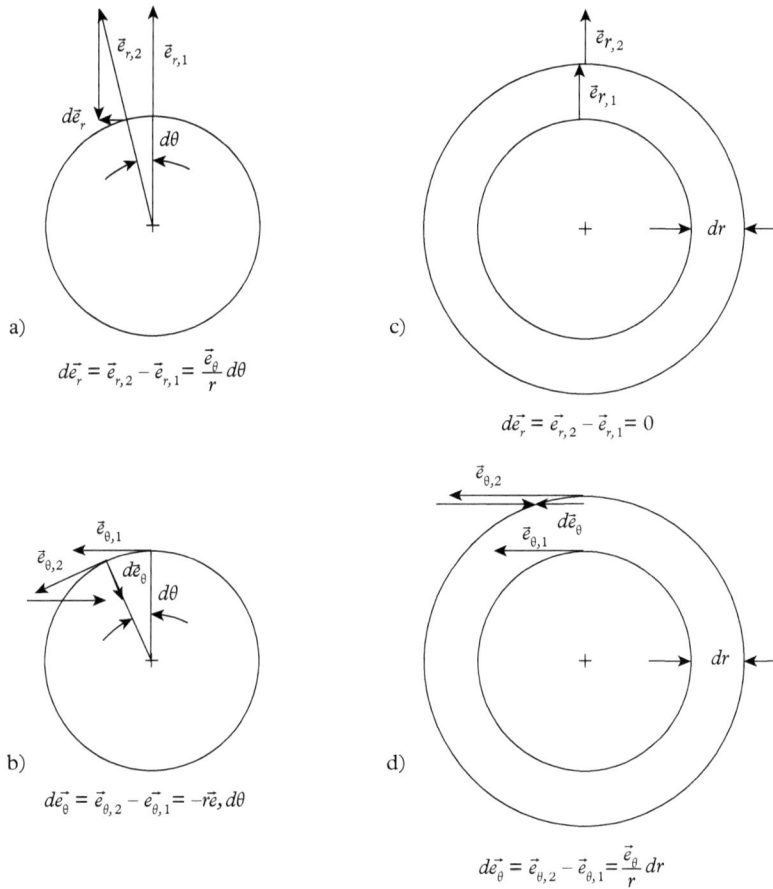

Figure 9.9 *Differentials of basis vectors in polar coordinates.*

Example 9.10 Contraction of the covariant derivative in (r, θ)

The two indices of the covariant derivative of Eq. (9.72) can be contracted

$$\nabla_\alpha V^\alpha = \frac{\partial V^\alpha}{\partial x^\alpha} + V^\mu \Gamma^\alpha_{\mu\alpha},$$
(9.76)

where

$$\Gamma^\alpha_{r\alpha} = \Gamma^r_{rr} + \Gamma^\theta_{r\theta} = \frac{1}{r}$$
$$\Gamma^\alpha_{\theta\alpha} = \Gamma^r_{\theta r} + \Gamma^\theta_{\theta\theta} = 0.$$
(9.77)

Therefore

$$\nabla_\alpha V^\alpha = \frac{\partial V^r}{\partial r} + \frac{\partial V^\theta}{\partial \theta} + \frac{1}{r} V^r,$$
(9.78)

which simplifies to

$$\nabla_\alpha V^\alpha = \frac{1}{r}\frac{\partial (rV^r)}{\partial r} + \frac{\partial V^\theta}{\partial \theta} = \vec{\nabla} \cdot \vec{V}.$$
(9.79)

This is the expression for divergence in polar coordinates.

9.3.3 Derivative notations

Textbooks on differential geometry use many differing forms for the derivative operation, which can be confusing. To help make the connection between the notation in this book and others, here are some examples.

Partial derivative of a vector field:

$$\frac{\partial V^a}{\partial x^b} = \partial_b V^a$$
$$= V^a_{,b}$$
(9.80)

Covariant derivative of a vector field:

$$\nabla_c V^a = V^a_{;c}$$
$$= \partial_c V^a + \Gamma^a_{bc} V^b$$
(9.81)

Covariant derivative of a covector field:

$$\nabla_c V_a = V_{a;c}$$
$$= \partial_c V_a - \Gamma^b_{ac} V_b$$
(9.82)

Derivative of a scalar field:

$$\nabla_a \phi = \partial_a \phi$$
(9.83)

Covariant derivatives of a rank-2 tensor:

$$\nabla_\beta V_{\mu\upsilon} = \partial_\beta V_{\mu\upsilon} - \Gamma^\alpha_{\mu\beta} V_{\alpha\upsilon} - \Gamma^\alpha_{\upsilon\beta} V_{\mu\alpha} \tag{9.84}$$

$$\nabla_\beta V^{\mu\upsilon} = \partial_\beta V^{\mu\upsilon} + \Gamma^\mu_{\alpha\beta} V^{\alpha\upsilon} + \Gamma^\upsilon_{\alpha\beta} V^{\mu\alpha} \tag{9.85}$$

Total derivative of a vector field

$$dV^a = (\partial_b V^a)\, dx^b \tag{9.86}$$

Total covariant derivative of a vector field:

$$\begin{aligned} DV^a &= (\nabla_b V^a)\, dx^b \\ &= dV^a + \Gamma^a_{cb} V^c dx^b \end{aligned} \tag{9.87}$$

Note the range of notations, some using commas, other using semi-colons, plus the differences between partial derivatives and covariant derivatives and to-tal derivatives. The distinctions among these different types of derivatives are important.

9.3.4 The connection of Christoffel symbols to the metric

The differential properties of a manifold are defined by its metric tensor. Therefore, the Christoffel symbols must be derivable from the metric tensor. The covariant derivative of the metric tensor is

$$\nabla_\mu g_{\alpha\beta} = \partial_\mu g_{\alpha\beta} - \Gamma^\upsilon_{\alpha\mu} g_{\upsilon\beta} - \Gamma^\upsilon_{\beta\mu} g_{\alpha\upsilon} = 0. \tag{9.88}$$

This identity is zero in all frames (an invariant). Rewriting with three index permutations

$$\begin{aligned} \partial_\mu g_{\alpha\beta} &= \Gamma^\upsilon_{\alpha\mu} g_{\upsilon\beta} + \Gamma^\upsilon_{\beta\mu} g_{\alpha\upsilon} \\ \partial_\beta g_{\alpha\mu} &= \Gamma^\upsilon_{\alpha\beta} g_{\upsilon\mu} + \Gamma^\upsilon_{\mu\beta} g_{\alpha\upsilon} \\ -\partial_\alpha g_{\beta\mu} &= -\Gamma^\upsilon_{\beta\alpha} g_{\upsilon\mu} - \Gamma^\upsilon_{\mu\alpha} g_{\beta\upsilon} \end{aligned} \tag{9.89}$$

adding, and using the fact that the Christoffel symbol is symmetric in the lower indices, gives

$$\partial_\mu g_{\alpha\beta} + \partial_\beta g_{\alpha\mu} - \partial_\alpha g_{\beta\mu} = 2 g_{\alpha\upsilon} \Gamma^\upsilon_{\beta\mu}. \tag{9.90}$$

This is inverted to give the Christoffel symbol of the second kind

$$\text{Christoffel symbol:} \qquad \Gamma^{\upsilon}_{\beta\mu} = \frac{1}{2} g^{\alpha\upsilon} \left(\partial_{\mu} g_{\alpha\beta} + \partial_{\beta} g_{\alpha\mu} - \partial_{\alpha} g_{\beta\mu} \right). \tag{9.91}$$

The $\Gamma^{\mu}_{\alpha\beta}$ are uniquely defined by the choice of metric tensor. In the special case of a *diagonal metric tensor*, the following simplifying expressions are helpful

$$\Gamma^{\alpha}_{\alpha\beta} = \Gamma^{\alpha}_{\beta\alpha} = \partial_{\beta} \left(\frac{1}{2} \ln |g_{\alpha\alpha}| \right),$$

$$\Gamma^{\alpha}_{\beta\beta} = -\frac{1}{2} \frac{1}{g_{\alpha\alpha}} \partial_{\alpha} g_{\beta\beta} \qquad \text{for } \alpha \neq \beta, \tag{9.92}$$

where all other $\Gamma^{\alpha}_{\beta\gamma}$ vanish. These last expressions can be used for orthogonal coordinate frames commonly encountered in general relativity.

9.4 Geodesic curves in configuration space

An element of key importance in the geometry of manifolds is the geodesic curve. A geodesic is the shortest arc between two points, and it is also the straightest. The geodesic equation can be derived in several different, but related ways. The shortest (or extremum) length of the geodesic curve is derived from variational principles (Euler equations), while the straightest path character is derived from the idea of parallel transport, in which a tangent vector is transported parallel to itself along the geodesic. A third approach to the geodesic takes the view of force-free motion of a test particle moving freely in a curved manifold. This third view ties in closely with other topics of this book, in which motion in a gravitational field is force-free motion in a warped space–time.

9.4.1 Variational approach to the geodesic curve

A geodesic is a curve of extremum length, either a maximum or a minimum, and hence it is derived directly from a variational approach and the Euler equations. The length of a curve in a manifold is defined by

$$s = \int ds = \int \sqrt{\left(\frac{ds}{d\lambda} \right)^2} \, d\lambda = \int L d\lambda, \tag{9.93}$$

where L is an effective Lagrangian and ds is the line element along a trajectory $x^a(\lambda)$ that is a function of the continuous-valued parameter λ. There are several options for what this parameter λ can be. For instance, it might be taken

as the time, or it may be taken proportional to the arc length itself.[5] The effective
"Lagrangian" of the variational problem is

$$L(x^a, \dot{x}^a) = \sqrt{g_{ab}\dot{x}^a\dot{x}^b} = \sqrt{w}, \tag{9.94}$$

where the "dot" denotes a derivative with respect to λ, and the metric tensor is a
function of the coordinates $g_{ab}(x^a)$. The parameter w is

$$w = g_{ab}\dot{x}^a\dot{x}^b = g_{ab}\frac{dx^a}{d\lambda}\frac{dx^b}{d\lambda} = \left(\frac{ds}{d\lambda}\right)^2. \tag{9.95}$$

With the Lagrangian defined in Eq. (9.94), the Euler equations are immediately
applied as

$$\frac{d}{d\lambda}\frac{\partial L}{\partial \dot{x}^c} - \frac{\partial L}{\partial x^c} = 0. \tag{9.96}$$

The components of the Euler equations are

$$\frac{\partial L}{\partial \dot{x}^c} = \frac{1}{L}g_{cb}\dot{x}^b \tag{9.97}$$

and

$$\frac{\partial L}{\partial x^c} = \frac{1}{2L}\dot{x}^a\dot{x}^b\frac{\partial g_{ab}}{\partial x^c} = \frac{1}{2L}\frac{\partial g_{de}}{\partial x^c}\dot{x}^d\dot{x}^e. \tag{9.98}$$

Inserting these into the Euler equations, Eq. (9.96) gives

$$\frac{d}{d\lambda}g_{cb}\dot{x}^b - \frac{1}{2}\frac{\partial g_{de}}{\partial x^c}\dot{x}^d\dot{x}^e = \frac{1}{2w}\frac{dw}{d\lambda}\frac{\partial w}{\partial \dot{x}^c}. \tag{9.99}$$

The dependence of g_{ab} on λ is through $x^a(\lambda)$, so the first term in the equation is
expanded to give

$$g_{cb}\frac{d^2x^b}{d\lambda^2} + \frac{\partial g_{cb}}{\partial x^d}\frac{dx^d}{d\lambda}\frac{dx^b}{d\lambda} - \frac{1}{2}\frac{\partial g_{de}}{\partial x^c}\frac{dx^d}{d\lambda}\frac{dx^e}{d\lambda} = \frac{1}{2w}\frac{dw}{d\lambda}\frac{\partial w}{\partial \dot{x}^c}. \tag{9.100}$$

The term on the right-hand side of Eq. (9.100) retains the dependence of w on
λ. When it is chosen to be the arc length $\lambda = s$, then

$$w = \left(\frac{ds}{d\lambda}\right)^2 = 1 \tag{9.101}$$

and the right-hand side of Eq. (9.100) is equal to zero. Alternatively, in the case of
a moving test particle under force-free motion, it is common to take the parameter
λ as the time t for which

$$w = \left(\frac{ds}{dt}\right)^2 = v^2 = const. \tag{9.102}$$

[5] In the case of space–time, a photon
trajectory is a null geodesic for which
$ds = 0$, and a different parameterization is
required. See Chapters 10 and 11.

with a constant speed, and again the right-hand side of Eq. (9.100) is equal to zero.[6] Equation (9.100) can then be rewritten as

$$\frac{d^2x^b}{d\lambda^2} + g^{cb}\left[\frac{\partial g_{cb}}{\partial x^d}\dot{x}^d\dot{x}^b - \frac{1}{2}\frac{\partial g_{de}}{\partial x^c}\dot{x}^d\dot{x}^e\right] = 0. \tag{9.103}$$

The second term of Eq. (9.103) can be rewritten using the symmetric relation for the derivative of the metric tensor

$$\frac{\partial g_{cb}}{\partial x^d} = \frac{1}{2}\left(\frac{\partial g_{cb}}{\partial x^d} + \frac{\partial g_{cd}}{\partial x^b}\right) \tag{9.104}$$

to yield

$$\frac{d^2x^b}{d\lambda^2} + g^{cb}\frac{1}{2}\left[\frac{\partial g_{cb}}{\partial x^d} + \frac{\partial g_{cd}}{\partial x^b} - \frac{\partial g_{db}}{\partial x^c}\right]\dot{x}^d\dot{x}^b = 0 \tag{9.105}$$

and the geodesic equation, expressed in terms of arc length s, is given by

$$\text{Geodesic equation:} \qquad \frac{d^2x^a}{ds^2} + \Gamma^a_{bc}\frac{dx^b}{ds}\frac{dx^c}{ds} = 0, \tag{9.106}$$

where Γ^a_{bc} is the Christoffel symbol of the second kind. The geodesic equation is a second-order ordinary differential equation (ODE) for $x^a(s)$ that can be solved either analytically or numerically if the Christoffel symbols are known (or equivalently if the metric tensor is known). It is a direct consequence of minimizing the length of a curve in a way that satisfies the Euler equations. The geodesic equation carries the constraint (a first integral)

$$g_{ab}\frac{dx^a}{ds}\frac{dx^b}{ds} = 1 \tag{9.107}$$

that guarantees a constant speed along the geodesic curve.

The geodesic equation, Eq. (9.106), is an ordinary second-order differential equation that can be converted to a set of coupled first-order equations that define a geodesic flow. The geodesic flow equations are

$$\text{Geodesic flow:} \qquad \begin{aligned} \frac{dv^a}{ds} &= -\Gamma^a_{bc}v^bv^c \\ \frac{dx^a}{ds} &= v^a, \end{aligned} \tag{9.108}$$

which generally are nonlinear equations that are solved using the techniques discussed in Chapter 3 for nonlinear flows. All that is needed is to define the Christoffel symbols that are connected to the metric of the underlying space through Eq. (9.91).

[6] On the other hand, if the speed is not constant, then see Section 9.5.

9.4.2   Parallel transport

A different approach for deriving the geodesic equation uses the nature of tangents on curved manifolds. This is the concept of parallel transport which shows that a geodesic is not only the shortest distance between two points, it is also the straightest. To start, consider transporting a vector $\vec{V}$ along a curve $\vec{x}(s)$ in a flat metric. The equation describing this process is

$$\frac{dV^a}{ds} = \frac{dx^b}{ds}\frac{\partial}{\partial x^b}V^a$$

$$= U^b \partial_b V^a, \tag{9.109}$$

where the tangent vector is

$$U^b = \frac{dx^b}{ds}. \tag{9.110}$$

Because the Γ^a_{bc} are zero in a flat space, the covariant derivative and the partial derivative are equal, giving

$$U^b \partial_b V^a = U^b \nabla_b V^a = \left(\vec{U} \cdot \vec{\nabla}\right) \vec{V}. \tag{9.111}$$

If the vector is transported parallel to itself, then there is no change in V along the curve, so that

$$\frac{dV^a}{ds} = 0. \tag{9.112}$$

Now in curved space, the basic question is: What curve transports its tangent vector onto itself? The expression for this is still Eq. (9.111), but now

$$U^b \nabla_b U^a = U^b \partial_b U^a + \Gamma^a_{bc} U^b U^c = 0, \tag{9.113}$$

where, as before, the tangent vector is $U^a = \frac{dx^a}{ds}$, which leads to

$$U^\beta \frac{\partial}{\partial x^\beta} = \frac{d}{ds}. \tag{9.114}$$

Substituting these into Eq. (9.113) gives

$$\frac{d}{ds}\left(\frac{dx^a}{ds}\right) + \Gamma^a_{bc}\frac{dx^b}{ds}\frac{dx^c}{ds} = 0, \tag{9.115}$$

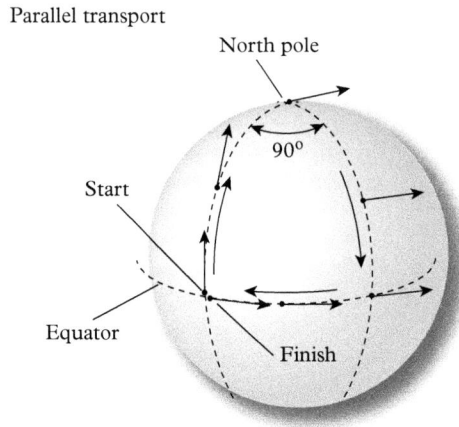

Figure 9.10 *Parallel transport on the surface of the Earth.*

which is the same geodesic equation as Eq. (9.106). It is an ODE that defines a curve that carries its own tangent vector onto itself. The curve is parameterized by a parameter s that can be identified with path length, but can be other parameters related linearly to s, such as the linear relation $s = vt + s_0$ that would make the geodesic equation a function of time (but only if the speed v is constant).

An example of parallel transport is shown in Fig. 9.10 for a vector transported across great circle routes (geodesics) on the surface of the Earth. A vector on the Equator pointing north is transported parallel to itself along a line of longitude to the North Pole. At the North Pole, the vector is transported parallel to itself along a southward longitude (in this example at 90° from the original longitude) to the Equator. It is then transported parallel to itself to the original starting point. Despite the fact that the vector was never "turned" during its journey, it has been rotated by 90° from its original orientation along this path. It can be shown that the finite rotation of the vector that occurs during parallel transport is equal to the solid angle enclosed by the path. In this case the solid angle is 1/8 of 4π sterradians, which equals $\pi/2$, and hence the 90° rotation.

The derivation of the geodesic equation using parallel transport did not rely on minimization nor on the Euler equations, yet leads to the same result. Therefore, parallel transport (straightest path) and the variational approach (shortest path) are consistent and equivalent.

9.4.3   Examples of geodesic curves

There are many examples of geodesic curves in various types of flat and curved manifolds. Here are some of the most common examples.

Example 9.11 Cartesian geodesic

The line element and metric tensor are

$$ds^2 = dx^2 + dy^2 \quad g_{ab} = \begin{pmatrix} 1 & 0 \\ 0 & 1 \end{pmatrix} \quad \Gamma^a_{bc} = 0$$

and the geodesic equation simplifies to

$$\frac{d^2 x^a}{ds^2} = 0$$

with parameterized solutions

$$x = as + b$$
$$y = cs + d$$

with a, b, c, and d constants. Substituting one into the other gives

$$y = \frac{c}{a}x + d - bc,$$

which are straight lines.

Example 9.12 Polar geodesic

The line element and metric tensor are

$$ds^2 = dr^2 + r^2 d\theta^2 \quad g_{ab} = \begin{pmatrix} 1 & 0 \\ 0 & r^2 \end{pmatrix}$$

$$\Gamma^\theta_{r\theta} = \frac{1}{r} \quad \Gamma^\theta_{\theta r} = \frac{1}{r} \quad \Gamma^r_{\theta\theta} = -r.$$

The geodesic equations are

$$\frac{d^2 r}{ds^2} + \Gamma^r_{\theta\theta}\left(\frac{d\theta}{ds}\right)^2 = 0$$

$$\frac{d^2\theta}{ds^2} + \Gamma^\theta_{r\theta}\frac{dr}{ds}\frac{d\theta}{ds} + \Gamma^\theta_{\theta r}\frac{d\theta}{ds}\frac{dr}{ds} = 0$$

and specifically

$$\frac{d^2 r}{ds^2} - r\left(\frac{d\theta}{ds}\right)^2 = 0$$

$$\frac{d^2\theta}{ds^2} + \frac{2}{r}\frac{dr}{ds}\frac{d\theta}{ds} = 0.$$

continued

Example 9.12 *continued*

These are solved by multiplying the first by $\sin\theta$, the second by $r\cos\theta$ and adding them. The resulting equation is expressed as

$$\frac{d}{ds}\left(r\cos\theta\frac{d\theta}{ds}+\sin\theta\frac{dr}{ds}\right)=0,$$

which is solved to give

$$s=r\sin(\theta-\theta_0),$$

which is the equation of a straight line in polar coordinates.

This example illustrates that a change in coordinates does not change the nature of the space. The flat plane remains flat with straight lines as geodesics. Even though some Christoffel symbols were not zero, the space is flat. A more general test of flatness is needed to distinguish flat spaces from curved spaces. This requires the concept of curvature.

Example 9.13 Geodesic on the surface of a sphere

The metric tensor on the surface of a sphere is (compare to Example 9.4)

$$g_{ab}=\begin{pmatrix}\cos^2 q^2 & 0\\ 0 & 1\end{pmatrix}$$

with

$$\frac{\partial g_{11}}{\partial q^2}=-2\sin q^2\cos q^2.$$

Only Christoffel symbols with indices of 1,1,2 in some order will be nonzero.
For the Christoffel symbols of the second kind

$$\Gamma^2_{11}=\sin q^2\cos q^2$$

$$\Gamma^1_{21}=\Gamma^1_{12}=\frac{-\sin q^2\cos q^2}{\cos^2 q^2}=-\tan q^2.$$

The geodesic equations are

$$\frac{d^2 q^1}{ds^2}-2\tan q^2\frac{dq^1}{ds}\frac{dq^2}{ds}=0$$

$$\frac{d^2 q^2}{ds^2}+\sin q^2\cos q^2\left(\frac{dq^1}{ds}\right)^2=0.$$

Solving these coupled second-order equations is not straightforward, although the solutions are simple: great-circle routes on the sphere. Instead of solving for these explicitly, we can try simple test cases.

Meridians of longitude: Choose $q^1 = a$ for $0 < a < 2\pi$ and $q^2 = t$ for $-\pi/2 < t < \pi/2$. In this case a is the longitude. Both equations are indeed zero, so meridians of longitude are geodesics.

Parallels of latitude: Choose $q^1 = t$ for $0 < t < 2\pi$ and $q^2 = b$, where b is the latitude. Again the equations sum to zero and parallels of latitude are geodesics.

Great circles: Choose

$$q^1 = \tan^{-1}\left(\frac{\tan s}{\sqrt{1+\alpha^2}}\right)$$

$$q^2 = \tan^{-1}\left(\frac{\alpha \tan s}{\sqrt{\alpha^2 + \sec^2 s}}\right),$$

where α is any real value. These can be proven to be geodesics. However, not all great circles are geodesics. An alternative, non-constant speed, parameterization of a great circle is $q^1 = t$ and $q^2 = \tan^{-1}(\alpha \sin t)$. Though it describes a great circle, this parameterization with time is not a geodesic. (See page 272 of James J. Callahan, *The Geometry of Spacetime: An Introduction to Special and General Relativity* (Springer, 2000)).

9.4.4 Optics, ray equation, and light "orbits"

There is a parallel between optics and mechanics in the use of minimum principles. In optics, it is Fermat's principle of least time, and in mechanics it is Jacobi's principle of least action. Furthermore, Hamilton's principle was originally motivated by optical phenomena. In this section, minimizing the optical path length between two points provides a ray equation that describes how a ray of light is refracted and bends in an inhomogeneous optical medium. The ray equation is a form of geodesic equation, and will be enlisted in Chapter 11 on gravitation to derive the deflection of light by gravitating bodies.

The optical path length between two points in a medium of refractive index n is

$$\int_{P_1}^{P_2} n\,ds = \int_{\lambda_1}^{\lambda_2} n\sqrt{dx^2 + dy^2 + dz^2} = \int_{\lambda_1}^{\lambda_2} n\sqrt{\frac{dx^2}{d\lambda^2} + \frac{dy^2}{d\lambda^2} + \frac{dz^2}{d\lambda^2}}\,d\lambda, \qquad (9.116)$$

where the trajectory is parametrized with the parameter λ using

$$ds = \sqrt{\dot{x}^2 + \dot{y}^2 + \dot{z}^2}\,d\lambda \qquad (9.117)$$

and

$$\dot{x} = \frac{dx}{d\lambda}, \text{ etc.} \qquad (9.118)$$

In the variational calculus, the extremum of the optical path length is found as

$$\delta \int_{P_1}^{P_2} n\, ds = \delta \int_{P_1}^{P_2} L\left(x, \dot{x}\right) d\lambda = 0, \tag{9.119}$$

where the effective Lagrangian is

$$L\left(x, \dot{x}\right) = n\left(x, y\right) \sqrt{\dot{x}^2 + \dot{y}^2 + \dot{z}^2}. \tag{9.120}$$

The Lagrangian is inserted into the Euler equations

$$\frac{d}{d\lambda} \frac{\partial L}{\partial \dot{x}^a} - \frac{\partial L}{\partial x^a} = 0 \tag{9.121}$$

to yield

$$\frac{\partial n}{\partial x^a} \sqrt{\dot{x}^2 + \dot{y}^2 + \dot{z}^2} - \frac{d}{d\lambda} \frac{n\dot{x}}{\sqrt{\dot{x}^2 + \dot{y}^2 + \dot{z}^2}} = 0 \tag{9.122}$$

or equivalently

$$\frac{\partial n}{\partial x^a} \sqrt{\dot{x}_b \dot{x}^b} - \frac{d}{d\lambda} \frac{n\dot{x}^a}{\sqrt{\dot{x}_b \dot{x}^b}} = 0. \tag{9.123}$$

These can be expressed as

$$\begin{aligned} \frac{d}{d\lambda} \frac{n\dot{x}^a}{\sqrt{\dot{x}_b \dot{x}^b}} &= \frac{\partial n}{\partial x^a} \sqrt{\dot{x}_b \dot{x}^b} \\ \frac{1}{\sqrt{\dot{x}_b \dot{x}^b}} \frac{d}{d\lambda} \frac{n\dot{x}^a}{\sqrt{\dot{x}_b \dot{x}^b}} &= \frac{\partial n}{\partial x^a} \\ \frac{d}{ds} \frac{n\dot{x}^a}{\sqrt{\dot{x}_b \dot{x}^b}} &= \frac{\partial n}{\partial x^a} \end{aligned} \tag{9.124}$$

using Eq. (9.117), and the final line is recast into its final form as the ray equation:

$$\textit{The ray equation:} \qquad \frac{d}{ds}\left(n \frac{dx^a}{ds}\right) = \frac{\partial n}{\partial x^a}, \tag{9.125}$$

which, in vector notation, is

$$\frac{d}{ds}\left(n\left(\vec{r}\right) \frac{d\vec{r}}{ds}\right) = \vec{\nabla} n\left(\vec{r}\right). \tag{9.126}$$

In a 3D inhomogeneous optical medium, the explicit equations are

$$\frac{d}{ds}\left(n(x,y,z)\frac{dx}{ds}\right) = \frac{\partial n(x,y,z)}{\partial x}$$

$$\frac{d}{ds}\left(n(x,y,z)\frac{dy}{ds}\right) = \frac{\partial n(x,y,z)}{\partial y} \qquad (9.127)$$

$$\frac{d}{ds}\left(n(x,y,z)\frac{dz}{ds}\right) = \frac{\partial n(x,y,z)}{\partial z}.$$

The ray equation is a second-order ordinary differential equation whose solution describes the path of a light ray through an inhomogeneous dielectric medium. As an example, if we consider only ray paths in the x–y plane, this is converted to a four-dimensional flow as

$$\dot{x} = \frac{q_1}{n(x,y)}$$

$$\dot{y} = \frac{q_2}{n(x,y)}$$

$$\dot{q}_1 = \frac{\partial n(x,y)}{\partial x} \qquad (9.128)$$

$$\dot{q}_2 = \frac{\partial n(x,y)}{\partial y}.$$

Note that

$$ds^2 = dx^2 + dy^2$$

$$\left(\frac{dx}{ds}\right)^2 + \left(\frac{dy}{ds}\right)^2 = 1, \qquad (9.129)$$

which yields

$$\frac{q_1^2}{n^2} + \frac{q_2^2}{n^2} = 1, \qquad (9.130)$$

which is expressed as

$$(\nabla S)^2 = q_1^2 + q_2^2 = n^2, \qquad (9.131)$$

where $S(x,y) = $ const. is the eikonal function. The k-vectors are perpendicular to the surfaces of constant S.

Example 9.14 GRIN lens

A graded index (GRIN) lens is often used to couple lasers into fiber optics. The lens is constructed by introducing a radial variation in the refractive index from high in the center to low towards the outside. For a parabolic refractive index profile, the entering rays execute periodic orbits across the optic axis. Furthermore, the travel time for all rays is the same, independent of their inclination to the optic axis (in the approximation of small angles).

The refractive index profile is

$$n(r) = n_0 - \frac{1}{2}n_2 r^2$$

and the gradient is

$$\vec{\nabla} n(r) = -n_2 \vec{r}.$$

The paraxial approximation is used for small angles for which $ds \approx dz$. The ray equation in the paraxial approximation becomes

$$\frac{d}{dz}\left(n(r)\frac{dr}{dz}\right) = -n_2 r.$$

The index is independent of z, and is approximately a constant n_0 for small angle deflections. The ray equation becomes

$$\frac{d^2 r}{dz^2} = -\frac{n_2}{n_0}r,$$

which has a simple oscillating solution

$$r(z) = r_0 \cos\left(\sqrt{\frac{n_2}{n_0}}z\right) + \theta_0 \sqrt{\frac{n_0}{n_2}}\sin\left(\sqrt{\frac{n_2}{n_0}}z\right)$$

for an initial ray position r_0 entering the lens at an initial angle θ_0. These solutions are periodic in the length

$$z_0 = \sqrt{\frac{n_0}{n_2}}.$$

The solutions are shown in Fig. 9.11 for three different initial conditions. The rays oscillate across the optic axis with the same period independent of the amplitude. The repeat length is z_0. If the incident wave is a plane wave, and if the lens has a length $z_0/4$, it focuses the rays within a spot on the optical axis on the output face. In a fiber coupling application, the center of a fiber optic cable would be located at this point, and the GRIN would focus the laser into the fiber.

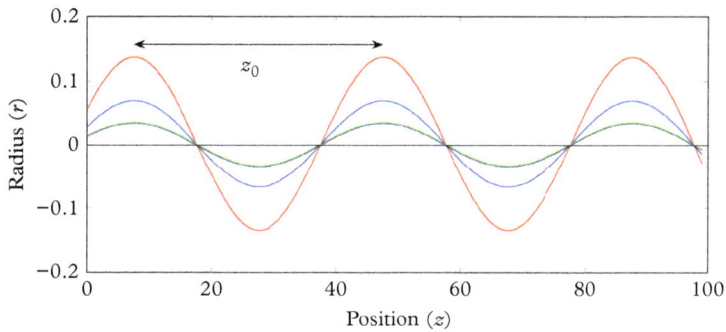

Figure 9.11 *Solutions for the ray equation for a graded-index (GRIN) lens for three initial conditions with light entering at different locations and at different angles.*

9.5   Geodesic motion

A third approach to the geodesic equation considers force-free motion. This section begins with the Euler–Lagrange equations in the case of zero potential, then generalizes to systems with nonzero potential. By redefining the metric tensor, motion in a potential can be described as a geodesic curve. In this way, forces are removed from dynamics, and all trajectories in such cases become force-free geodesics. This viewpoint is known as the "geometrization of mechanics," which has a rich history and important consequences for general relativity.

9.5.1   Force-free motion

In the case of force-free motion, when there is no potential function, the Lagrangian is

$$L = T = \frac{1}{2} g_{ab} \dot{q}^a \dot{q}^b. \qquad (9.132)$$

The generalized coordinates q^a do not need to be Cartesian coordinates, nor single-particle coordinates. They can represent collective modes of a multi-particle system, and imposed constraints on the system can give the configuration space nonzero curvature. In this general case, the g_{ab} have the units of mass, but are not necessarily identified as the individual masses of the particles comprising the system. The g_{ab} furthermore can have position dependence defined by the constrained motion in configuration space.

Applying the Euler–Lagrange equations to the Lagrangian

$$\frac{d}{dt}\left(\frac{\partial L}{\partial \dot{q}^c}\right) - \frac{\partial L}{\partial q^c} = 0 \tag{9.133}$$

yields

$$\frac{d}{dt}\left(2g_{cb}\dot{q}^b\right) - \frac{\partial g_{bd}}{\partial q^c}\dot{q}^b\dot{q}^d = 0. \tag{9.134}$$

The first time derivative is

$$\frac{d}{dt}\left(g_{cb}\dot{q}^b\right) = g_{cb}\ddot{q}^b + \frac{\partial g_{cb}}{\partial q^d}\dot{q}^d\dot{q}^b \tag{9.135}$$

to give

$$g_{cb}\ddot{q}^b + \frac{\partial g_{cb}}{\partial q^d}\dot{q}^b\dot{q}^d - \frac{1}{2}\frac{\partial g_{bd}}{\partial q^c}\dot{q}^b\dot{q}^d = 0. \tag{9.136}$$

Multiplying on the left with the inverse metric yields

$$\ddot{q}^a + g^{ad}\left(\frac{\partial g_{db}}{\partial q^c} - \frac{1}{2}\frac{\partial g_{bc}}{\partial q^d}\right)\dot{q}^b\dot{q}^c = 0, \tag{9.137}$$

which can be rewritten using the symmetry of the metric tensor to give

$$\ddot{q}^a + \frac{1}{2}g^{ad}\left(\frac{\partial g_{db}}{\partial q^c} + \frac{\partial g_{dc}}{\partial q^b} - \frac{\partial g_{bc}}{\partial q^d}\right)\dot{q}^b\dot{q}^c = 0. \tag{9.138}$$

This equation, re-expressed using the Christoffel symbol, is

$$\ddot{q}^a + \Gamma^a_{bc}\dot{q}^b\dot{q}^c = 0, \tag{9.139}$$

which is the geodesic equation that was derived before. Therefore, the multi-particle trajectory of force-free motion is a geodesic curve that minimizes the "distance" travelled on the action-space manifold, where the "distance" along the geodesic curve can be expressed as

$$s = \int_{t_1}^{t_2} dt\sqrt{g_{ab}\dot{q}^a\dot{q}^b} = \int_{t_1}^{t_2} dt\sqrt{2T}. \tag{9.140}$$

9.5.2   Geodesic motion through potential energy landscapes

Geodesics are closely related to Jacobi's principle in mechanics. In each case, a property is minimized (action for dynamics and distance for geodesics). The question is whether this analogy can be made explicit even in the case when forces act on the system. This approach defines an appropriate "metric" for the configuration space for which the minimum "distance" yields the minimum action.

Define a position-dependent Riemannian metric tensor as

$$\gamma_{ab}\left(x^a\right) = 2\left(E - U\left(x^a\right)\right) g_{ab} \tag{9.141}$$

with a kinetic energy

$$T = \frac{1}{2}\left(\frac{ds}{dt}\right)^2, \tag{9.142}$$

where g_{ab} is the configuration-space metric, U is the potential, and the total energy E is constant. The action line element is

$$
\begin{aligned}
d\sigma = \sqrt{d\sigma^2} &= \sqrt{\gamma_{ab}dx^a dx^b} \\
&= \sqrt{2\left(E - U\right) g_{ab}dx^a dx^b} \\
&= \sqrt{2\left(E - U\right)}ds
\end{aligned}
\tag{9.143}
$$

and we have the correspondence (from Chapter 2)

$$\frac{\partial L}{\partial \dot{q}^a} \dot{q}^a = 2T = \left(\frac{ds}{dt}\right)^2 = 2\left(E - U\right). \tag{9.144}$$

The time element is then

$$dt = \frac{ds}{\sqrt{2\left(E - U\right)}} \tag{9.145}$$

and the action integral is

$$\int_\gamma \frac{\partial L}{\partial \dot{q}^a} \dot{q}^a \, dt = \int_\gamma \sqrt{2\left(E - U\right)}ds = \int_\gamma d\sigma. \tag{9.146}$$

The path that minimizes the action is also the path that minimizes the distance on the Riemannian action manifold.

There is neither time nor velocity dependence in the final integral. The dynamics are determined entirely from the geometry of the Riemannian space. This strongly parallels the concept of trajectories in general relativity, in which energy is the source of space–time curvature, and trajectories are geodesics on the space–time manifold.[7]

To derive the geodesic equation for trajectories through a potential landscape, begin with the action metric

$$\gamma_{ab} = (E - U)\, g_{ab} \tag{9.147}$$

and its inverse

$$\gamma^{ab} = \frac{g^{ab}}{(E - U)}. \tag{9.148}$$

Applying Eq. (9.91) to the action metric yields the relationship between the Christoffel symbols of the configuration space and the action space as

$$\Gamma^a_{bc(\gamma)} = \Gamma^a_{bc(g)} - \frac{1}{2\,(E - U)} \left(\delta^a_b \frac{\partial U}{\partial x^c} + \delta^a_c \frac{\partial U}{\partial x^b} - g_{bc} g^{ad} \frac{\partial U}{\partial x^d} \right). \tag{9.149}$$

The equation of motion in configuration space is

$$\frac{d^2 x^a}{dt^2} + \Gamma^a_{bc(g)} \frac{dx^b}{dt} \frac{dx^c}{dt} = -g^{ad} \frac{\partial U}{\partial x^d} \tag{9.150}$$

and substituting in the Christoffel symbol for the dynamical action metric gives

$$\begin{aligned} \frac{d^2 x^a}{dt^2} + \Gamma^a_{bc(\gamma)} \frac{dx^b}{dt} \frac{dx^c}{dt} &+ \frac{1}{(E - U)} \frac{\partial U}{\partial x^c} \frac{dx^c}{dt} \frac{dx^a}{dt} \\ &= -g^{ad} \frac{\partial U}{\partial x^d} \left[1 - \frac{g_{bc}}{2\,(E - U)} \frac{dx^b}{dt} \frac{dx^c}{dt} \right]. \end{aligned} \tag{9.151}$$

However,

$$\frac{g_{bc}}{2\,(E - U)} \frac{dx^b}{dt} \frac{dx^c}{dt} = 1 \tag{9.152}$$

and the term on the right vanishes, so the equation of motion becomes

$$\frac{d^2 x^a}{dt^2} + \Gamma^a_{bc(\gamma)} \frac{dx^b}{dt} \frac{dx^c}{dt} = -\frac{1}{(E - U)} \frac{\partial U}{\partial x^c} \frac{dx^c}{dt} \frac{dx^a}{dt}. \tag{9.153}$$

This equation can be rewritten in terms of the action element $d\sigma = 2T dt$ as

$$2T \frac{d}{d\sigma} \left(2T \frac{dx^a}{d\sigma} \right) + \Gamma^a_{bc(\gamma)} (2T)^2 \frac{dx^b}{d\sigma} \frac{dx^c}{d\sigma} = -\frac{2T}{(E - U)} \frac{dU}{dt} \frac{dx^a}{d\sigma} \tag{9.154}$$

[7] From a historical perspective, the geometrization of mechanics did not originate with Einstein, but had earlier origins with Hertz and with Darboux in the 1890s, predating even special relativity. An earlier priority for a geometric view of mechanics can be attributed to Lipshitz in the 1870s, shortly after Reimann's lectures became widely known when they were published in 1866. For a full account of the history of the geometrization of physics, see J. Lutzen, *Mechanistic Images in Geometric Form: Heinrich Hertz's Principle of Mechanics* (Oxford University Press, 2005).

and carrying through the derivatives

$$\left(2T\frac{d^2x^a}{d\sigma^2} - 2\frac{dx^a}{d\sigma}\frac{dU}{d\sigma}\right) + \Gamma^a_{bc(\gamma)}2T\frac{dx^b}{d\sigma}\frac{dx^c}{d\sigma} = -\frac{1}{(E-U)}\frac{dU}{dt}\frac{dx^a}{d\sigma} \tag{9.155}$$

and rearranging

$$\frac{d^2x^a}{d^2\sigma} + \Gamma^a_{bc(\gamma)}\frac{dx^b}{d\sigma}\frac{dx^c}{d\sigma} = -\frac{1}{2(E-U)^2}\frac{dU}{dt}\frac{dx^a}{d\sigma} + \frac{1}{2T^2}\frac{dx^a}{d\sigma}\frac{dU}{dt}. \tag{9.156}$$

The right-hand side vanishes identically, yielding the geodesic equation in terms of the action element and the action Christoffel symbols

$$\frac{d^2x^a}{d\sigma^2} + \Gamma^a_{bc(\gamma)}\frac{dx^b}{d\sigma}\frac{dx^c}{d\sigma} = 0, \tag{9.157}$$

which is the geodesic equation. Therefore, the action metric of Eq. (9.147) leads to a geodesic equation that is equivalent to the equations of force-free motion.

The action metric defines distances on the configuration space. The geodesic path is the shortest (or straightest) path through that space. Therefore, the description of the trajectory is purely geometric, without resort to forces. It can be said that the potential U modifies the geometry of the space, and trajectories are force-free curves through the modified geometry. This same language will be used in the theory of general relativity when speaking of mass modifying (or warping) space–time, and trajectories are force-free motions through the modified space.

Example 9.15 Geodesic motion in a constant field

Use the geodesic equation to analyze the planar trajectory of a mass m in a constant gravitational potential $U(y)$. Consider a Cartesian configuration space with Cartesian g_{ab}. All of the Christoffel symbols of the configuration are zero. However, the action space has nonzero Christoffel symbols. These are

$$\Gamma^a_{bc(\gamma)} = \frac{-1}{2(E-U)}\left(\delta^a_b\frac{\partial U}{\partial x^c} + \delta^a_c\frac{\partial U}{\partial x^b} - g_{bc}g^{ad}\frac{\partial U}{\partial x^d}\right)$$

$$\Gamma^0_{01} = \Gamma^0_{10} = \frac{-1}{2(E-U)}\frac{\partial U}{\partial y}$$

$$\Gamma^1_{00} = \frac{1}{2(E-U)}\frac{\partial U}{\partial y}$$

$$\Gamma^1_{11} = \frac{-1}{2(E-U)}\frac{\partial U}{\partial y}$$

continued

Example 9.15 *continued*

---

and all other Christoffel symbols are zero. Starting with Eq. (9.153), the equations of motion are

$$\ddot{x} + \Gamma^0_{01}\dot{x}\dot{y} + \Gamma^0_{10}\dot{y}\dot{x} = -\frac{1}{(E-U)}\frac{\partial U}{\partial y}\dot{y}\dot{x}$$

$$\ddot{x} + -\frac{1}{(E-U)}\frac{\partial U}{\partial y}\dot{y}\dot{x} = -\frac{1}{(E-U)}\frac{\partial U}{\partial y}\dot{y}\dot{x}$$

$$\ddot{x} = 0$$

and

$$\ddot{y} + \Gamma^1_{00}\dot{x}^2 + \Gamma^1_{11}\dot{y}^2 = -\frac{1}{(E-V)}\frac{\partial U}{\partial y}\dot{y}^2$$

$$\ddot{y} + \frac{1}{2\,(E-U)}\frac{\partial U}{\partial y}\dot{x}^2 - \frac{1}{2\,(E-U)}\frac{\partial U}{\partial y}\dot{y}^2 = -\frac{1}{(E-U)}\frac{\partial U}{\partial y}\dot{y}^2$$

$$\ddot{y} + \frac{1}{m}\frac{\partial U}{\partial y} = 0.$$

This last equation is recognized as the equation for a free trajectory, and for $U(y) = mgy$, the trajectory is a parabola.

Admittedly, using the geodesic equation for this simple problem is far more complicated than solving for the motion directly, but for more complex problems in configuration spaces with complex constraints or with curvature, the geodesic equation can provide a safe approach when physical intuition may not be clear. Of course, a Lagrangian approach is also appropriate in this case. The importance of geodesic motion is primarily conceptual, providing a common perspective from which to see many different types of motions, whether simple trajectories in flat spaces, or complex trajectories in warped space time of general relativity. In Chapter 11, relativistic orbits of planets and photon paths past the Sun will be seen to follow geodesics.

9.6 Summary

The *metric tensor* uniquely defines the geometric properties of a *metric space* which supports vectors and covectors. The metric tensor is used to convert between these two forms and participates in the *inner product* that generates a scalar quantity that defines lengths and distances. *Differential geometry* is concerned with the derivatives of vectors and tensors within a metric space. Because basis vectors can change during displacement, derivatives of vector quantities include the derivatives of basis vectors which lead to *Christoffel symbols* that are directly connected to the metric tensor. This connection between derivatives and the metric tensor provides the tools necessary to define *geodesic curves* in the metric space. Two separate approaches lead to the *geodesic equation*. One is based on variational calculus and demonstrates that geodesic curves are the shortest curves. The other is based on *parallel transport* and demonstrates that geodesic curves are the straightest curves. *Geodesic motion* is the trajectory of a particle through a metric space defined by an *action metric* that includes a potential function. In this way, dynamics converts to geometry as a trajectory in a potential converts to a geodesic curve through an appropriately-defined metric space.

Line element

The line element is defined in terms of the metric tensor as

$$ds^2 = g_{ab}dx^a dx^b. \tag{9.4}$$

Metric tensor

The metric tensor determines the properties of a metric space

$$g_{ab} = \frac{\partial x}{\partial q^a}\frac{\partial x}{\partial q^b} + \frac{\partial y}{\partial q^a}\frac{\partial y}{\partial q^b} + \frac{\partial z}{\partial q^a}\frac{\partial z}{\partial q^b}. \tag{9.6}$$

The metric tensor components are obtained through the inner product between basis vectors

$$g_{ab} = \vec{e}_a \cdot \vec{e}_b. \tag{9.26}$$

Basis vectors

Basis vector components derive from the coordinate transformation between general coordinates and Cartesian coordinates through

$$\vec{e}_a = \left(\frac{\partial x}{\partial q^a}, \frac{\partial y}{\partial q^a}, \frac{\partial z}{\partial q^a}\right). \tag{9.24}$$

Basis vectors transform inversely to vector components.

Transformation properties

Contravariant and covariant vectors transform inversely to each other. Covariant vectors transform as basis vectors.

$$\begin{aligned} A^{\bar{b}} &= R^{\bar{b}}_a A^a \\ A_{\bar{b}} &= R^a_{\bar{b}} A_a \\ \vec{e}_{\bar{b}} &= R^a_{\bar{b}} \vec{e}_a. \end{aligned} \tag{9.34}$$

Derivative of a vector

The derivative of a vector includes the derivative of the basis vector using the chain rule

$$\frac{\partial \vec{V}}{\partial x^\beta} = \frac{\partial V^\alpha}{\partial x^\beta}\vec{e}_\alpha + V^\alpha \frac{\partial \vec{e}_\alpha}{\partial x^\beta}. \tag{9.69}$$

Christoffel symbol

The Christoffel symbol of the second kind arises from the covariant derivative of a basis vector

$$\frac{\partial \vec{e}_\alpha}{\partial x^\beta} = \Gamma^\mu_{\alpha\beta}\vec{e}_\mu, \tag{9.70}$$

where the Christoffel symbol is related to the metric tensor through

$$\Gamma^\gamma_{\beta\mu} = \frac{1}{2}g^{\alpha\gamma}\left(\partial_\mu g_{\alpha\beta} + \partial_\beta g_{\alpha\mu} - \partial_\alpha g_{\beta\mu}\right). \tag{9.91}$$

Geodesic equation

The geodesic equation defines a straightest path of shortest length through a metric space

$$\frac{d^2 x^a}{ds^2} + \Gamma^a_{bc}\frac{dx^b}{ds}\frac{dx^c}{ds} = 0. \tag{9.106}$$

The ray equation

The ray equation is an analogy to the geodesic equation. Light follows a null geodesic that is defined by the ray equation

$$\frac{d}{ds}\left(n\frac{dx^a}{ds}\right) = \frac{\partial n}{\partial x^a}, \tag{9.125}$$

and in vector notation

$$\frac{d}{ds}\left(n\,(\vec{r})\,\frac{d\vec{r}}{ds}\right) = \vec{\nabla}n\,(\vec{r}). \tag{9.126}$$

9.7   Bibliography

G. Arfkin,*Mathematical Methods for Physicists* (Academic Press, 1985). 3rd ed.

J. Lützen, "The geometrization of analytical mechanics: A pioneering contribution by J. Liouville (ca. 1850)," in *History of Modern Mathematics*, Vol. 2, David E. Rowe and John McCleary (Eds.) (Academic Press, Boston, 1989), p. 77.

B. F. Schutz, *A First Course in General Relativity* (Cambridge University Press, 1985).

J. L. Synge and A. Schild, *Tensor Calculus* (Dover, 1949).

R. Talman, *Geometric Mechanics* (Wiley, 2000).

9.8   Homework exercises

1. **Parabolic trajectory:** Calculate the total path length as a function of time of a general parabolic trajectory with arbitrary initial velocity (v_x, v_y) in a constant gravitational field. Evaluate (expand) the short-time and long-time behavior and interpret as simple limits.

2. **Rigid pendulum:** A pendulum is constructed as a mass on a rigid massless rod. The mass is dropped from rest at a height y above the center of rotation. Find the path length as a function of time. Find the unit tangent vector as a function of time.

3. **Rotation:** Prove $\vec{e}_a = R_a^{\bar{b}}\vec{e}_{\bar{b}}$ for a general 2D rotation starting from $A^a\left(R_a^{\bar{b}}\vec{e}_{\bar{b}} - \vec{e}_a\right) = 0$ by explicitly carrying out the implicit summation.

4. **Invariants:** Prove that $\vec{A} = A^1\vec{e}_1 + A^2\vec{e}_2 + A^3\vec{e}_3$ is invariant under any linear coordinate transformation.

5. **Lorentz transformation:** Find the affine parameters in Eq. (1.53) for the variables (x, t) that match the Lorentz coordinate transformations (Eq. (10.3)).

6. **Coordinate transformation:** Consider the coordinate system sometimes used in electrostatics and hydrodynamics

$$xy = u$$
$$x^2 - y^2 = v$$
$$z = z$$

   Find (a) the Jacobian matrix and the Jacobian; and (b) the basis vectors.

7. **Transformation:** A possible transformation of the 2D vector B^k by the operator (matrix) R_j^i is $A^i = R_j^i B^k R_k^j$. Write out the explicit expressions (explicit summation) for (A^x, A^y). (Example: $A^x = R_x^x B^x R_x^x + R_x^x B^y R_y^x + R_y^x B^x R_x^y + \ldots$ etc.)

8. **Non-cummutativity:** Show explicitly that $A_b^a B_c^b \neq A_c^b B_b^a$, which is the statement of non-cummutativity of two matrices (two transformations).

9. **Invariant metric tensor:** Show that the metric tensor is independent of coordinate transformation in the case of Cartesian coordinates and a rotation.

10. **Torus:** Calculate the distance on a flat 2D torus and on an embedded 3D torus for the curve $\theta = \phi$ between point $(\theta, \phi) = (0, 0)$ and $(\theta, \phi) = (\pi/2, \pi/2)$ for $R = 10$ and $r = 1$. How easy is the embedded integral to perform?

11. **Lorentz transformation:** How would you define the Lorentz transformation as a pseudo-Riemannian affine transformation? (Hint: Define appropriate basis vectors for the transformation.)

12. **Affine coordinates:** Find the basis vectors of an affine coordinate system

13. **Covariant derivative:** Derive the covariant derivative of a covariant vector (Eq. (9.72)).

14. **Christoffel symbols:** The contraction of the Christoffel symbol of the second kind with the inverse metric tensor

$$\Gamma^{\upsilon}_{\beta\mu} = g^{\alpha\upsilon}\Gamma_{\alpha,\beta\mu}$$

defines the Christoffel symbol of the first kind

$$\Gamma_{\alpha,\beta\mu} = \frac{1}{2}\left(\partial_{\mu}g_{\alpha\beta} + \partial_{\beta}g_{\alpha\mu} - \partial_{\alpha}g_{\beta\mu}\right).$$

Derive all the Christoffel symbols of the first and second kind for spherical coordinates.

15. **Geodesic equation derivation:** Show that

$$\frac{\partial}{\partial\dot{x}^c}\left(g_{ab}\dot{x}^a\dot{x}^b\right) = 2g_{cb}\dot{x}^b$$

16. **Great circle routes:** Prove that great circle routes on the Earth are geodesics.

17. **Helix:** Using a metric tensor approach, show that a geodesic on a flat cylinder is a helix.

18. **3D torus:** Derive the geodesic equation for a geodesic on the surface of a curved torus embedded in three dimensions.

19. **Unbound orbit:** Derive the geodesic equation for geodesic motion of an unbounded orbit (a parabolic orbit) in a $1/r^2$ potential.

20. **Anisotropic oscillator:** Use the geodesic equation for geodesic motion to find the trajectory of a 2D anisotropic harmonic oscillator.

21. **Light orbits:** A refractive index profile that has cylindrical symmetry has a shell of high index surrounding a core of low index. The refractive index function is

$$n(r) = 1 + n_2 r \exp\left(-r^2/2\sigma^2\right).$$

Solve the geodesic flow equations (9.128) for ray optics for initial conditions on the rays (location and direction) that give rise to light "orbits" in this refractive index landscape.

Relativistic Dynamics

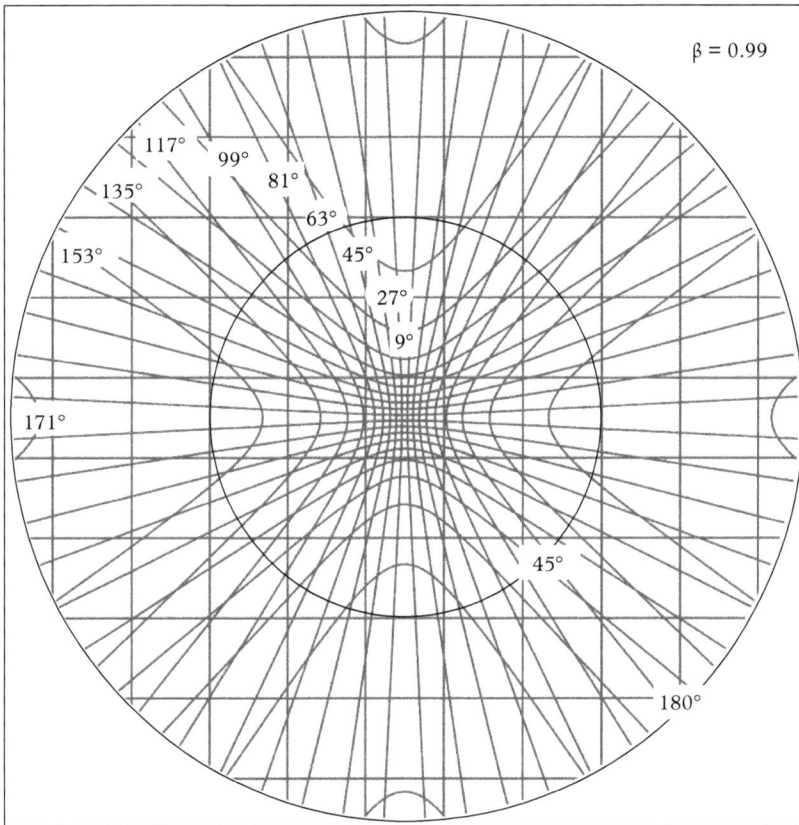

$\beta = 0.99$

117°
99°
135°
81°
63°
153°
45°
27°
9°
171°
45°
180°

The view, looking forward, while moving at 99% of the speed of light.

Relativistic dynamics takes place in four-dimensional (4D) spaces that have non-Euclidean measures of distance. In both the special and the general theory of relativity, a major emphasis is placed on geometric interpretation of distance, focusing on the "length" of vectors on one hand, and arc lengths on the other. In both cases, coordinate transformations play a dominant role. In the special theory, the Lorentz transformation takes an observer from one inertial frame to another. In general relativity, it is the transformation from local "flat" coordinates to global "curvilinear" coordinates that define curvature in space–time.

The theory of special relativity describes what different observers "see" in their own inertial frames as they pass each other at speeds approaching the speed of light. The Lorentz[1] transformation is a linear transformation that skews and dilates the coordinate axes of one frame relative to the other and destroys the concept of simultaneity. Within and between these frames, 4D vectors called four-vectors reside in non-Euclidean Minkowski space.[2] The metric tensor defines a mapping of a vector and a covector onto the space of real numbers—to scalars that are invariant under all coordinate transformations. These invariants are the immutable quantities of reality. In this sense, the study of "relativity" is misnamed, because it is not that all things are relative, and that nothing is absolute, but rather that many properties *are* invariant to coordinate change. One of the fundamental postulates of relativity is the equivalence of all observers. This is simply stating that reality is coordinate-free. You cannot change reality (i.e., physics) simply by changing your coordinates. The physics remains absolute and real, and there is nothing relative about it.

10.1 The special theory

The special theory pertains to observers who are moving in inertial frames and what they measure. An inertial frame is one that moves at constant velocity. Envision that there is some physical process that takes place, let's say a particle executing a trajectory, and that it is observed by two different observers, each in their own inertial frame. A pertinent question to ask is how each observer measures the physical process. One possibility is that each observer is positioned at the origin of their respective coordinate systems holding a very accurate clock, and they "watch" the physical process unfold from their standpoint. But this causes problems, because it takes time for the information to travel (at the speed of light) from one part of the trajectory, called an event, to the observers at the different origins. The find the actual path of the particle, the observers would need to propagate their information back to the event to know when and where it occurred. Therefore, to ask what an observer "sees" is not easily converted into what happened.

An alternative is to replace each observer at their origin with an array of clocks that fills space, as shown in Fig. 10.1. These clocks are globally synchronized so that all clocks in a single frame read the same time. An event is recorded at the location in space where it occurs and is timed against the reading of the local clock. This takes away any need for a signal to propagate some distance in order to record the event. A trajectory is not "observed" as it happens by an observer at the origin. Each local clock records an event, and a trajectory is a string of events that occur at a succession of positions at a succession of times. Once the physical process is over, the string of measurements can be relayed back to the observers for them to analyze.

[1] The transformation properties of space and time were developed by Hendrik Lorentz (1853–1928) between the years 1892 and 1905, starting long before Einstein. He introduced these relations to explain the transformation properties of electromagnetic waves that generate the famous null result of the Michelson-Morley experiment.

[2] Hermann Minkowski (1864–1909) was a German theoretical physicist. The concept of space-time was introduced by Minkowski in an article published in 1908 titled Raum und Zeit. He showed that space-time constituted a pseudo-Riemann manifold.

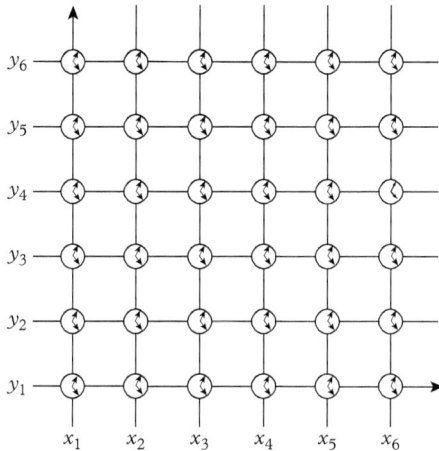

Figure 10.1 *Space is filled with synchronized clocks that record local events. A trajectory is a series of recorded local events as a particle visits a sequence of locations in succession. The trajectory is reconstructed only after all the recorded events have been transmitted back to the origin. This construction of distributed clocks as measurement devices removes the difficulty of deciding what an observer in a certain frame "sees" by removing delays caused by the finite speed of light.*

Now that we have answered *how* the observers measure a physical process, a second question presents itself, which is *what* each observer observes with their distributed sensor network. This will be different for the two observers. For instance, a ball that is stationary in one frame will appear to be moving in another frame that is moving relative to the first. In fact, as the relative speed approaches the speed of light, nothing will look the same from the two different frames. The array of clocks will tick differently, and meter sticks will shrink differently, depending on who is watching what. At this level of relativity theory everything looks relative, and it may not be obvious how to define the things that stay the same, that are invariant, for both observers. This leads to a third question, which is: What do all observers have *in common*, independently of the details of what they observe? To begin to answer this question, Einstein stated two simple postulates from which all the complexities and invariants of relativity theory may be derived.

The two postulates upon which relativity is based

Postulate I: All inertial observers are equivalent.

This means that it is impossible to make local experiments to ascertain whether one's own frame is in motion. It also means that a straight trajectory in one frame (a geodesic in an inertial frame) must be a straight trajectory in another relatively-moving inertial frame. Any allowable coordinate transformations between the frames therefore must take straight lines into straight lines.

Postulate II: The speed of light is constant in all frames.

This means that a wave front observed to move at the speed of light in one frame must look the same in the other frame: a spherical wave front will look spherical to both.

The first postulate is not new to special relativity, but applies as well to Galilean relativity. It is a statement that physics cannot depend on the choice of coordinate system. This postulate is one of the cornerstones in the foundation upon which all physics is built. The second postulate is more specific, and not immediately obvious. It places the speed of light as a fundamental invariant in all physical theories. Despite the obviousness of the first postulate, and the simplicity of the second postulate, they begin a line of fairly simple mathematical derivations that lead to consequences that are increasingly bazaar and unintuitive. These consequences are the topics of this chapter, but at their heart are simple invariants that anchor them all.

10.2   Lorentz transformations

It has been the consistent theme, throughout this textbook, to consider trajectories as the primary objects of interest. A conventional trajectory is a single-parameter curve in three-space, consisting of successive positions occupied at successive times. However, in the theory of special relativity, positions and times coalesce into a single space–time, and trajectories are curves within a metric space. The geometric properties of these curves are explored within the different coordinate frames of observers moving relative to each other. A coordinate transformation must be found that converts the coordinates of one frame to the other while satisfying the two postulates of relativity.

10.2.1   Standard configuration (SC)

Consider two reference frames (Fig. 10.2), one primed and one unprimed. Frame O' is in constant motion along the x-axis with speed v relative to the unprimed frame O.

The standard configuration (SC) is the configuration when the x-axes of each frame are oriented along the direction of the relative velocity. In addition, when the origins O and O' coincide (as O' passes O), the distributed clocks within each frame are reset. Now consider a light flash that is emitted from the origin when the origins coincide. By postulate II the location of the spherical light front along the x-axis is at

$$x = ct$$
$$x' = ct'. \tag{10.1}$$

By Postulate I, lines must transform into lines, mandating a linear coordinate transformation, also known as an affine transformation. The simplest form for a general linear transformation is

$$x' = \gamma\,(x - vt)$$
$$x = \gamma\,(x' + vt'), \tag{10.2}$$

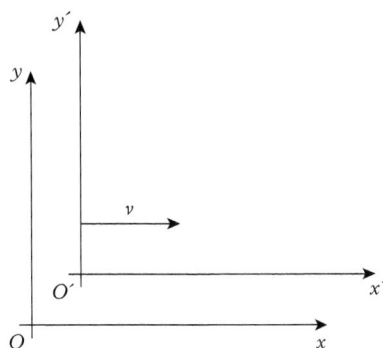

Figure 10.2 *Two relativistic inertial frames. The origin O is the fixed frame. O′ is moving with a velocity v relative to O. In the standard configuration the clocks are synchronized in each frame at the instant O′ passes O (they are shown offset here for visibility).*

where the terms in parentheses are simply the Galilean relativity transformations, and γ is a proportionality factor. Applying Eq. (10.1) to Eq. (10.2) leads directly to the Lorentz transformations

Lorentz transformations:
$$
\begin{aligned}
t' &= \gamma\left(t - \frac{v}{c^2}x\right) & t &= \gamma\left(t' + \frac{v}{c^2}x'\right) \\
x' &= \gamma\,(x - vt) & x &= \gamma\,(x' + vt') \\
y' &= y & y &= y' \\
z' &= z & z &= z',
\end{aligned}
\tag{10.3}
$$

where

$$
\gamma = \frac{1}{\sqrt{1-\beta^2}}
$$
$$
\beta = \frac{v}{c}.
\tag{10.4}
$$

The Lorentz factor γ is the key parameter in all special relativity problems. It is nearly unity for velocities less than 1% of the speed of light, which is the regime of Newtonian mechanics. Even for speeds around 10% of the speed of light, the factor is only equal to 1.005. However, it diverges rapidly as β increases above 80%, as shown in Fig. 10.3.

It is common to consider the "rest" frame of some object. This is the frame in which the object is stationary, or at rest. For instance, a fundamental particle like a muon might be generated in a high-energy collision in a laboratory, and the muon travels in the lab frame at a speed approaching the speed of light. This chapter

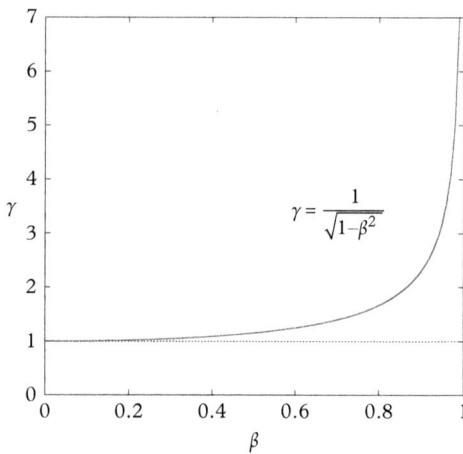

Figure 10.3 *The Lorentz factor γ as a function of β. It remains near unity even up to $\beta = 0.5$, but then diverges rapidly as β approaches unity.*

takes the convention that the primed frame is moving with the muon, so that the rest frame of the muon is the primed frame, and the laboratory frame (sometime also called the "fixed" frame) is unprimed. The properties of the muon can be considered in its rest frame, such as the lifetime of the muon before it decays to an electron and a neutrino. The mean time it takes the muon to decay as measured in the laboratory frame is different, dictated by the Lorentz transformation. However, the mean lifetime of a muon is a physical property of the fundamental particle and should not be some relative property. Therefore, the rest frame of an object is a special frame in which the properties of a physical object are defined. A meter stick is one meter long in its rest frame. A clock ticks once per second in its rest frame. These are also known as the *proper* length and the *proper* time. The existence of a special frame does not cause any trouble with the two postulates of relativity. The Lorentz transformations point to this special frame, because this is the frame with the maximum measured length and smallest measured times among all possible relatively moving frames.

10.2.2   Minkowski space

The trajectory of an object in special relativity is plotted with one time axis against three space axes, in other words, a four-space. In the standard configuration, only the x-axis is plotted. To give the axes equal dimensions, the time axis is multiplied by the speed of light. An example of a space–time diagram is shown in Fig. 10.4a. A photon travels on a trajectory that makes a 45° angle called the light line. If the y-axis were plotted coming out of the page, light would propagate on a "light cone." The oblique lines in Fig. 10.4b are how axes of frame O' with relativistic velocity $\beta = 0.333$ look to us in O when O' moves to the right. The O' coordinate system, as seen from O, is no longer orthogonal. Vertical lines are now skewed. When O' is moving to the right, the angle θ of the skew is

$$\tan \theta = \frac{\gamma v t'}{ct} = \frac{\gamma v t'}{c \gamma t'} = \beta. \tag{10.5}$$

Similarly, the axes are skewed in the opposite direction in Fig. 10.4c for $\beta = -0.333$.

One of the important consequences of the Lorentz transformation is the loss of the absolute notion of simultaneity. In any given frame, a set of events may be simultaneous, as in Fig. 10.5a. These are events at different locations that occur at the same time in the primed frame. However, these same events, as measured in the unprimed frame, appear as in Fig. 10.5b. They are no longer simultaneous when observed in this other relatively moving frame. Many of the apparent paradoxes of relativity theory are resolved once the notion of simultaneity is removed.

Rest (proper) frame

a)

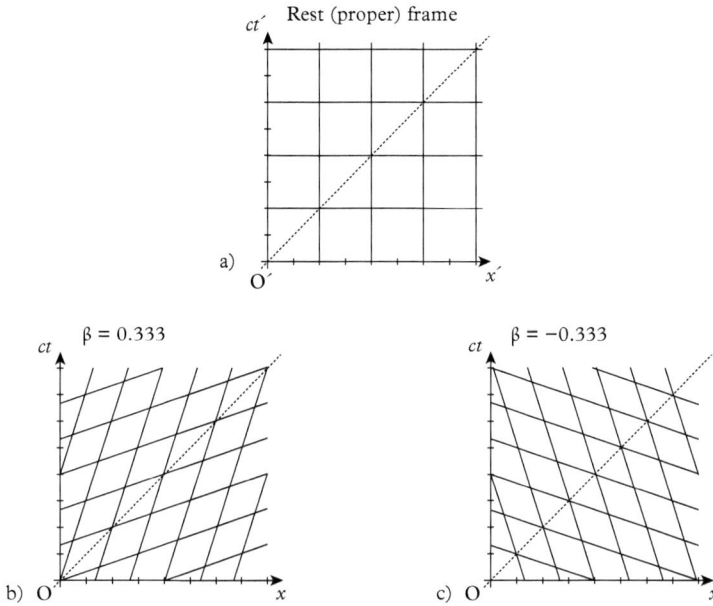

Figure 10.4 *a) Space–time diagram with the ct-axis vertical and the x-axis horizontal. The trajectory of a photon is shown as the 45° line. b) The oblique lines represent the axes of a frame moving to the right at a velocity β = 0.333, or to the left at a velocity β = −0.333 when viewed from the lab frame. The photon makes a diagonal trajectory in all frames.*

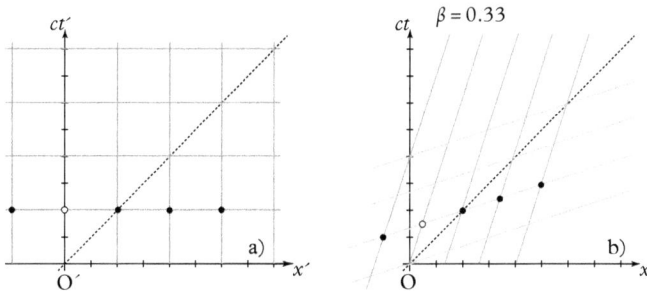

β = 0.333

β = −0.333

b) O

c) O

Figure 10.5 *Example of the relativity of simultaneity. On the left is the O′ rest-frame with events occurring simultaneously at different locations along the x-axis. On the right is the O′ frame viewed relative to a stationary frame O. The events are not seen as simultaneous in the frame O.*

β = 0.33

a)

b)

Minkowski space is a 4D space that requires vectors with four components to specify an event. The position four-vector is

$$x^a = \begin{pmatrix} ct \\ x \\ y \\ z \end{pmatrix}. \tag{10.6}$$

The numbering convention is $a \in \{ \begin{array}{cccc} 0 & 1 & 2 & 3 \end{array} \}$, where the zeroth element is the time-like coordinate. The product of speed of light times time is a distance, keeping the units consistent among all components.

The four-vector is transformed by the Lorentz transformation (Einstein summation convention assumed) as

$$x^{\bar{b}} = \Lambda^{\bar{b}}_a x^a,$$ (10.7)

where the transformation matrix in the standard configuration is

$$\Lambda^{\bar{b}}_a = \begin{pmatrix} \gamma & -\beta\gamma & 0 & 0 \\ -\beta\gamma & \gamma & 0 & 0 \\ 0 & 0 & 1 & 0 \\ 0 & 0 & 0 & 1 \end{pmatrix}.$$ (10.8)

The inverse transformation is

$$x^a = \Lambda^a_{\bar{b}} x^{\bar{b}},$$ (10.9)

where

$$\Lambda^a_{\bar{b}} = \left(\Lambda^{\bar{b}}_a\right)^{-1}$$ (10.10)

and

$$\Lambda^a_{\bar{b}} = \begin{pmatrix} \gamma & \beta\gamma & 0 & 0 \\ \beta\gamma & \gamma & 0 & 0 \\ 0 & 0 & 1 & 0 \\ 0 & 0 & 0 & 1 \end{pmatrix}.$$ (10.11)

The Lorentz transformation is an affine transformation in which the x'- and t'-axes are not orthogonal when viewed from the lab frame (shown in Fig. 10.4), but it is a linear transformation. Therefore, the Jacobian matrix of the transformation $\mathcal{J}^a_b$ is equal to the transformation itself in Eq. (10.8) with a Jacobian determinant

$$\mathcal{J} = \gamma^2 - \beta^2\gamma^2 = 1$$ (10.12)

and hence the transformation conserves space–time volume.

10.2.3   Kinematic consequences of the Lorentz transformation

The Lorentz transformation is a simple consequence of the two postulates of relativity, and this transformation is the starting point for mathematical derivations of the kinematic consequences of the postulates. There are many unexpected relativistic phenomena. Here, we consider time dilation, length contraction, velocity addition, the Lorentz boost, and the Doppler effect.

10.2.3.1 Time dilation

Consider a ticking clock at the origin of the O' frame at $x' = 0$. If it ticks exactly when it passes the origin O when $t' = t = 0$, then it will tick again at time $t = \Delta t$ for which the first line in Eq. (10.3) becomes

$$\Delta t = \gamma \, \Delta t'. \tag{10.13}$$

Because $\gamma > 1$, the time duration observed in O is longer than what is observed in O' (the rest frame of the clock), and hence *moving clocks tick slowly.*

One way to visualize moving clocks running slowly is with a "light clock." A light clock in its rest frame is shown in Fig. 10.6a. A light pulse originates at the bottom, travels a distance L to the top mirror, where it reflects and returns to the bottom. The time for the round trip is $\Delta t' = 2L/c$. When the same clock is viewed moving past in the lab frame in Fig. 10.6b, the light pulse must travel a longer distance. It is still traveling at the speed of light, so it takes longer to perform the round trip. The total time is

$$\Delta t = \frac{2}{c}\sqrt{L^2 + (v\Delta t/2)^2}. \tag{10.14}$$

Solving this for Δt gives

$$\Delta t^2 = \frac{4}{c^2}\left(L^2 + (v\Delta t/2)^2\right)$$
$$\Delta t^2 \left(1 - \beta^2\right) = \left(\frac{2L}{c}\right)^2 \tag{10.15}$$
$$\Delta t = \gamma \, \Delta t'$$

which is the time dilation.

10.2.3.2 Length contraction

Another consequence of the Lorentz transformation is that objects moving at relativistic speeds appear shortened in the direction of motion. This is derived from Eq. (10.3) in the standard configuration in which a moving meter stick is

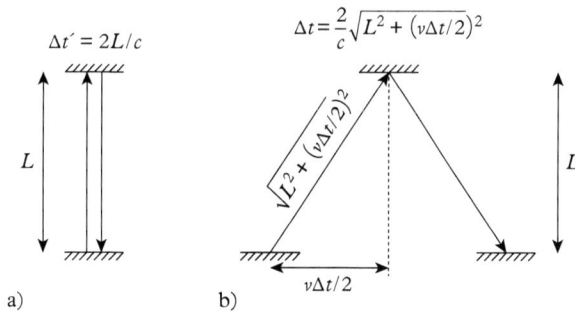

Figure 10.6 (a) *A light clock with a round trip $\Delta t' = 2L/c$ in the clock rest frame. (b) When viewed in the lab frame, the light path is longer, so the clock ticks slower.*

oriented along the *x*-axis. The primed frame is the rest frame of the meter stick that has a proper length L'. The unprimed frame is the laboratory frame relative to which the meter stick is moving with speed v. The meter stick is measured at $t = 0$ when the trailing edge of the meter stick is at the origin of both frames at $x_{Trail} = x'_{Trail} = 0$ and $t_{Trail} = t'_{Trail} = 0$. Using the first equation in Eq. (10.3) for the leading edge of the meter stick

$$t = \gamma \left(t'_{Lead} + \frac{v}{c^2} x'_{Lead} \right) = 0$$
$$t'_{Lead} = -\frac{v}{c^2} x'_{Lead}$$

(10.16)

and inserting the time of the measurement in the moving frame into the second equation of Eq. (10.3) gives

$$x_{Lead} = \gamma \left(x'_{Lead} - \frac{v^2}{c^2} x'_{Lead} \right)$$
$$= \gamma x'_{Lead} \left(1 - \frac{v^2}{c^2} \right)$$
$$= \frac{x'_{Lead}}{\gamma}.$$

(10.17)

Therefore, the measured length in the fixed frame is

$$L = x_{Lead} - x_{Trail}$$
$$= \frac{x'_{Lead}}{\gamma} = \frac{L'}{\gamma}$$

(10.18)

and the meter stick is observed to be shorter by a factor of γ. Hence, *moving objects appear shortened in the direction of motion.*

10.2.3.3 *Velocity addition*

Velocity addition arises from the problem of describing how two observers see the same moving object. For instance, in the primed frame an observer sees a speed u'_x, while in the unprimed frame an observer see the same object moving with a speed u_x. The speed of the primed frame relative to the unprimed frame is v. The speed u'_x is obtained using the differential expressions of Eq. (10.3).

$$\frac{dx'}{dt'} = \frac{\gamma (dx - v dt)}{\gamma \left(dt - \frac{v}{c^2} dx \right)}.$$

(10.19)

Dividing the numerator and denomenator on the right-hand-side by dt gives

$$\frac{dx'}{dt'} = \frac{(dx/dt - v)}{\left(1 - \frac{v}{c^2} dx/dt \right)}.$$

(10.20)

With $u_x = dx/dt$, this is simply

$$u'_x = \frac{u_x - v}{1 - \frac{u_x v}{c^2}}. \tag{10.21}$$

The same procedure applies to transverse velocities:

$$\frac{dy'}{dt'} = \frac{dy}{\gamma\left(dt - \frac{v}{c^2}dx\right)}$$
$$= \frac{dy/dt}{\gamma\left(1 - \frac{v}{c^2}\frac{dx}{dt}\right)} \tag{10.22}$$

and the full set of equations are

Velocity Addition:
$$u'_x = \frac{u_x - v}{1 - \frac{u_x v}{c^2}}$$
$$u'_y = \frac{u_y}{\gamma\left(1 - \frac{u_x v}{c^2}\right)} \tag{10.23}$$
$$u'_z = \frac{u_z}{\gamma\left(1 - \frac{u_x v}{c^2}\right)}.$$

The inverse relationships between O and O' simply reverse the sign of the relative speed v. Examples of velocity addition are shown in Fig. 10.7. The resultant observed speeds never exceed the speed of light.

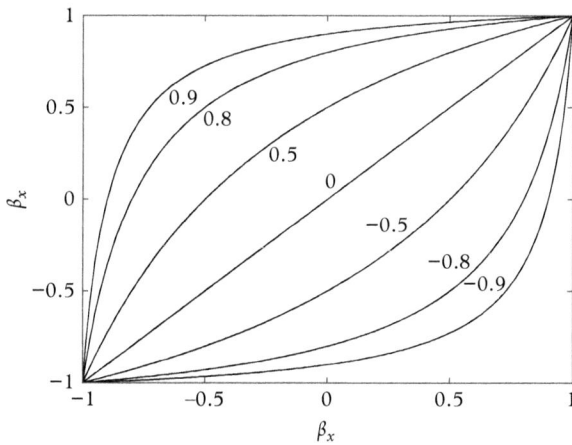

Figure 10.7 *Observed speed β_x in the O frame for an object moving along the x'-axis with a speed β_x' in the O' frame. The different curves are marked with the relative speed v between the frames.*

10.2.3.4 Lorentz boost

Relativistic velocity addition leads to the Lorentz boost, also known as the headlight effect (Fig. 10.8). When two relative observers see a light ray inclined at an oblique angle relative to the direction of frame motion, they observe the light ray making different angles. If the angle observed in the primed frame is given by

$$\cos \theta' = \frac{\Delta x'}{c\Delta t'}, \tag{10.24}$$

then the angle in the unprimed frame is

$$\cos \theta = \frac{\Delta x}{c\Delta t} = \frac{\gamma \left(\Delta x' + v\Delta t'\right)}{c\gamma \left(\Delta t' + v\Delta x'/c^2\right)}$$

$$= \frac{\gamma \left(\Delta x'/\Delta t' + v\right)}{c\gamma \left(1 + v\Delta x'/\Delta t'c^2\right)} \tag{10.25}$$

$$= \frac{\left(\Delta x'/\left(c\Delta t'\right) + \beta\right)}{\left(1 + \dfrac{v}{c}\Delta x'/\left(c\Delta t'\right)\right)}.$$

Re-expressing this in terms of the primed-frame angle gives

Lorentz Boost: $$\cos \theta = \frac{\cos \theta' + \beta}{1 + \beta \cos \theta'}. \tag{10.26}$$

[3] This principle is used in X-ray light sources in so-called wrigglers and undulators that cause transverse accelerations and radiation of X-rays that experience a Lorentz boost to form strongly collimated X-ray beams.

This is the transformation rule to transform a light ray angle from one frame to another. The reason it is called the headlight effect is because an object that emits isotropically in its rest frame will appear as a forward-directed headlight in the unprimed frame if it is moving at highly relativistic speeds. As the relative speed approaches the speed of light, $\cos \theta$ approaches unity for any angle θ', and the rays are compressed into the forward direction along the line of motion.[3]

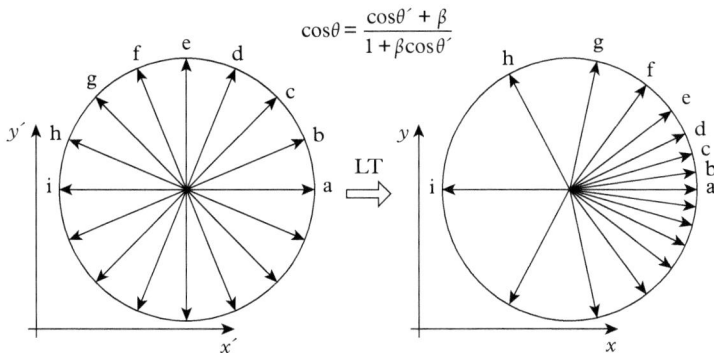

Figure 10.8 *The headlight effect for* $\beta = 0.8$. *In the rest frame, an isotropic light source emits uniformly in all directions. The same light source, observed in the lab frame, has a stronger emission in the forward direction—like a headlight.*

10.2.3.5 *Doppler effect*

The Doppler effect is an important effect in astrophysics, and it also leads to effects that are easily measured in the laboratory even for very low speeds that are far from the speed of light. For instance, green light reflected from a mirror moving at only 1 cm s^{-1} is shifted in frequency by 20 kHz, which is easily measured on an oscilloscope using an interferometer.

The Doppler effect is derived from the Lorentz transformation by considering the wavelength measured by two relative observers (Fig. 10.9). In the primed frame, which is taken as the rest frame of a light emitter, the measured wavelength is

$$\lambda' = c\Delta t'. \tag{10.27}$$

But in the unprimed frame, the emitter moves with a speed u during the emission of a single wavelength, and the resulting wavelength that is observed during an emission time is

$$\lambda = c\Delta t \mp u\Delta t$$
$$= (c \mp u)\,\Delta t, \tag{10.28}$$

where the minus sign is when the source is moving in the same direction as the emission direction and vice versa. The emission time is related to the emission time in the rest frame by time dilation, which gives

$$\lambda = (c \mp u)\,\gamma\,\Delta t'$$
$$= \gamma\,(1 \mp \beta)\,\lambda' \tag{10.29}$$

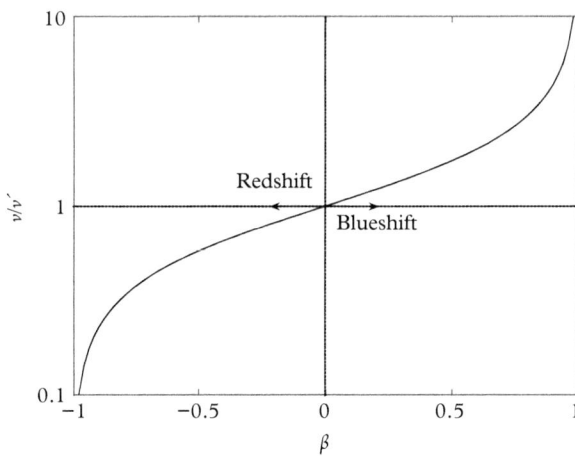

Figure 10.9 *Relative Doppler ratio showing redshifts and blueshifts for a source moving away from, or toward, the observer.*

to give the change in wavelength

$$\lambda = \sqrt{\frac{(1 \mp \beta)}{(1 \pm \beta)}} \lambda' \qquad (10.30)$$

or the change in frequency

Doppler Effect: $\nu = \sqrt{\dfrac{(1 \pm \beta)}{(1 \mp \beta)}} \nu'.$ $\qquad (10.31)$

The plus sign in the last equation is for blue-shifted light, and the minus sign is for red-shifted light.

10.2.4   Classic paradoxes

Because of the non-intuitive nature of relativity, there are many physical phenomena that appear at first to lead to contradictions and hence are paradoxes that defy "common sense." However, none of these apparent paradoxes are true paradoxes, because each can be explained fully within the framework of the theory of relativity. The process of explaining why these are *not* paradoxes gives deep insight into the physics of relativity and can help build intuition in this unintuitive science. There are many classic paradoxes, of which three are described here. These are the *muon paradox*, the *twin paradox*, and the *pole and barn paradox.*

10.2.4.1   *Muon paradox*

A muon is a fundamental particle (a lepton) that has some of the properties of an electron, but it is much heavier, with a rest mass that is about 200 times heavier than the electron. In its rest frame, a muon decays into an electron (plus a neutrino) with a mean decay time of $\tau = 2.2$ μs. Muons are common constituents of the flux of cosmic rays that bombard the Earth's surface. They travel close to the speed of light, and hence the mean decay length of the muon would be $c\tau = 660$ m. The muon flux can be measured at the peak of a mountain 660 m high, and measured also at sea level. The flux ratio in this case might be expected to be $\exp(-660/c\tau) = 0.37$. However, in actual experiments, the flux ratio is about 82%. In this paradox, the resolution is obvious—the muon's lifetime is time dilated. Muons traveling at 0.98 times the speed of light, with $\gamma = 5$, have a lifetime that is about five times longer than the rest decay time, and the flux reaching sea level is $\exp(-660/c5\tau) = 0.82$. An alternative way to look at this problem is from the point of view of the muon. In its frame, it decays in 2.2 μs, but the height of the mountain is length-contracted to $660/5 = 132$ m and hence fewer muons decay between the measurement at the top of the mountain and sea level.

10.2.4.2   *Twin paradox*

This is perhaps the most famous of the relativity paradoxes and requires considerable more effort to explain compared to the muon paradox. Consider a pair of twins of exactly the same age: Bob and Alice. Alice is on Earth, while Bob is on a space ship traveling at $\beta = 0.8$ with a Lorentz factor $\gamma = 5/3$. From Alice's viewpoint, Bob travels for 5 years and then turns around and returns to Earth after a total trip time of 10 years. Because she sees Bob traveling near the speed of light, she sees his clock running slowly and hence she believes he has not aged as much as she has. Conversely, from Bob's frame, it is Alice who is moving at relativistic speeds. He sees her clock running slowly and hence he believes that she has not aged as much as he has. In other words, each sees the other's clock running slowly and each thinks the other has not aged as much. The question is, when they get back together, who has actually aged the least?

To settle this question, let's put some numbers to the problem. First, it is clear that Alice ages 10 years. Hence

$$\Delta t_A = 10 \text{ yrs}.$$

Alice sees Bob's clock running slowly by a factor of 5/3, and hence she thinks

$$\Delta t_B = \frac{3}{5} 10 \text{ yrs} = 6 \text{ yrs}.$$

In Alice's frame, Bob travels a distance of $0.8c \cdot 5$ yrs = 4 lyrs outward bound for a total distance of 8 light years round-trip. From his point of view, this distance is length-contracted by a factor of 3/5, so he only measures a total round-trip distance of 4.8 light years, which takes him 6 years to cover. Therefore, Bob's elapsed time is

$$\Delta t'_B = \frac{\frac{3}{5} 8 \text{ lyrs}}{0.8c} = 6 \text{ yrs}.$$

But he sees Alice's clock running slowly by the factor of 5/3, so he thinks her clock has elapsed

$$\Delta t'_A = \frac{3}{5} 6 \text{ yrs} = 3.6 \text{ yrs}.$$

This is then the paradox. Alice has aged 10 years but thinks Bob has aged 6 years. Bob has aged 6 years, but thinks Alice has aged 3.6 years. Each thinks the other is younger, although they both agree on Bob's age. As we put numbers to this paradox, we were careful to include both time dilation and length contraction effects, yet we are still stuck with the paradox.

To resolve the paradox, we need to do better "bookkeeping" on the elapsed times. To accomplish this, let both Bob and Alice emit light pulses once per year during the trip, and each measures how many times they receive light flashes from the other. By counting light flashes, they know for certain how much time has elapsed on each other's clocks. Counting light pulses emitted by a clock is equivalent to measuring a Doppler frequency shift. The situation is calculated in two parts—one for the outbound journey and one for the inbound journey. In both cases, Alice's and Bob's frequencies in their rest frames are

$$f_A = 1/\text{yr}$$
$$f_B' = 1/\text{yr}.$$

For the outbound journey, the frequency ratios are

$$\frac{f_B}{f_B'} = \sqrt{\frac{1-\beta}{1+\beta}} = \frac{1}{3}$$

$$\frac{f_A'}{f_A} = \sqrt{\frac{1-\beta}{1+\beta}} = \frac{1}{3}$$

from which

$$f_B = \frac{1}{3}\ \text{yr}^{-1}$$

$$f_A' = \frac{1}{3}\ \text{yr}^{-1}.$$

These are the Doppler-shifted frequencies that Alice sees for Bob, and Bob sees for Alice. Each sees the same slow rate of pulse arrival, but each sees this rate for different amounts of time. The durations of the outward trip are

$$\Delta t_{out} = 5\ \text{yrs} + 0.8 \cdot 5\ \text{yrs} = 9\ \text{yrs}$$
$$\Delta t_{out}' = 3\ \text{yrs}$$

for Alice and Bob, respectively. Therefore, Alice receives three flashes from Bob's outward journey, and Bob receives only one flash from Alice during his outbound journey.

For in the inbound journey, the frequency ratios are

$$\frac{f_B}{f_B'} = \sqrt{\frac{1+\beta}{1-\beta}} = 3$$

$$\frac{f_A'}{f_A} = \sqrt{\frac{1+\beta}{1-\beta}} = 3$$

from which

$$f_B = 3\ \text{yr}^{-1}$$
$$f_A' = 3\ \text{yr}^{-1}.$$

But the durations of the inward journey are

$$\Delta t_{in} = 1\ \text{yr}$$
$$\Delta t_{in}' = 3\ \text{yrs}.$$

Therefore, Alice receives three flashes from Bob, but Bob receives nine flashes from Alice.

In total, for the round-trip, Alice receives 3 + 3 = 6 flashes from Bob. Bob receives 1 + 9 = 10 flashes from Alice. Therefore, Alice is 10 years older after the trip, but Bob is only 6 years older, so Bob is younger than Alice, and each now agrees on the correctness of the calculations. The space–time diagram from Alice's frame is shown in Fig. 10.10. The diagonal lines are light pulses emitted by Alice (emitted from the ct-axis) and the other diagonal lines are the light pulses emitted by Bob. Bob emits 6 pulses in total (the final pulse is emitted just as he arrives back at Earth), and Alice emits 10 pulses. Hence, Alice has aged 10 years to Bob's 6 years.

This explanation of the twin paradox relies on accurate bookkeeping—in other words, just arithmetic. But a qualitative explanation is still needed. The original paradox was expressed as each seeing the other's clock running slowly during the travel. This put a false emphasis on an apparent symmetry between Bob and Alice. However, this paradox has a fundamental asymmetry—because Bob must turn around. During his turnaround, he would experience deceleration and then acceleration forces that Alice would never experience. Therefore, Bob must know that he had altered his motion during the trip, while Alice did not. In fact, when Bob decelerates and then reaccelerates to return home, he is no longer in an inertial reference frame. This puts the problem into the realm of the theory of general relativity, which treats non-inertial frames and the physics of gravity which is beyond the scope of this textbook. Nonetheless, this Doppler approach does give the correct resolution to the twin paradox if Bob's turnaround is instantaneous.

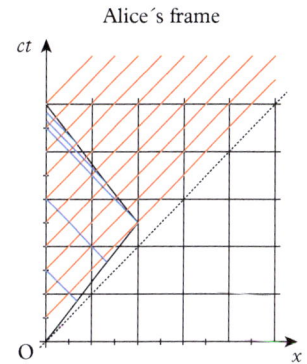

Figure 10.10 *Space–time diagram of the twin paradox from Alice's frame. The diagonal lines emerging from the ct-axis are the photon trajectories of each of her pulses. The other diagonal lines are the photon trajectories of each of Bob's pulses. Alice emits 10 light pulses in ten years, while Bob emits only 6. Therefore, when he returns, Bob is younger than Alice.*

10.2.4.3 Pole and barn paradox

In this paradox, Alice is a pole-vaulter carrying a pole that she measures to be 22 m long in her rest frame. She runs at a relativistic speed of $\beta = 0.99$ ($\gamma = 7$) at a barn that is 20 m long (in its rest frame) with a front and back door. In the barn rest frame, the pole is length-contracted to $22/7 = 3.14$ m. Hence the pole can be completely inside the barn—in fact, both the barn doors can be closed while the pole is inside. But from Alice's point of view, the barn is length-contracted to a length $20/7 = 2.86$ m, so in fact the pole sticks out of both the front and back doors as she passes through. The paradox is: How can the barn doors be closed in the barn frame (completely enclosing the pole), when from Alice's frame the barn is so short that the pole can never fit inside and the doors cannot be closed?

The resolution of this paradox is in simultaneity (or the lack of it). In the barn frame the front and back doors close simultaneously, completely enclosing the pole. But in Alice's frame, the doors do not close simultaneously. In fact, the back door opens to allow the front of Alice's pole to enter, then the front door opens to allow the front of Alice's pole to leave the barn, but the back of her pole has yet to enter. Once the back of the pole enters the barn, the back door closes. Finally, as the back of the pole leaves the front of the barn, the front door closes. At no time were both the front and back door closed, so the pole never smashes into a door. The pole was never entirely inside the barn in Alice's frame.

10.3   Metric structure of Minkowski space

Minkowski space is a flat space without curvature, but it is not a Euclidean space. It is pseudo-Riemannian and has some unusual properties, such as distances that can be negative. Its geometric properties are contained in its metric tensor.

10.3.1   Invariant interval

For two events, consider the quantity

$$(\Delta s)^2 = \sum_{j=1}^{3} \Delta x_j'^2 - c^2 \Delta t'^2. \tag{10.32}$$

How does it transform? Individually

$$\begin{aligned}
\left(\Delta x'\right)^2 &= \gamma^2 (\Delta x - v \Delta t)^2 \\
c^2 \Delta t'^2 &= c^2 \gamma^2 \left(\Delta t - \frac{v}{c^2} \Delta x\right)^2
\end{aligned} \tag{10.33}$$

and combining these gives

$$\begin{aligned}
\left(\Delta x'\right)^2 - c^2 \Delta t'^2 &= \gamma^2 \left[\Delta x^2 - 2v\Delta t \Delta x + v^2 \Delta t^2 - c^2 \Delta t^2 + 2v\Delta t \Delta x - \frac{v^2}{c^2}\Delta x^2 \right] \\
&= \gamma^2 \left(\Delta x^2 - c^2 \Delta t^2\right) + \gamma^2 \beta^2 \left(c^2 \Delta t^2 - \Delta x^2\right) \\
&= \gamma^2 \left(1 - \beta^2\right)\left(\Delta x^2 - c^2 \Delta t^2\right) \\
&= \Delta x^2 - c^2 \Delta t^2.
\end{aligned} \tag{10.34}$$

Therefore,

$$\Delta s^2 = \left(\Delta x'\right)^2 - c^2 \Delta t'^2 = \Delta x^2 - c^2 \Delta t^2, \tag{10.35}$$

where Δs^2 between the same two events takes on the same value in all reference frames, no matter what the relative speeds are. The quantities Δx and Δt need not be small, but can have any magnitude.

The invariant interval leads to the closely-associated idea of an invariant hyperbola. For a given Δs, the equation

$$\Delta s^2 = x'^2 - c^2 t'^2 = x^2 - c^2 t^2 \tag{10.36}$$

describes hyperbola in Minkowski space. All points on a given hyperbola are "equidistant" from the origin. Lorentz transformations keep an event always on the same hyperbola. In Fig. 10.11, a set of events (the solid circles) that are simultaneous in the rest frame of a moving system is transformed into coordinates observed by a fixed observer (the open circles). The Lorentz transformation

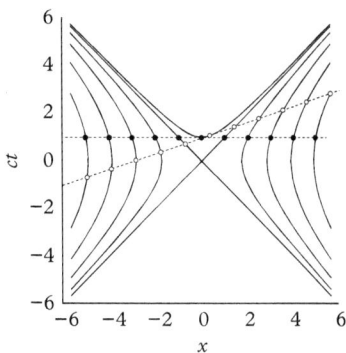

Figure 10.11 *Graph showing invariant hyperbola. All points on the same hyperbolic curve are the same "distance" from the origin. The solid points are simultaneous events in the primed frame. The open circles are the same events viewed in the unprimed frame. A given event, when transformed between frames, remains always on the same hyperbola.*

transforms each event onto a new point on its own hyperbola because the metric "distance" Δs between two events is an invariant property, unaffected by transformation. The hyperbola associated with each event plot on top of each other in both the fixed and moving frames—indeed, it does not matter what frame is used to view the hyperbola, but the individual event locations on the hyperbola shift as the frame of view changes.

The invariant interval suggests the choice for a differential line element in Minkowski space (see Chapter 9) as

$$ds^2 = -c^2 dt^2 + dx^2 + dy^2 + dz^2$$
$$= g_{ab} dx^a dx^b, \tag{10.37}$$

where

$$g_{ab} = \begin{pmatrix} -1 & 0 & 0 & 0 \\ 0 & 1 & 0 & 0 \\ 0 & 0 & 1 & 0 \\ 0 & 0 & 0 & 1 \end{pmatrix} \tag{10.38}$$

is the Minkowski metric expressed in matrix form.[4] The determinant of the Minkowski metric equals -1, so this metric is known as pseudo-Riemannian. It defines a metric "distance" on the space, but the distance can take on positive, zero or negative values. Therefore, although Minkowski space is "flat," it is not Euclidean with Cartesian distances.

In spite of its unusual geometry, the metric tensor plays all the same roles in Minkowski space that it does in Riemannian geometry. The inner product is defined in terms of g_{ab} through Eq. (10.37). Similarly, a position four-covector is obtained from a vector as

$$x_b = g_{ab} x^a \tag{10.39}$$

or

$$\begin{pmatrix} -ct & x & y & z \end{pmatrix} = \begin{pmatrix} -1 & 0 & 0 & 0 \\ 0 & 1 & 0 & 0 \\ 0 & 0 & 1 & 0 \\ 0 & 0 & 0 & 1 \end{pmatrix} \begin{pmatrix} ct \\ x \\ y \\ z \end{pmatrix}. \tag{10.40}$$

The Minkowski matrix in Eq. (10.38) converts a column vector into another column vector, but it is represented here as a row vector (with a change in sign on the first element of the position four-vector) as a reminder that it is a covector.

10.3.2 Proper time and dynamic invariants

The invariant interval leads to an important quantity called proper time. The word "proper" is from the French "propre" which means "one's own." It is also

[4] The metric tensor is not strictly a matrix because it has two covariant indices. However, it is convenient to give it matrix form as an operator that can multiply four-vectors.

known as the rest-frame time, such as the mean decay lifetime for a radioactive particle, or the ticking of a well-made clock. These are physical properties of the particles or the apparatus and therefore are invariant.

When two events are separated in time, but occur at the same location in space, the invariant interval is

$$\Delta s^2 = -c^2 \Delta t'^2 = -c^2 d\tau^2. \tag{10.41}$$

The proper time is then

$$d\tau = \sqrt{-\frac{\Delta s^2}{c^2}}, \tag{10.42}$$

where only the positive square root is taken. By time dilation it is also

Proper time interval: $\qquad d\tau = \dfrac{1}{\gamma} dt. \tag{10.43}$

The proper time plays an important role when defining other four-vectors beyond the position four-vector. For instance, the velocity four-vector is obtained from the position four-vector by taking the derivative with respect to the proper time:

$$v^a = \frac{dx^a}{d\tau} = \gamma \begin{pmatrix} c \\ u^x \\ u^y \\ u^z \end{pmatrix}. \tag{10.44}$$

The derivative with respect to proper time produces an object with the same transformation properties as the position four-vector. If the derivative had been taken with respect to time, the corresponding four-velocity would not have transformed the same way as a four-vector because the time in the denominator would also be transformed.

When the four-velocity is defined in this way by using the invariant interval, the inner product of the four-velocity with itself is

$$\begin{aligned} g_{ab}v^a v^b &= -\gamma^2 c^2 + \gamma^2 u^2 \\ &= -\gamma^2 \left(c^2 - u^2 \right) \\ &= -c^2 \end{aligned} \tag{10.45}$$

which is clearly an invariant quantity. This result is dictated by the metric properties of any physical space. The length of a vector is an intrinsic scalar property and cannot be dependent on the choice of coordinate system. Scalars are just numbers, and must be the same in any frame.

The four-momentum of a particle is obtained readily from the four-velocity by multiplying by the rest mass of the particle

$$p^a = mv^a$$

$$= \gamma m \begin{pmatrix} c \\ u^x \\ u^y \\ u^z \end{pmatrix}. \tag{10.46}$$

The inner product of the four-momentum with itself is

$$g_{ab}p^a p^b = -\gamma^2 m^2 c^2 + \gamma^2 m^2 u^2$$
$$= -m^2 c^2, \tag{10.47}$$

where m is the rest mass of the particle, and the inner product is again an invariant.

10.3.3 Null geodesics

In relativity, both special and general, the trajectories of photons hold a special place as a family of invariant curves. They are geodesics, but of a special kind called null geodesics. Once you have mapped out the light curves of a space–time, you have a fairly complete picture of how the space–time behaves—even in the case of strongly warped space–time around black holes.

A metric space with the property $g_{ab}V^a V^b \geq 0$ is positive definite, and distances are calculated as we intuitively expect in two and three dimensions. Conversely, a metric space with property $g_{ab}V^a V^b < 0$ is indefinite (also known as pseudo-Riemannian). Indefinite metric spaces can have curves with zero arc-length. Recall that in Minkowski space, which is pseudo-Riemannian, every point on an invariant hyperbola is equidistant from the origin. Similarly, light curves have zero distance to the origin. These null geodesics have the property

$$ds^2 = 0 \tag{10.48}$$

along the curve. Because the path length is zero, null geodesics require parameterization by the variable t rather than s:

$$\frac{d^2 x^a}{dt^2} + \Gamma^a_{bc} \frac{dx^b}{dt} \frac{dx^c}{dt} = 0 \tag{10.49}$$

with the further constraint

$$g_{rs} \frac{dx^r}{dt} \frac{dx^s}{dt} = \left(\frac{ds}{dt} \right)^2 = 0, \tag{10.50}$$

which is the constraint of Eq. (9.107), with $ds = 0$.

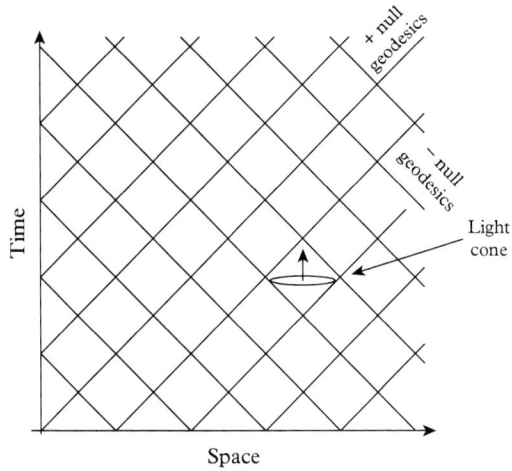

⁵ For a mathematical description of a relativistic anharmonic oscillator, see W. Moreau, R. Easther, and R. Neutze, *Am. J. Phys.* vol. 62, pp. 531–535 (1994).

Figure 10.12 *Null geodesics in Minkowski space. Photons have trajectories (world lines) that are oriented at 45°.*

Null geodesics in Minkowski space–time are found by considering the Christoffel symbols $\Gamma^a_{bc} = 0$ that lead to the linear equation $x^a = x^a_0 + v^a t$. Furthermore, $g_{ab}V^aV^b = 0$ leads to

$$g_{ab}\left(x^a - x^a_0\right)\left(x^b - x^b_0\right) = 0, \qquad (10.51)$$

which gives

$$(ct - ct_0)^2 = (x - x_o)^2 + (y - y_o)^2 + (z - z_o)^2. \qquad (10.52)$$

In the (ct, x) plane, these are lines at 45°, as shown in Fig. 10.12. The null geodesics in Minkowski space are the trajectories of photons.

10.3.4 Trajectories as world lines

Trajectories of massive particles are drawn on Minkowski space–time diagrams as "world lines." The particles can be subject to varying forces that cause them to accelerate or decelerate, but their speed always remains less than the speed of light and, hence, on a space–time diagram they always have slopes greater than 45°. An example of a world line is shown in Fig. 10.13 for a relativistic harmonic oscillator. The trajectory is oscillatory, but the equation for this oscillator is not a simple linear equation.⁵

All pairs of points on a single connected world line have $\Delta s^2 < 0$ and hence have a "time-like" interval. These points with time-like intervals can be causally connected such that events unambiguously can be defined to occur in the past or in the future relative to one another. A world line is constrained by a "light cone." On a space–time diagram, there can be events that are not causally connected, and hence also cannot unambiguously be defined to be in the future or the past relative to each other. Such descriptions as future and past are relative for these types of events and depend on the frame of reference. These events are separated by an interval $\Delta s^2 > 0$ and are said to have a "space-like" interval.

10.4 Relativistic dynamics

Just as classical mechanics is traditionally divided into kinematics (how things move) and dynamics (why things move), relativistic mechanics likewise has this natural division. Relativistic dynamics is concerned with momenta and forces under relativistic speeds. In relativity, forces and accelerations do not share transformation properties like position and velocity. Even in Galilean relativity, positions and velocities transform, but forces and accelerations are frame-invariant. This is a result of the fundamental principle of relativity: that physics is invariant to the choice of inertial frame. Since Newtonian physics is defined by forces and accelerations, these are consequently invariant to the choice of inertial reference frame. On the other hand, in special relativity, this simple invariance of acceleration and force vectors is *not* preserved, but must be replaced with invariant products of vectors and covectors. For this reason, forces and accelerations *do* have relativistic transformations, but their invariant products are the main objects of interest.

10.4.1 Relativistic energies

Kinetic energy is derived by considering the three-force acting on a particle that is moving with a speed $\boldsymbol{u}$ relative to a fixed observer. Newton's second law is still related to the momentum of the particle

$$\mathbf{F} = \frac{d\mathbf{p}}{dt} = \frac{d}{dt}\left(\gamma m \mathbf{u}\right), \tag{10.53}$$

as observed in the fixed lab frame. By integrating the mechanical power, the work done on the particle is found to be

$$\begin{aligned} W = T &= \int \frac{d}{dt}\left(\gamma m \mathbf{u}\right) \cdot \mathbf{u}\,dt \\ &= m \int_0^u u\, d\left(\gamma u\right). \end{aligned} \tag{10.54}$$

Integrating by parts gives

$$\begin{aligned} T &= \gamma m u^2 - m \int_0^u \frac{u\,du}{\sqrt{1 - u^2/c^2}} \\ &= \gamma m u^2 + m c^2 \sqrt{1 - u^2/c^2} - m c^2 \\ &= \gamma m c^2 - m c^2 \end{aligned} \tag{10.55}$$

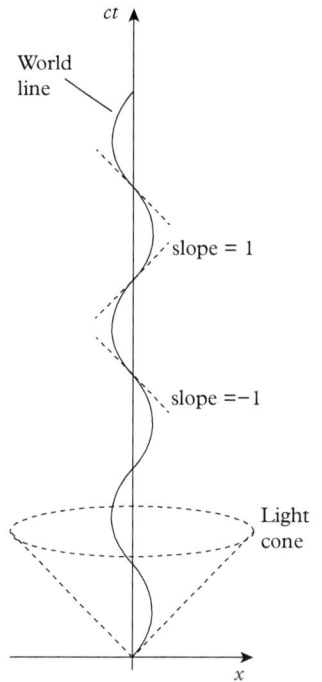

Figure 10.13 *World line of a massive particle oscillating relativistically. The slope of the space–time trajectory always has a magnitude greater than unity.*

From this result it is simple to make the assignment

$$T = E - E_0 \tag{10.56}$$

for the kinetic energy of the particle, where the total energy is

$$E = \gamma mc^2 \tag{10.57}$$

and the rest energy of the particle is

$$E_0 = mc^2. \tag{10.58}$$

The kinetic energy is therefore also expressed as

$$T = (\gamma - 1)\, mc^2. \tag{10.59}$$

These last three expressions are very useful when working with total, rest, and kinetic energy as well as for mass–energy conversions.

10.4.2 Application: mass–energy conversions

The equivalence between mass and energy enables the conversion from one form to the other. In other words, energy can be converted to mass, and mass can be converted to energy—most notably in the form of fusion or fission energy. As an example of energy conversion to mass, consider the binding of a neutron and a proton in the nucleus of a deuterium atom. The energy of the separated constituents is

$$E_\infty = m\left(^1_1\mathrm{H}\right) + m_n c^2, \tag{10.60}$$

where m_n is the mass of the neutron and $m\left(^1_1\mathrm{H}\right)$ is the mass of hydrogen that includes the electron and the binding energy of the electron to the proton. The energy of a deuterium atom is similarly

$$E_D = m\left(^2_1\mathrm{H}\right) c^2. \tag{10.61}$$

Therefore, the binding energy of deuterium is given by

$$\begin{aligned} \Delta E_B &= E - E_D \\ &= m\left(^1_1\mathrm{H}\right) c^2 + m_n c^2 - m\left(^1_1\mathrm{H}\right) c^2. \end{aligned} \tag{10.62}$$

The conversion from mass units to energy is

$$c^2 = 931.5\,\frac{\mathrm{MeV}}{u}, \tag{10.63}$$

where the mass unit, denoted by u, is defined by ^{12}C which has 12 mass units for 12 nucleons. Numerically, the energy difference for the deuteron is

$$\Delta E_B = (1.007825u + 1.008665u - 2.014102u)\left(931.5\frac{\text{MeV}}{u}\right)$$

$$= 0.002388u\left(931.5\frac{\text{MeV}}{u}\right) \qquad (10.64)$$

$$= 2.224422\,\text{MeV}.$$

It is common to quote this result as the binding energy per nucleon, which is 1.12 MeV per nucleon in this case.

In this example, the deuterium atom is heavier than the mass of its separated components. This extra mass comes from the binding energy of the neutron to the proton. It is a direct example of the conversion energy (in this case binding energy) to mass.

10.4.3   Application: antimatter pair production

One of the consequences of mass–energy equivalence is the possibility to create new matter out of energy. The creation of matter is governed by conservation laws that go beyond simple energy conservation, which are topics in high-energy particle physics that go beyond the scope of this book. However, one of the conservation laws is the conservation of lepton number.[6] If kinetic energy is to be converted to the creation of an electron, then the conservation of lepton number requires that an anti-electron (known as a positron) also must be created.

As an example, consider a relativistic electron that collides head-on with an electron in a thin metal foil in the laboratory shown in Fig. 10.14. The kinetic

[6] Leptons are one of the fundamental classes of massive particles that include electrons and positrons as well as neutrinos. Electrons have positive lepton number +1, while positrons have negative lepton number −1. Hence, pair production conserves lepton number.

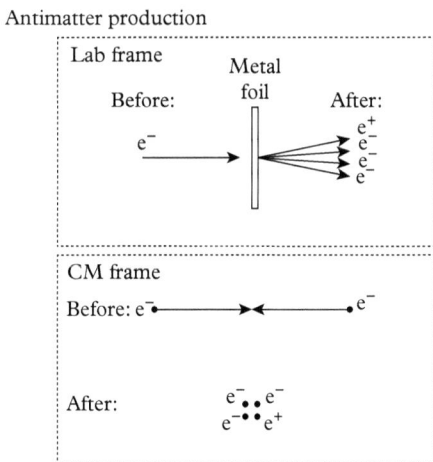

Figure 10.14 *Antimatter   (positron) production through the energetic collision of two electrons. The upper panel shows in incident electron on a metal foil in the laboratory frame. The bottom panel shows the same process from the point of view of the center of mass.*

energy of the incident electron can be converted into the creation of an electron-positron pair in addition to the original two electrons (the incident electron and the electron in the foil). The pair-production reaction is written as

$$2e^- \rightarrow 3e^- + e^+. \tag{10.65}$$

This reaction is most easily solved by considering the reaction from the point of view of the center-of-mass (CM) frame. The initial energy in the CM frame is

$$E_i' = 2\gamma' m_e c^2, \tag{10.66}$$

where the primes relate to the CM frame. The lowest energy for which this reaction can take place is set by the final state when all of the particles are stationary in the CM frame. In this case, the only energy is mass energy, and the final energy is

$$E_f' = 4m_e c^2. \tag{10.67}$$

Equating the initial with the final energy gives

$$2\gamma' m_e c^2 = 4m_e c^2$$
$$\gamma' = 2 \tag{10.68}$$

which establishes the Lorentz factor. For the original two electrons in the CM frame, the kinetic energy is

$$K' = (\gamma' - 1) m_e c^2 = 0.511 \, \text{MeV}/c^2 \tag{10.69}$$

for each one. To find the speeds of the initial two electrons in the CM frame, solve for v' as

$$2 = \frac{1}{\sqrt{1 - \dfrac{v'^2}{c^2}}}$$
$$v'^2 = \frac{3}{4}c^2 \tag{10.70}$$
$$v' = 0.866c.$$

These are transformed back into the lab frame where the electron in the metal foil is stationary, so that

$$u = v' \tag{10.71}$$

and solving for the speed,

$$v = \frac{v' + u}{1 + \dfrac{uv'}{c^2}} = 0.9897c, \tag{10.72}$$

gives the Lorentz factor in the lab frame,

$$\gamma = \frac{1}{\sqrt{1 - \dfrac{v^2}{c^2}}} \approx 7, \tag{10.73}$$

for the incident electron. The original energy in the lab frame is

$$E_i = 7 m_e c^2 + m_e c^2 = 8 m_e c^2 \tag{10.74}$$

and the incident electron has a kinetic energy

$$K_i = 6 m_e c^2. \tag{10.75}$$

The final total energy is shared equally among all four particles moving with speed v'

$$E_f = 4 \frac{1}{\sqrt{1 - \dfrac{v'^2}{c^2}}} m_e c^2 = 8 m_e c^2 \tag{10.76}$$

and the final kinetic energy is then

$$K_f = 8 m_e c^2 - 4 m_e c^2 = 4 m_e c^2. \tag{10.77}$$

Therefore, positron production at threshold also has the unique property that all of the produced particles have the same velocity.

10.4.4   Momentum transformation

By using the identifications

$$p = \gamma m u \quad \text{and} \quad E = \gamma m c^2, \tag{10.78}$$

the four-momentum inner product is also expressed as

$$p^2 = p_a p^a = m^2 c^2 + \gamma^2 m^2 u^2 = \gamma^2 m^2 c^2$$
$$m^2 c^2 + p^2 = \frac{E^2}{c^2} \tag{10.79}$$

which yields the important result

Energy, mass, and momentum: $E^2 = \left(m c^2 \right)^2 + p^2 c^2. \tag{10.80}$

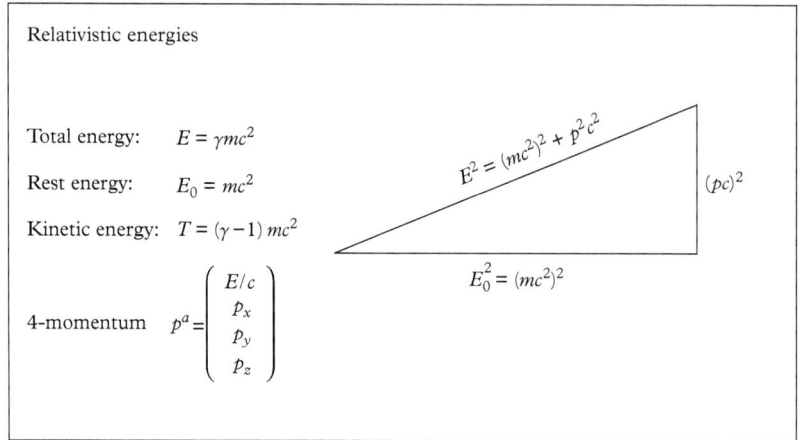

Figure 10.15 *Relativistic energies (total, rest, and kinetic) and their relationship to momentum, derived by constructing the invariant product of the four-momentum.*

This relates total energy and rest energy to the particle momentum. For a photon that has no rest mass, the simple (and Einstein's famous) relationship is

$$E = pc. \tag{10.81}$$

The relativistic energies are summarized in Fig. 10.15, showing the geometric relationship between total energy, rest energy, and momentum.

The four-momentum of Eq. (10.46) can be expressed as

$$p^a = \begin{pmatrix} E/c \\ p_x \\ p_y \\ p_z \end{pmatrix}, \tag{10.82}$$

where the invariant momentum product is given in Eq. (10.80). The four-momentum is a four-vector with the usual Lorentz transformation properties,

$$
\begin{aligned}
E' &= \gamma \left(E - v p_x\right) & E &= \gamma \left(E' + v p'_x\right) \\
p'_x &= \gamma \left(p_x - \frac{v}{c^2}E\right) & p_x &= \gamma \left(p'_x + \frac{v}{c^2}E'\right) \\
p'_y &= p_y & p_y &= p'_y \\
p'_z &= p_z & p_z &= p'_z
\end{aligned}
\tag{10.83}
$$

that couple the time-like component (energy) to the space-like components (momentum). The conservation of three-momentum plays a key role in high-energy scattering processes like Compton scattering.

10.4.5   Application: Compton scattering

Compton scattering is the process by which a photon scatters off a stationary electron. The Feynman diagram for Compton scattering is shown in Fig. 10.16.

Compton scattering

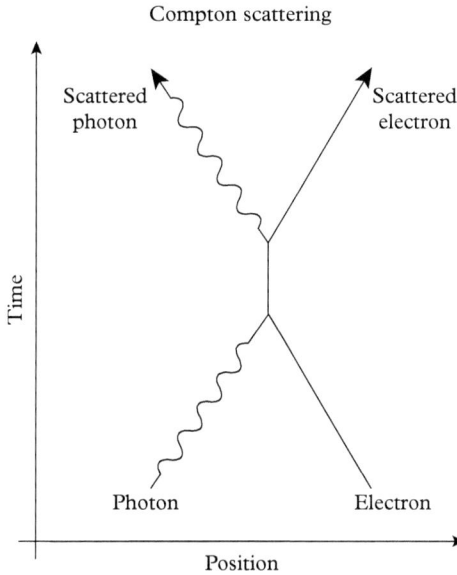

Figure 10.16 *Feynman diagram of Compton scattering where a photon scatters off of an electron.*

Feynman diagrams are routinely used to illustrate the interactions among fundamental particles. They are like cartoons of space–time diagrams with position along the horizontal axis and time increasing along the vertical axis. In this Feynman diagram, the electron first absorbs the photon, and then emits a new one. In the process, the individual energies and momenta of the photons and the electron change, but both energy and momentum are conserved between the initial and final states.

The energy of the scattered electron and photon can be derived as a function of the scattering angle. Energy conservation establishes

$$E_{init} = E_{final}$$
$$E + mc^2 = E' + E_e, \tag{10.84}$$

where m is the electron mass, E is the incident photon energy, E' is the energy of the scattered photon, and E_e is the energy of the scattered electron. The three-momentum conservation gives

$$p = p_e \cos\phi + p' \cos\theta$$
$$0 = p_e \sin\phi - p' \sin\theta, \tag{10.85}$$

where θ and φ are the scattering angles that define the directions of the scattered photon and electron, respectively. There are four unknowns (θ, ϕ, E_e, E') but only three equations. It is possible to eliminate E_e and φ by asking only about properties of the photon. The momentum equations are

$$p_e \cos \phi = p - p' \cos \theta$$
$$p_e \sin \phi = p' \sin \theta. \tag{10.86}$$

By squaring each equation and adding them,

$$p_e^2 = p^2 - 2pp' \cos \theta + p'^2 \cos^2 \theta + p'^2 \sin^2 \theta$$
$$p_e^2 = p^2 - 2pp' \cos \theta + p'^2 \tag{10.87}$$

the energy of the electron is

$$E_e^2 = c^2 p_e^2 + m^2 c^4. \tag{10.88}$$

Combining these equations gives

$$\left(E + mc^2 - E' \right)^2 = c^2 \left(p^2 - 2pp' \cos \theta + p'^2 \right) + m^2 c^4,$$
$$Emc^2 - mc^2 E' = EE' \left(1 - \cos \theta \right), \tag{10.89}$$
$$\frac{E - E'}{EE'} = \frac{1}{mc^2} \left(1 - \cos \theta \right),$$

which finally gives the *Compton scattering* formula

$$\frac{1}{E'} - \frac{1}{E} = \frac{1}{mc^2} \left(1 - \cos \theta \right). \tag{10.90}$$

This last equation is often written in terms of the photon wavelength using

$$E = \frac{hc}{\lambda}, \tag{10.91}$$

which is one of the few places in this textbook that makes a direct connection to quantum physics. The relationship between the initial and final wavelength of the photon is

$$\lambda' - \lambda = \lambda_e \left(1 - \cos \theta \right), \tag{10.92}$$

where the *Compton wavelength* for photon scattering from an electron is

$$\lambda_e = \frac{h}{m_e c} = 2.426 \times 10^{-12} \text{m}. \tag{10.93}$$

For forward scattering ($\theta = 0$), there is no shift in the photon wavelength. For backscattering ($\theta = \pi$), the change in the photon wavelength is a maximum, with the value

$$\Delta\lambda = 2\lambda_e, \tag{10.94}$$

which is about 5 picometers (5 pm). Obviously, Compton scattering is negligible for visible light (wavelength around 0.5 microns), but it becomes sizeable for X-rays.

10.4.6 Force transformation

The relativistic transformation of forces provides a first step towards the general theory of relativity. Forces on free objects in an inertial frame produce non-inertial frames (the frames attached to the accelerating objects). But before we look at accelerating frames, we first must find how forces transform between two inertial frames. The transformation of forces between the frames leads to an important equivalence between f_x and f_x' along the direction of motion.

From Newton's second law, three-forces (ordinary forces) are derived from the time-derivative of the momentum. The four-force, by analogy, is derived as

$$f^a = \frac{dp^a}{d\tau} = \frac{d}{dt'}\begin{pmatrix} E/c \\ p_x \\ p_y \\ p_z \end{pmatrix} = \gamma \frac{d}{dt}\begin{pmatrix} E/c \\ p_x \\ p_y \\ p_z \end{pmatrix} = \gamma \begin{pmatrix} P/c \\ f_x \\ f_y \\ f_z \end{pmatrix}, \qquad (10.95)$$

where f_x, f_y, and f_z are the "ordinary" three-forces, and P is the power transferred by the force to the object. Now consider two inertial frames as in Fig. 10.17, with a point mass that experiences a force. The point mass is stationary in the primed frame, and is moving along the x-axis with velocity v in the unprimed frame. The four-force in the particle rest frame (the primed frame) is

$$f^{\bar{b}} = \begin{pmatrix} 0 \\ f_x' \\ f_y' \\ f_z' \end{pmatrix}, \qquad (10.96)$$

where the time component is zero because the particle is at rest (power equals force times velocity). Now consider the same object from the unprimed frame in which the point mass is moving with β oriented along the x-axis. Then

$$f^a = \begin{pmatrix} \beta\gamma f_x \\ \gamma f_x \\ \gamma f_y \\ \gamma f_z \end{pmatrix}, \qquad (10.97)$$

where the time component uses the power $P = vf_x$. But the inverse Lorentz transformation of the four-force is

$$f^a = \Lambda^a_{\bar{b}} f^{\bar{b}} = \begin{pmatrix} \gamma & \beta\gamma & 0 & 0 \\ \beta\gamma & \gamma & 0 & 0 \\ 0 & 0 & 1 & 0 \\ 0 & 0 & 0 & 1 \end{pmatrix} \begin{pmatrix} 0 \\ f_x' \\ f_y' \\ f_z' \end{pmatrix}$$

$$= \begin{pmatrix} \beta\gamma f_x' \\ \gamma f_x' \\ f_y' \\ f_z' \end{pmatrix}. \qquad (10.98)$$

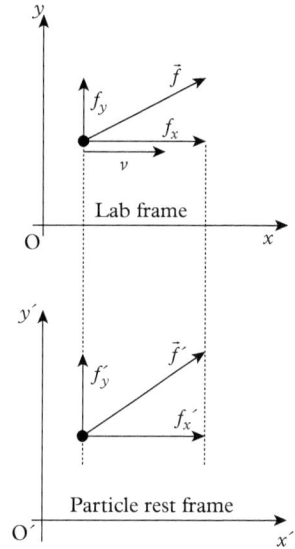

Figure 10.17 *A mass moving at velocity v along the x-axis experiences a force f in the fixed frame. In the rest frame of the mass, the force along the direction of motion remains unchanged, but the transverse forces are transformed.*

Equating the components in the two four-force expressions in Eqs. (10.97) and (10.98) gives the relationships among the three-force components in the two frames as

$$f_x = f'_x$$
$$f_y = f'_y/\gamma \qquad (10.99)$$
$$f_z = f'_z/\gamma.$$

Therefore, the components of the force in the direction of relative motion are equal, which is the same result for that component that was obtained from Galilean relativity. In this sense, acceleration (in the direction of motion) is independent of the inertial frame. However, the forces transverse to the direction of motion do transform with a factor of γ, which is clearly non-Galilean.

This result has importance for the relative view of accelerating objects. If a force $f = mg$ acts on an object in one frame to accelerate it, then in the frame of the object (in which it is instantaneously at rest) it experiences the same force. In this way, it is possible to consider a rocket ship that exerts a constant thrust to give it a constant acceleration of g. The astronaut inside feels his weight $m_{astro}g$ (which is what he would feel if the rocket ship were at rest at the surface of the Earth). And when viewed from the fixed frame, the rocket ship also experiences a constant force $F_{rocket} = m_{rocket}g$. However, the acceleration observed from the rest frame will decrease as the rocket moves more relativistically and approaches the speed of light. Furthermore, an accelerating object is only instantaneously in one inertial frame—it moves through a succession of inertial frames. This leads to a more complicated description of accelerated motion, but takes the first step towards a general theory of relativity that applies to *all* frames, not only inertial ones.

10.5 Linearly accelerating frames (relativistic)

The last part of this chapter on special relativity bridges to the general theory by considering a frame with constant linear acceleration relative to an inertial frame. We treated this case in Chapter 1 as a non-relativistic extension of Galilean relativity. Here we include relativistic effects. The linearly accelerating frame is obviously not inertial, and hence we may expect to find new and counterintuitive phenomena in this case, especially with regards to the behavior of light rays. The reason that this section is included here at the end of the chapter on special relativity instead of at the beginning of General Relativity is because it has a closer correspondence with special relativity. We are dealing with frames that are instantaneously inertial relative to each other. This allows us to apply Lorentz transformations locally, even though the global properties of the two frames are considerably different.

To begin to explore this connection, consider two identical ticking clocks at rest in the accelerating frame of a spaceship that is accelerating at the rate g. The location of the forward clock is x'_1, and the location of the aft clock is x'_2. The question is whether these clocks are ticking at the same rate. The challenge is to relate the properties of these two clocks to each other. This cannot be done as easily as in an inertial frame where photons can be exchanged to synchronize and compare two clocks, because the photons themselves are modified by the emission and absorption processes at the clocks (see gravitational redshift in the next chapter). However, there is a simple approach to make this comparison that employs a third clock that is falling freely in the frame of the spaceship. The third clock passes the first clock, and because they are located at the same place in space, their times can be compared directly. Assume that the speed of the falling clock when it passes clock 1 is given by β_1. Then when the falling clock passes clock 2 its speed is given by β_2. The time on the falling clock is dilated by the amounts

$$dt_1 = \gamma_1 dt_1'$$
$$dt_2 = \gamma_2 dt_2',$$

(10.100)

where the clock's inertial rest frame is the unprimed frame. But in its own rest frame, it is obvious that $dt_1 = dt_2$, and therefore $\gamma_1 dt_1' = \gamma_2 dt_2'$. This leads to

$$\frac{dt_1'}{dt_2'} = \frac{\gamma_2}{\gamma_1} = \left(\frac{1 - u_1^2/c^2}{1 - u_2^2/c^2}\right)^{1/2}$$

$$\approx 1 - \frac{1}{2}\frac{u_1^2 - u_2^2}{c^2},$$

(10.101)

where a low-velocity expansion has been used. Clearly the two clocks, one forward and one aft, are ticking at different rates in the accelerating frame of the spaceship.

An astronaut on the spaceship is at rest in the spaceship frame, and the acceleration of the ship at the rate g is experienced as a phenomenon equivalent to a gravitation field of strength g. Therefore, the astronaut would perform a simple calculation on the speed of the falling clock that re-expresses it in terms of gravitational potential as

$$\frac{dt_1'}{dt_2'} \approx 1 - \frac{g}{c^2}\left(x_2' - x_1'\right).$$

(10.102)

This result suggests that the invariant interval, using Eq. (10.102), is

$$ds^2 = -\left(1 + \frac{g}{c^2}x'\right)^2 c^2 dt'^2.$$

(10.103)

If we take "coordinate time" to be set by a clock at $x' = 0$, then a clock at a position x' has the time interval

$$dt' = \frac{d\tau}{\left(1 + \frac{g}{c^2}x'\right)}$$

(10.104)

relative to the proper time interval of the reference clock. Therefore, clocks that are along the positive x'-axis (higher in potential energy in the pseudo-gravitational field experienced as the acceleration g) have a shorter time interval and run faster. Clocks with lower potential energy (at negative x') run slower. Therefore a clock at the tip of the accelerating rocket ship runs fast relative to a clock at the base of the rocket ship. This phenomenon provides an initial glimpse into one of the effects of general relativity.

Constructing constantly accelerating frames is fraught with difficulties that lead to paradoxes related to the elastic deformation of rigid frames. To avoid these difficulties, a subterfuge is used that begins with a constantly accelerating point mass. This point mass passes continuously through successive inertial frames. If an extended collection of such point masses could be constructed, then they could represent masses in such an accelerated frame. The reason for taking pains to construct such a frame is to obtain a frame whose metric is time independent. This makes it possible to compare such a frame with a gravitational field whose metric also is independent of time and hence to make the connection to the principle of equivalence, which was one of Einstein's cornerstones in the construction of the general theory. The equivalence principle (EP) states that there is an equivalence between a uniformly accelerating frame and a gravitational field. In other words, if you are in an elevator (without windows to look out) you cannot tell whether you are far out in space and accelerating at a rate g, or whether you are moving at constant velocity in a constant gravitational field.

To begin this construction, consider a point mass that experiences a constant force $f_x' = m_0 g$ when viewed from the rest frame of the mass. However, from Eq. (10.99) we have $f_x' = f_x$ and hence in the unprimed frame

$$\frac{dp_x}{dt} = m_0 g \tag{10.105}$$

or

$$\frac{d\,(\gamma m_0 \beta)}{c dt} = m_0 \frac{g}{c^2}, \tag{10.106}$$

which yields the simple differential equation

$$d\,(\gamma\beta) = \frac{g}{c^2}\,(cdt)\,. \tag{10.107}$$

This integrates immediately to

$$\gamma\beta = \kappa ct, \tag{10.108}$$

where the constant is

$$\kappa = \frac{g}{c^2}. \tag{10.109}$$

Solving for $\beta = \kappa ct/\gamma$ and $\gamma = 1/\sqrt{1-\beta^2}$ gives

$$\beta = \frac{\kappa ct}{\sqrt{1 + (\kappa ct)^2}} \quad \text{and} \quad \gamma = \sqrt{1 + (\kappa ct)^2}. \tag{10.110}$$

These are the instantaneous velocity and Lorentz factor of the particle in O' seen from the fixed lab frame O.

Substituting β into the definition of velocity in the fixed frame

$$dx = \beta c dt \qquad (10.111)$$

and integrating gives

$$\kappa x = \int \frac{\kappa ct}{\sqrt{1 + (\kappa ct)^2}} d(\kappa ct) = \sqrt{1 + (\kappa ct)^2} + \kappa x_P. \qquad (10.112)$$

Rearranging this gives

$$\kappa^2 (x - x_P)^2 - (\kappa ct)^2 = 1. \qquad (10.113)$$

This is the equation for a hyperbola where x_P represents the location of an event P that is the focus of the hyperbola. The space–time diagram is shown in Fig. 10.18. The event associated with the particle at $t = 0$ is A. The focal event P is located at $x_P = x_0 - 1/\kappa$. The x'-axis of the instantaneously inertial frame co-moving with the particle is the line through PB. As B evolves in time, the x'-axis transforms continuously towards the asymptotic line of the hyperbola.

The event P has unexpected properties because the world line that starts at A asymptotes to the light line that emanates from P. Therefore, P behaves like the *event horizon of a black hole*. A clock at P would appear infinitely time dilated (stopped) to the observer in O' accelerating with the particle. A photon emitted from P at the instant that the mass began accelerating would never catch up with it. Therefore, in this problem of constant acceleration, a phenomenon

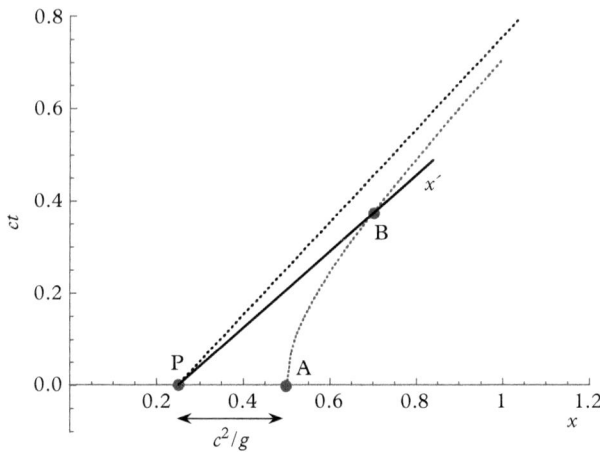

Figure 10.18 *The trajectory of a uniformly accelerating point mass is a hyperbola. The particle begins accelerating at $t' = t = 0$ at event A. At a later event B a co-moving inertial frame with the same instantaneous velocity has an x'-axis that passes through B and the event P, which acts as the focus to the hyperbolic world line of the particle.*

has appeared that we usually associate with massive bodies (black holes). The behavior of constant acceleration in special relativity has some of the properties of gravitation, as we shall see in the final chapter.

10.6 Summary

The invariance of the *speed of light* with respect to any inertial observational frame leads to a surprisingly large number of unusual results that defy common intuition. Chief among these are *time dilation*, *length contraction*, and the loss of *simultaneity*. The *Lorentz transformation* intermixes space and time, but an overarching structure is provided by the metric tensor of *Minkowski space–time*. The pseudo-Riemannian metric supports *four-vectors* whose norms are invariants, independent of any observational frame. These invariants constitute the *proper* objects of reality studied in the special theory of relativity. Relativistic dynamics defines the *equivalence of mass and energy* that has many applications in nuclear energy and particle physics. Forces have transformation properties between relatively moving frames that set the stage for a more general theory of relativity that describes physical phenomena in non-inertial frames.

Lorentz transformations

Lorentz transformations are affine transformations that transform positions and time between relative frames

$$
\begin{aligned}
t' &= \gamma\left(t - \frac{v}{c^2}x\right) & t &= \gamma\left(t' + \frac{v}{c^2}x'\right) \\
x' &= \gamma\left(x - vt\right) & x &= \gamma\left(x' + vt'\right) \\
y' &= y & y &= y' \\
z' &= z & z &= z'
\end{aligned}
\tag{10.3}
$$

The four-vector is transformed by the Lorentz transformation (Einstein summation convention assumed) as

$$
x^{\bar{b}} = \Lambda^{\bar{b}}_a x^a,
\tag{10.7}
$$

where the transformation matrix is

$$
\Lambda^{\bar{b}}_a = \begin{pmatrix} \gamma & -\beta\gamma & 0 & 0 \\ -\beta\gamma & \gamma & 0 & 0 \\ 0 & 0 & 1 & 0 \\ 0 & 0 & 0 & 1 \end{pmatrix}.
\tag{10.8}
$$

Velocity addition

Velocity components observed between two frames moving along the *x*-axis at a relative speed v are

$$u'_x = \frac{u_x - v}{1 - \dfrac{u_x v}{c^2}}$$

$$u'_y = \frac{u_y}{\gamma \left(1 - \dfrac{u_x v}{c^2}\right)} \tag{10.23}$$

$$u'_z = \frac{u_z}{\gamma \left(1 - \dfrac{u_x v}{c^2}\right)},$$

where u_a are the velocity components in the unprimed frame.

Invariant interval

The invariant interval is a differential line element in Minkowski space that has the same value in all frames. It is defined as

$$ds^2 = -c^2 dt^2 + dx^2 + dy^2 + dz^2$$
$$= g_{ab} dx^a dx^b, \tag{10.37}$$

where the Minkowski metric tensor expressed as a matrix is

$$g_{ab} = \begin{pmatrix} -1 & 0 & 0 & 0 \\ 0 & 1 & 0 & 0 \\ 0 & 0 & 1 & 0 \\ 0 & 0 & 0 & 1 \end{pmatrix}. \tag{10.38}$$

Four-momentum

The four-momentum is

$$p^a = m v^a$$

$$= \gamma m \begin{pmatrix} c \\ u^x \\ u^y \\ u^z \end{pmatrix} \tag{10.46}$$

and the inner product with itself is

$$g_{ab} p^a p^b = -\gamma^2 m^2 c^2 + \gamma^2 m^2 u^2$$
$$= -m^2 c^2 \tag{10.47}$$

which is an invariant.

Energy–momentum

The contributions of momentum and mass to total energy are

$$E^2 = \left(mc^2\right)^2 + p^2 c^2, \tag{10.80}$$

where the total energy is

$$E = \gamma mc^2 \tag{10.57}$$

and the rest energy of the particle is

$$E_0 = mc^2. \tag{10.58}$$

The kinetic energy is therefore

$$T = (\gamma - 1)\, mc^2. \tag{10.59}$$

Force transformation

Forces transform between frames moving relatively along the x-axis as

$$\begin{aligned} f_x &= f'_x \\ f_y &= f'_y/\gamma \\ f_z &= f'_z/\gamma, \end{aligned} \tag{10.99}$$

where the longitudinal component of the force is the same in both frames.

10.7　Bibliography

J. J. Callahan, *The Geometry of Space–time: An Introduction to Special and General Relativity* (Springer, 2000).

R. D'Inverno, *Introducing Einstein's Relativity* (Oxford University Press, 1992).

T. A. Moore, *A Traveler's Guide to Space–time: An Introduction to the Special Theory of Relativity* (McGraw-Hill, 1995).

R. A. Mould, *Basic Relativity* (Springer, 1994).

V. Petkov, *Relativity and the Nature of Space–time* (Springer, 2005).

R. Talman, *Geometric Mechanics* (John Wiley & Sons, 2000).

10.8　Homework exercises

1. **Lorentz transformation:** Derive Eqs. (10.3) and (10.4) from Eqs. (10.1) and (10.2).
2. **Synchronization:** Two clocks at the origins of the K and K' frames (which have a relative speed v) are synchronized when the origins coincide. After a time t, an observer at the origin of the K system observes the K' clock by means of a telescope. What does the K' clock read?
3. **Light clock:** Take the light clock in Fig. 10.6a and place it so the light path is along the direction of motion. Rederive the time dilation of the clock in this configuration.

4. **Length contraction:** A stick of length l is fixed at an angle θ from its x-axis in its own rest frame K. What is the length and orientation of the stick as measured by an observer moving along x with speed v?

5. **Headlight effect:** Consider a light point-source that emits isotropically with a constant intensity output in its own frame. If it is moving with speed v in a lab frame, derive the intensity as a function of angle from the forward direction that is observed in the lab frame.

6. **Doppler effect:** A light source moving with speed β along the x-axis emits a photon that has an angular frequency ω' in its rest frame. The photon is observed in the fixed frame to travel at an angle θ_0 relative to the direction of motion. (a) Calculate the Doppler shift as a function of β and θ_0. What is the shift for $\theta_0 = 90°$? Is it blue-shifted or red-shifted? (b) Calculate the Doppler shift as a function of β and θ_1, where θ_1 is the angle from the detection point to the position of the light source at the moment the photon is detected. What is the shift for $\theta_1 = 90°$? Is it blue-shifted or red-shifted? (c) What is the shift for $\theta_0 + \theta_1 = 180°$? Is it blue-shifted or red-shifted? This last answer is the transverse Doppler shift observed in transverse laser scattering experiments.

7. **Invariant product:** Show that the inner product between any two four-vectors necessarily is an invariant, even if the two four-vectors are not of the same type. For instance if one is the position four-vector and the other is the momentum four-vector.

8. **Pair production:** Derive the threshold energy for a photon to scatter from a stationary electron to produce an electron–positron pair. Assume that all kinetic energy after the interaction is shared equally among the three leptons (original electron and the electron–positron pair).

9. **Relatistic force:** Show that the relativistic form of Newton's second law becomes

$$F = \frac{m}{\left(1 - \beta^2\right)^{3/2}} \frac{du}{dt}.$$

Is force a vector or a covector? Why? Can you construct a four-force that has an invariant contraction?

10. **Euler–Lagrange:** Show that the relativistic Lagrangian provides the appropriate relativistic dynamics from the Euler–Lagrange equations.

$$L = -m_0 c \sqrt{-\dot{x}_a \dot{x}^a} \sqrt{1 - \beta^2} - U\left(x^a\right)$$

11. **Constant acceleration:** You are on a space ship that begins accelerating at $t' = 0$ with a constant acceleration $a' = g$ (measured by the force you experience locally). After one year passes on your clock, how far away is the Earth in your frame? How far away are you from the Earth in the Earth's frame? Could you get to Alpha Centauri by this time?

11

The General Theory of Relativity and Gravitation

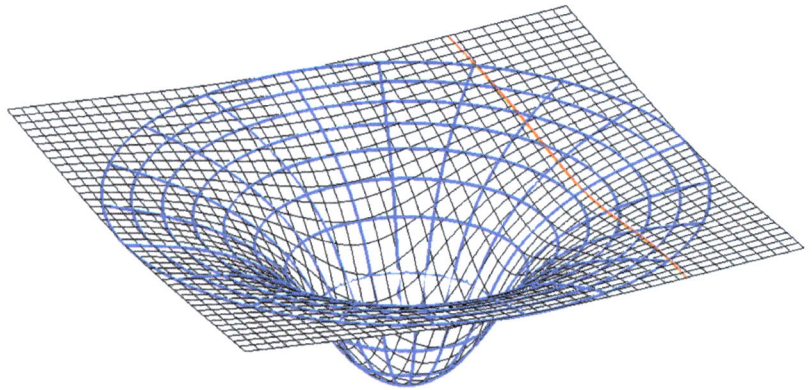

The central postulate of general relativity (GR) holds that all observers are equivalent, *even if they are not in inertial frames.* This means that observers in accelerating frames and observers in gravitational fields must measure the same local properties of physical objects. This chapter introduces the concept of curved spaces and their tensor metrics and tensor curvatures. By defining the effect of mass–energy on space–time curvature, simple consequences of the general theory are derived, such as gravitational acceleration and the deflection of light by gravity.

11.1 Riemann curvature tensor

Curvature of surfaces in three-dimensional (3D) Euclidean space is easily visualized. A sphere, for instance, has curvature which makes it impossible to map the surface of the Earth onto a flat sheet. It is also impossible, on a globe, to draw a triangle, composed of sections of three great-circle routes, that has angles that add to 180°. In fact, the sum of the angles is always greater than 180°. Furthermore, parallel lines can and do meet on the surface of a sphere: think of two lines of longitude that are parallel at the Equator but that meet at the North

Pole. Conversely, the surface of a cylinder has no curvature, because it can be cut along any "straight" line and laid flat.

These intuitive properties of surfaces embedded in Euclidean spaces do not go very far when one begins to ask questions about the curvature of a metric space, especially when the metric space is four-dimensional and pseudo-Riemannian. To tackle questions of curvature in these cases, it is necessary to consider the actions of covariant gradients on vectors on the metric space. For instance, consider the commutator relation

$$\nabla_c \nabla_d V^a - \nabla_d \nabla_c V^a. \tag{11.1}$$

Does this commutator equal zero? This simple question opens up fundamental aspects of curved spaces. In flat Euclidean spaces, the quantity does commute. Even on manifolds that have global curvature, such as the surface of the cylinder, the quantity commutes. But on the surface of a sphere it does not. This is because the surface of the sphere has intrinsic curvature that destroys the concept of translating a vector in a "parallel" manner. As we saw in Chapter 9, this translation, or transport, of a vector on a manifold is at the heart of curvature. In this chapter, we find that curvature of space–time is caused by energy density, and the curvature determines the geometry of trajectories near gravitating objects.

To evaluate the commutator, take the first term

$$\nabla_d \left(\nabla_c V^a\right) = \nabla_d \left(\partial_c V^a + \Gamma^a_{bc} V^b\right)$$
$$= \partial_d \left(\partial_c V^a + \Gamma^a_{bc} V^b\right) + \Gamma^a_{ed} \left(\partial_c V^e + \Gamma^e_{bc} V^b\right) - \Gamma^e_{cd} \left(\partial_e V^a + \Gamma^a_{be} V^b\right) \tag{11.2}$$

then the second

$$\nabla_c \left(\nabla_d V^a\right) = \nabla_c \left(\partial_d V^a + \Gamma^a_{bd} V^b\right)$$
$$= \partial_c \left(\partial_d V^a + \Gamma^a_{bd} V^b\right) + \Gamma^a_{ec} \left(\partial_d V^e + \Gamma^e_{bd} V^b\right) - \Gamma^e_{dc} \left(\partial_e V^a + \Gamma^a_{be} V^b\right). \tag{11.3}$$

Subtract these two and use $\partial_d \partial_c V^a = \partial_c \partial_d V^a$. The result is

$$\nabla_c \nabla_d V^a - \nabla_d \nabla_c V^a = R^a_{bcd} V^b + \left(\Gamma^e_{cd} - \Gamma^e_{dc}\right) \nabla_e V^a, \tag{11.4}$$

where a new tensor has been defined as

Riemann curvature tensor: $\quad R^a_{bcd} = \partial_c \Gamma^a_{bd} - \partial_d \Gamma^a_{bc} + \Gamma^e_{bd} \Gamma^a_{ec} - \Gamma^e_{bc} \Gamma^a_{ed}, \tag{11.5}$

which is the Riemann[1] curvature tensor. An all-covariant version of the tensor is

$$R_{abcd} = g_{ar} R^r_{bcd}. \tag{11.6}$$

[1] Bernard Riemann (1826–1866) was a German mathematician who made fundamental contributions to differential geometry with the introduction of metric spaces in his Habilitationsschrift delivered at Göttingen in 1854 with the title On the Hypotheses which lie at the Foundation of Geometry.

In broad terms, the Riemann curvature tensor is like a generalized second derivative of the metric tensor, while the Christoffel symbols are like generalized first derivatives of the metric tensor. The complexity of these tensors is mainly bookkeeping to keep track of the interdependence of basis vectors in general curvilinear coordinates.

The Riemann curvature tensor in general has 81 elements in 3D space, although it possesses many symmetries that reduce the number of independent elements to six. Contractions of the Riemann tensor lead to additional useful tensors, such as the Ricci[2] tensor

$$\text{Ricci tensor:} \quad R_{ab} = R^c_{acb} = g^{cd} R_{dacb}, \tag{11.7}$$

where the tensor R_{dacb} is the Riemann tensor of the first kind, while with the single raised index R^a_{bcd} it is the Riemann tensor of the second kind. Further contraction leads to the Ricci scalar

$$\text{Ricci scalar:} \quad R = g^{ab} R_{ab}. \tag{11.8}$$

These tensors and scalars are just consequences of the commutator of the second derivatives of Eq. (11.4). However, they are important measures of curvature of a metric space. To gain some insight into curved spaces, it is helpful to consider a few examples.

Example 11.1 Surface of a cylinder (2D surface)

The surface of a cylinder is not curved. Start with the line element

$$ds^2 = dz^2 + r^2 d\theta^2$$

$$g_{ab} = \begin{pmatrix} r^2 & 0 \\ 0 & 1 \end{pmatrix} \qquad g^{ab} = r \begin{pmatrix} \frac{1}{r^2} & 0 \\ 0 & 1 \end{pmatrix}$$

$$\Gamma^1_{11} = \frac{1}{2}\left[g^{11}\frac{\partial g_{11}}{\partial \theta} + g^{11}\frac{\partial g_{11}}{\partial \theta} - g^{11}\frac{\partial g_{12}}{\partial \theta} + g^{12}\frac{\partial g_{22}}{\partial \theta} + g^{12}\frac{\partial g_{21}}{\partial \theta} - g^{12}\frac{\partial g_{11}}{\partial z} \right]$$
$$= 0$$

and likewise all the others are zero because g_{ab} is constant (remember that r is a constant). Therefore $R = 0$. A cylinder surface is *not* curved. This is because parallel lines remain parallel if we think of the cylinder surface being formed from a rolled up flat sheet of paper with parallel lines on it.

---

[2] Gregorio Ricci-Curbastro (1853–1925) was the inventor of tensor calculus. He published, with his student Levi-Civita, the theory of tensor calculus in 1900. This book made tensor analysis generally accessible and was a major reference for Albert Einstein as he applied tensor calculus to the general theory of relativity.

Example 11.2 Surface of a sphere (2D surface)

Find the Ricci scalar of a spherical surface.

$$x = r \sin \theta \cos \phi$$
$$y = r \sin \theta \sin \phi \qquad ds^2 = r^2 d\theta^2 + r^2 \sin^2\theta \, d\phi^2$$
$$z = r \cos \theta$$

$$g_{ab} = \begin{pmatrix} r^2 & 0 \\ 0 & r^2 \sin^2\theta \end{pmatrix} \qquad g^{ab} = \begin{pmatrix} \dfrac{1}{r^2} & 0 \\ 0 & \dfrac{1}{r^2 \sin^2\theta} \end{pmatrix}$$

$$\Gamma^1_{11} = \frac{1}{2}\left[g^{11}\frac{\partial g_{11}}{\partial \theta} + g^{11}\frac{\partial g_{11}}{\partial \theta} - g^{11}\frac{\partial g_{12}}{\partial \theta} + g^{12}\frac{\partial g_{22}}{\partial \theta} + g^{12}\frac{\partial g_{21}}{\partial \theta} - g^{12}\frac{\partial g_{11}}{\partial \phi} \right]$$

$$= 0$$

$$\Gamma^1_{22} = \frac{1}{2}\left[g^{11}\frac{\partial g_{12}}{\partial \phi} + g^{11}\frac{\partial g_{12}}{\partial \phi} - g^{11}\frac{\partial g_{22}}{\partial \theta} + g^{12}\frac{\partial g_{22}}{\partial \phi} + g^{12}\frac{\partial g_{22}}{\partial \phi} - g^{12}\frac{\partial g_{22}}{\partial \phi} \right]$$

$$= \frac{1}{2}(-1)\frac{1}{r^2}\frac{\partial}{\partial \theta}\left(r^2 \sin^2\theta \right) = \frac{-1}{2}\frac{1}{r^2}2r^2 \sin\theta \cos\theta \, \theta$$

$$= -\sin\theta \cos\theta$$

$\Gamma^2_{12} = \Gamma^2_{21} = \cot\theta$ and all others are zero.

Next, the Riemann curvature tensor is

$$R^1_{212} = \frac{\partial}{\partial \theta}\Gamma^1_{22} - \frac{\partial}{\partial \phi}\Gamma^1_{21} + \Gamma^1_{22}\Gamma^1_{11} + \Gamma^2_{22}\Gamma^1_{21} - \Gamma^1_{21}\Gamma^1_{12} - \Gamma^2_{21}\Gamma^1_{22}$$

$$= \frac{\partial}{\partial \theta}\left(-\sin\theta \cos\theta \right) - \cot\theta \left(-\sin\theta \cos\theta \right)$$

$$= -\left(-\sin^2\theta + \cos^2\theta \right) + \cot\theta \left(\sin\theta \cos\theta \right)$$

$$= \sin^2\theta - \cos^2\theta + \cos^2\theta$$

$$= \sin^2\theta$$

All other R^a_{bcd} are zero. Now define the Ricci tensor

$$R_{ab} = R^c_{acb} = g^{cd}R_{dacb}$$
$$R_{22} = \sin^2\theta$$

and

$$R_{ab} = \begin{pmatrix} 0 & 0 \\ 0 & \sin^2\theta \end{pmatrix},$$

continued

Example 11.2 *continued*

which leads to the Ricci scalar

$$R = g^{ab}R_{ab} = \begin{pmatrix} \dfrac{1}{r^2} & 0 \\ 0 & \dfrac{1}{r^2\sin^2\theta} \end{pmatrix} \begin{pmatrix} 0 & 0 \\ 0 & \sin^2\theta \end{pmatrix}$$

$$= \frac{1}{r^2}$$

This is recognized as the curvature of the surface of a sphere, as expected.

11.2 The Newtonian correspondence

A particle with no forces acting on it executes a geodesic trajectory. This fact was derived through a variational calculation on the Lagrangian of a free particle. The resulting trajectory in Euclidean three-space is a straight line. This is also easy to see in Minkowski space in which force-free relativistic particles, including photons, have world lines that are straight.

$$\text{Inertial frame: } \frac{d^2x^a}{d\tau^2} = 0. \tag{11.9}$$

The next step is to ask what trajectory a particle in free-fall executes when it is near a gravitating body. The answer is again a geodesic. However, the geodesic in this case is no longer a straight line. The geodesic also needs to be viewed in four-space, with time-like components that are affected by the curvature of space–time induced by matter (and energy). The equations of motion are then

$$\text{Non-inertial frame: } \frac{d^2x^a}{d\tau^2} + \Gamma^a_{bc}\frac{dx^b}{d\tau}\frac{dx^c}{d\tau} = 0. \tag{11.10}$$

The second term in the equation of motion for the non-inertial frame contains all the forces associated with the coordinate transformation from one frame to another, such as the centrifugal force and Coriolis force. In addition, it describes gravitational forces.

The trajectory of a particle in GR must approach the trajectory of a particle in Newtonian gravity when the field is weak and the particle velocity is small. Taking the non-inertial equation of motion in the Newtonian limit and making the correspondence to Newtonian gravity provides a means of finding the expression for $g_{oo}(M)$ as a function of mass M.

For a slow particle with $dx/dt \ll c$ and $dx/ds \ll cdt/ds$ then

$$\frac{d\tau}{dt} = 1 \tag{11.11}$$

for $\gamma \approx 1$. This approximation allows us to write

$$\Gamma^a_{bc} \frac{dx^c}{ds} \frac{dx^b}{ds} \approx \Gamma^a_{00} \left(\frac{cdt}{ds} \right)^2, \tag{11.12}$$

where the zeroth term has dominated the summation because c is so much larger than the other velocities. The equation of motion becomes

$$\frac{d^2x^a}{ds^2} + \Gamma^a_{00} \left(\frac{cdt}{ds} \right)^2 = 0, \tag{11.13}$$

where

$$\Gamma^a_{00} = -\frac{1}{2} g^{ab} \frac{\partial g_{00}}{\partial x^b}. \tag{11.14}$$

At this point, a second simplifying approximation is made (the first was the slow particle) that assumes that the metric is nearly that of Minkowski space, such that

$$g_{ab} = \eta_{ab} + h_{ab} = \begin{pmatrix} -1 & 0 & 0 & 0 \\ 0 & 1 & 0 & 0 \\ 0 & 0 & 1 & 0 \\ 0 & 0 & 0 & 1 \end{pmatrix} + \begin{pmatrix} h_{00} & 0 & 0 & 0 \\ 0 & h_{11} & 0 & 0 \\ 0 & 0 & h_{22} & 0 \\ 0 & 0 & 0 & h_{33} \end{pmatrix} \tag{11.15}$$

with the weak field approximation $|h_{ab}| \ll 1$. The connection to the metric is then

$$\Gamma^a_{00} = -\frac{1}{2} \eta^{ab} \frac{\partial h_{00}}{\partial x^b} \tag{11.16}$$

and the equation of motion becomes

$$\frac{d^2\vec{x}}{ds^2} = \frac{1}{2} \left(\frac{cdt}{ds} \right)^2 \vec{\nabla} h_{00}$$
$$\frac{d^2\vec{x}}{dt^2} = \frac{1}{2} c^2 \vec{\nabla} h_{00} \tag{11.17}$$

which relates the particle acceleration to the perturbation of the metric.

The next step is to make the connection between the geodesic trajectory and the equations of Newtonian gravity. This step is based on the equivalence principle, which states that the motion of a particle in a steadily accelerating (non-inertial) reference frame is equivalent to the motion of a particle in a spatially uniform gravitational field. Recall that the result from Newtonian gravity is

$$\frac{d^2\vec{x}}{dt^2} = -\vec{\nabla}\left(-\frac{GM}{r}\right). \tag{11.18}$$

Equating Eqs. (11.17) and (11.18) determines h_{00} to be

$$h_{00} = -\frac{2}{c^2}\left(-\frac{GM}{r}\right) \tag{11.19}$$

and this gives

$$g_{00} = -1 + \frac{2GM}{c^2 r}. \tag{11.20}$$

Therefore, the invariant interval in this weak-field and low-velocity limit is

$$ds^2 = -\left(1 - \frac{2GM}{c^2 r}\right)c^2 dt^2 + dx^2 + dy^2 + dz^2. \tag{11.21}$$

The time-like term is affected by mass and hence gravity. This equation is similar to Eq. (10.103) that was derived in special relativity for constant acceleration. There, too, only the time-like term was affected. One of the interesting aspects of this derivation is the absence of any gravitational force. Only the geometry of space–time entered into the derivation through the geodesic equation. From this weak-field geodesic approach (using the Equivalence Principle), it is only possible to derive g_{00} to lowest order, even though the space-component metric tensors g_{ab} are of the same magnitude in a spherically symmetric mass distribution.[3] The space-like components of the metric tensor are obtained from Einstein's field equation.

11.3 Einstein's field equations

Kinematics arising from the geodesic equation does not by itself describe the origin of gravitational fields. For this, it is necessary to define a source term in an equation that is analogous to Poisson's equation for Newtonian gravity. The source term should depend on the energy–momentum tensor

$$T^{ab} = \rho_0 U^a U^b, \tag{11.22}$$

where ρ_0 is the proper mass density and U^a is the four-velocity. In the energy–momentum tensor, both mass energy and kinetic energy densities contribute to the gravitational source term.

[3] See D'Inverno, p. 166.

11.3.1   Einstein tensor

To arrive at the desired differential equation for gravity, Einstein sought a rank-2 tensor operator that would be related to the rank-2 energy–momentum tensor. The most general rank-2 tensor that can be constructed from the metric tensor is

$$O^{ab} = R^{ab} + \mu g^{ab} R + \Lambda g^{ab} \tag{11.23}$$

that uses the Ricci tensor and scalar in addition to the metric tensor. Additional constraints must be satisfied (such as Bianchi identities), which determine the constant to be $\mu = -\dfrac{1}{2}$. This leads to the Einstein tensor

$$G^{ab} = R^{ab} - \frac{1}{2} g^{ab} R. \tag{11.24}$$

The field equations of GR then are

$$\text{\textit{Einstein field equations}:} \qquad G^{ab} + \Lambda g^{ab} = \frac{8\pi G}{c^4} T^{ab}, \tag{11.25}$$

where the constant of proportionality before the energy–momentum tensor makes the correspondence to Newtonian gravity and the gravitational constant G. The constant Λ is the so-called cosmological constant that has a long and interesting history and has reemerged in recent years as a feature of dark energy. The Einstein field equations are second-order nonlinear partial differential equations. Only for cases of special symmetry and boundary conditions are exact analytical solutions available. For instance, one of the simplest cases is for a non-rotating mass with spherical symmetry that leads to the Schwarzschild solution, to be discussed in Section 11.4. Before looking at the exact solution, it is instructive to look at the Newtonian limit and confirm that Newtonian gravity is incorporated by the field equations.

11.3.2   Newtonian limit

The Newtonian limit of the field equations is derived by recognizing that, for weak gravity, the magnitudes of the energy–momentum tensor satisfy the conditions

$$\left| T^{00} \right| \gg \left| T^{oa} \right| \gg \left| T^{bc} \right|. \tag{11.26}$$

This leads to approximations that extract the weak-field components of the metric tensor where gravity causes a perturbation of the Minkowski manifold through Eq. (11.15). To find the deviation h_{ab} from the Minkowski metric, the Riemann curvature tensor can be expressed as

$$R^{\alpha}_{\beta\mu\nu} = \frac{1}{2} g^{\alpha\sigma} \left[\partial_\beta \partial_\mu g_{\sigma\nu} - \partial_\beta \partial_\nu g_{\sigma\mu} + \partial_\sigma \partial_\nu g_{\beta\mu} - \partial_\sigma \partial_\mu g_{\beta\nu} \right] \tag{11.27}$$

and, in terms of h_{ab},

$$R_{\alpha\beta\mu\nu} = \frac{1}{2}\left[\partial_\beta\partial_\mu h_{\alpha\nu} - \partial_\beta\partial_\nu h_{\alpha\mu} + \partial_\alpha\partial_\nu h_{\beta\mu} - \partial_\alpha\partial_\mu h_{\beta\nu}\right],$$ (11.28)

which leads after considerable algebra[4] to an approximate expression for the Einstein tensor as

$$
\begin{aligned}
G^{ab} &= R^{ab} - \frac{1}{2}g^{ab}R \\
&= -\frac{1}{2}\left[-\frac{1}{c^2}\frac{\partial^2}{\partial t^2} + \nabla^2\right]\left[h^{ab} - \frac{1}{2}\eta^{ab}h\right] \\
&= -\frac{1}{2}\Box^2\bar{h}^{ab} = \frac{8\pi G}{c^4}T^{ab}.
\end{aligned}
$$ (11.29)

The trace h is

$$h_{xx} + h_{yy} + h_{zz} - h_{00} = h = h_a^a$$ (11.30)

and the trace-reverse of h^{ab} is defined as

$$\bar{h}^{ab} = h^{ab} - \frac{1}{2}\eta^{ab}h.$$ (11.31)

The correspondence between h^{ab} and R^{ab} in Eq. (11.29) is made through the differential operator (the D'Alembertian $\Box^2$).

Because of the weak-field conditions for the energy–momentum tensor in Eq. (11.26) and relating Eq. (11.29) to Eq. (11.25), the only trace-reverse that is sizable (in the Newtonian limit of low velocities and low fields) is $\bar{h}^{00}$. The dominant term in the energy–momentum tensor is

$$T_{00} = \rho c^2$$ (11.32)

and the field equations become

$$\Box^2\bar{h}^{00} = -\frac{16\pi G}{c^2}\rho.$$ (11.33)

It is noteworthy that Eq. (11.33) for vacuum, when $\rho = 0$, is a homogeneous wave equation. This is a first indication of the existence of gravity waves that propagate at the speed of light, although more conditions are needed to ensure their existence (which are indeed met).

For low particle velocities, only the gradient component of the D'Alembertian is sizable, yielding

$$\nabla^2\left[h^{00} - \frac{1}{2}\eta^{00}h\right] = -\frac{16\pi G}{c^2}\rho,$$ (11.34)

[4] And the application of gauge transformations; see Schultz, p. 205.

which looks suspiciously similar to Gauss' law for Newtonian gravity. The Newtonian potential is related to mass density through

$$\nabla^2 \Phi = 4\pi G\rho. \tag{11.35}$$

Therefore, it is an obvious choice to make the correspondence

$$h^{00} - \frac{1}{2}\eta^{00}h = -4\phi, \tag{11.36}$$

where

$$\phi = \frac{\Phi}{c^2} \tag{11.37}$$

is the gravitational potential (normalized by c^2).

The weak-field conditions of Eq. (11.26) applied to Eq. (11.29) lead to similar magnitudes for $\bar{h}^{ab}$ and T^{ab}, which are small for space components. This yields the space-component equations

$$h^{ab} - \frac{1}{2}h \approx 0 \tag{11.38}$$

for $[a, b] = 1, 2, 3$, which gives

$$h^{xx} = h^{yy} = h^{zz} = \frac{1}{2}h. \tag{11.39}$$

Finally, the trace h is

$$h = \eta_{ab}h^{ab} = -h^{00} + \frac{3}{2}h, \tag{11.40}$$

which is solved for h^{00} to give

$$h^{00} = \frac{1}{2}h. \tag{11.41}$$

From Eq. (11.36) this yields

$$h = -4\phi \tag{11.42}$$

and the metric is finally

$$ds^2 = -(1 + 2\phi)\, c^2 dt^2 + (1 - 2\phi)\left(dx^2 + dy^2 + dz^2\right). \tag{11.43}$$

This metric describes the spherically-symmetric, weak-field limit of the space–time geometry of the Einstein field equations. For the potential outside the spherically symmetric mass distribution, the potential is

$$\phi = \frac{\Phi}{c^2} = -\frac{GM}{c^2 r} \tag{11.44}$$

and the metric becomes

$$ds^2 = -\left(1 - \frac{2GM}{c^2 r}\right)c^2 dt^2 + \left(1 + \frac{2GM}{c^2 r}\right)\left(dx^2 + dy^2 + dz^2\right), \tag{11.45}$$

which is the metric outside of a planet or star. Previously, in Eq. (11.21), only the time component had been modified by the potential, where the effects of the curvature of space had been neglected. This previous result was derived from the equivalence principle in which a steadily accelerating reference frame was compared to a spatially uniform gravitational field. The form of Eq. (11.45) is now the complete result that includes general relativistic effects on both the space and the time components of the metric tensor (but still in the weak-field limit).

11.4   Schwarzschild space–time

The weak-field metric is only an approximate solution to Einstein's field equation. Within a year after the publication in 1915 of the general theory of relativity, Karl Schwarzschild[5] published the first exact solution for a non-rotating spherically-symmetric mass. The Schwarzschild metric solution to the field equation is

$$ds^2 = -\left(1 - \frac{2MG}{c^2 r}\right) c^2 dt^2 + \frac{dr^2}{\left(1 - \frac{2MG}{c^2 r}\right)} + r^2 d\Omega^2. \tag{11.46}$$

This metric holds for arbitrarily strong gravity, as long as it has spherical symmetry. The time component has a zero and the space term has a divergence at the Schwarzschild radius

$$R_S = 2GM/c^2. \tag{11.47}$$

This radius is the event horizon of a black hole where time stops and space compresses. Nothing at radii smaller than R_S can be known to the outside world, and nothing falling past R_S can ever return, even light.

Light is the perfect tool to use to map out the geometry of Schwarzschild space–time because it defines the null geodesics that constrain all possible world lines. Consider the radial motion of a photon with the line element

$$ds^2 = -\left(1 - \frac{R_S}{r}\right) c^2 dt^2 + \frac{dr^2}{\left(1 - \frac{R_S}{r}\right)} = 0, \tag{11.48}$$

from which it follows

$$cdt = \pm \frac{dr}{(1 - R_S/r)}. \tag{11.49}$$

This integrates to

$$ct = \pm \left[r + R_S \ln |r - R_S| + C\right], \tag{11.50}$$

[5] Karl Schwarzschild (1873–1916) was a German physicist and astronomer whose name is used to describe the event horizon of a black hole—the Schwarzschild radius. He received his PhD in 1896 on a topic posed by Poincaré. At the outbreak of World War I in 1914 he volunteered as an officer in the German army and saw action on both the western and eastern fronts. It was while he was on the eastern front in 1915 and early 1916 that he wrote his three famous papers on exact solutions to Einstein's field equations for spherical symmetry.

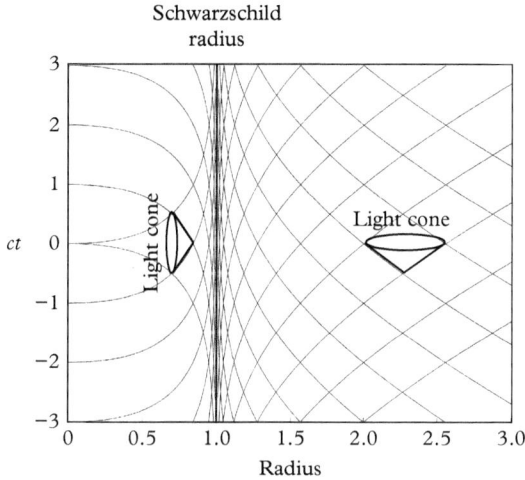

Figure 11.1 *Schwarzschild space–time showing the null geodesics. The coordinate singularity at $R/R_S = 1$ divides the space–time into two regions. Far from the Schwarzschild radius, the light lines are approximately 45°. Inside the Schwarzschild radius, the light lines all terminate on the singularity at the origin.*

where C is the integration constant, and the plus and minus signs stand for outgoing and infalling null geodesics (light lines).

The null geodesics of the Schwarzschild space–time are plotted in Fig. 11.1 for outgoing and incoming light curves. The coordinate singularity at $R/R_S = 1$ divides the space–time into two regions. In the region outside the Schwarzschild radius far from the gravitating body, the light lines make approximately 45° angles, as expected for Minkowski space–time. In the region inside the Schwarzschild radius, the light cones are tipped by 90° and point to the singularity at the origin. All massive particles have world lines that lie within the light cones and hence all trajectories, even light, eventually end on the singularity at the origin.

The singular behavior of the coordinate description of the Schwarzschild space time at $R = R_S$ is an artifact of the choice of the coordinate variables (t, r, θ, ϕ). Therefore, the Schwarzschild metric is not the most convenient choice for describing the properties of systems that are near the Schwarzschild radius or for objects that pass through it. In fact, the proper time and lengths of objects falling into a black hole experience no discontinuous behavior at the Schwarzschild radius. There are many other choices of coordinate variables that avoid the coordinate singularity at R_S. However, these go beyond the topics of the current textbook.[6]

11.5 Kinematic consequences of gravity

There are many important kinematic consequences of the Schwarzschild metric that may be derived for conditions that are outside the apparent singularity at the Schwarzschild radius. The proper time and distance of an observer approaching

[6] Alternate choices of coordinate variables include Eddington–Finkelstein coordinates and Kruskal coordinates.

a gravitating body remain well behaved, but an observer far from the gravitating body sees infalling clocks ticking slower, lengths contracting, and photons becoming red-shifted.

11.5.1 Time dilation

Consider a calibrated clock in a gravitational potential $\phi = -GM/c^2 r$ ticking with proper time $d\tau$

$$d\tau = \frac{1}{c}\sqrt{-ds^2}. \tag{11.51}$$

If it is stationary in its own frame, then the space terms vanish and the relation becomes

$$d\tau = dt\sqrt{1 - \frac{2GM}{c^2 r}}, \tag{11.52}$$

where dt is the coordinate time. It is clear that when r equals infinity then coordinate time and proper time are equal. Therefore the coordinates are referenced to infinite distance from the gravitating body. As the clock descends towards the gravitating body the factor in the square root decreases from unity, meaning that the coordinate time span is larger than the proper time span—clocks near a gravitating body slow down. As the clock approaches the Schwarzschild radius, it slows and stops as it disappears into the event horizon (the light reflecting from the clock (in order to "see" it) becomes infinitely red-shifted and so the clock would not be visible).

11.5.2 Length contraction

Consider a meter stick oriented parallel to the radius vector. The proper length of the stick is ds, which is related to the coordinate length by

$$ds = \sqrt{ds^2} = dr/\sqrt{1 - 2GM/c^2 r}. \tag{11.53}$$

Therefore, a yardstick approaching a gravitating body is contracted, and the length of the yardstick shrinks to zero as the stick passes the event horizon.

11.5.3 Redshifts

The shift in frequency of a photon emitted from an object near a gravitating body is most easily obtained by considering two clocks at two radial locations r_1 and r_2. These clocks are used to measure the frequency of light emitted from one and detected by the other. The ratio of measured frequencies is equal to the ratio of coordinate times for the clocks, giving the result

$$\frac{d\tau_1}{d\tau_2} = \frac{\nu_2}{\nu_1} = \sqrt{\frac{1 - 2GM/c^2 r_1}{1 - 2GM/c^2 r_2}}.$$

(11.54)

Therefore, a photon falling to Earth is blue-shifted. Conversely, a photon emitted to the sky from Earth is red-shifted. A photon emitted near an event horizon is red-shifted, with the redshift becoming infinite at the event horizon.

From these considerations, one concludes that a clock falling toward the event horizon of a black hole ticks more slowly, becomes squashed in the direction of fall, and turns progressively more red (the photon frequency moving into the infrared and then radio frequencies, and ultimately to zero frequency).

11.6 The deflection of light by gravity

In 1911, while at the University of Prague, Einstein made his famous prediction that the gravitational field of the Sun would deflect light.[7] The experimental proof of this theoretical prediction was delayed by World War I. However, in May 1919, Arthur Eddington lead an expedition to an island off Africa (and another expedition was undertaken to an island off Brazil) to observe stars during a total eclipse of the Sun. His observations confirmed Einstein's theory and launched Einstein to superstar status on the world stage.

The general relativistic calculation of the deflection of light by gravity can be based on null geodesics in warped space–time. A gravitating body creates a metric with curvature, and a photon trajectory follows a geodesic within this metric. The curvature is greatest near the massive body, with two straight-line asymptotes far from the gravitational source, as shown in Fig. 11.2, with an angular deflection that is a function of the impact parameter b.

11.6.1 Refractive index of GR

An alternative calculation that does not use the geodesic equation explicitly (but which is consistent with it implicitly) treats warped space–time as if it had a spatially varying refractive index.[8] The light trajectories then follow the ray equation

[7] Einstein's work in 1911 was still based on the equivalence principle, which ignores curvature of space. His full derivation of the deflection of light by the gravity of the Sun came in 1915, along with his derivation of the precession of the orbit of Mercury.

[8] See Cheng, p. 124.

Apparent
position

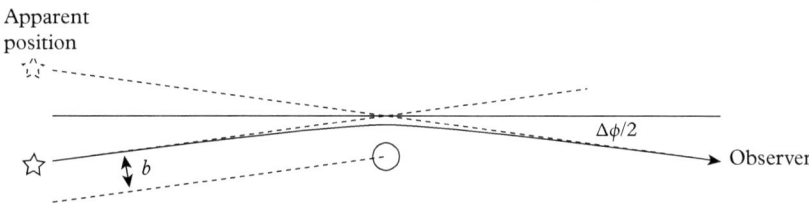

Figure 11.2 *Deflection of light by the Sun where the angular deflection is a function of the impact parameter b. The figure is not to scale; the angular deflection for a ray grazing the surface of the Sun is only 8 microradians.*

of Chapter 9. The fundamental postulate of special relativity states that the speed of light is a constant (and the same) for all observers. This statement needs to be modified in the case of GR. While the speed of light measured locally is always equal to c, the apparent speed of light observed by a distant observer (far from the gravitating body) is modified by time dilation and length contraction. This makes the apparent speed of light, as observed at a distance, vary as a function of position. Application of Fermat's principle for light leads to the ray equation, and hence to the equation for the deflection of light by gravity.

The invariant element for a light path in the Schwarzschild geometry with $d\theta = d\phi = 0$ is

$$ds^2 = g_{00}c^2 dt^2 + g_{rr}dr^2 = 0. \tag{11.55}$$

The apparent speed of light is then

$$\frac{dr}{dt} = c\sqrt{-\frac{g_{00}(r)}{g_{rr}(r)}} = c(r), \tag{11.56}$$

where $c(r)$ is always less than c, when observing it from flat space. The "refractive index" of space is defined, as for any optical material, as the ratio of the constant speed divided by the observed speed

$$n(r) = \frac{c}{c(r)} = \sqrt{-\frac{g_{rr}(r)}{g_{00}(r)}}. \tag{11.57}$$

The Schwarzschild metric has the property

$$g_{rr} = \frac{-1}{g_{00}}, \tag{11.58}$$

so the effective refractive index of warped space–time is

$$n(r) = \frac{1}{-g_{00}(r)} = \frac{1}{1 - \dfrac{2GM}{c^2 r}} \tag{11.59}$$

with a divergence at the Schwarzschild radius.

The refractive index of warped space–time in the limit of weak gravity can be used in the ray equation Eq. (9.125) from Chapter 9,

$$\frac{d}{ds}\left(n\frac{dx^a}{ds}\right) = \frac{\partial n}{\partial x^a}, \tag{11.60}$$

where the gradient is

$$\begin{aligned}
\frac{\partial n}{\partial x} &= -n^2 \frac{2GM}{c^2}\frac{x}{r^3} \\
\frac{\partial n}{\partial y} &= -n^2 \frac{2GM}{c^2}\frac{y}{r^3}.
\end{aligned} \tag{11.61}$$

The ray equation is a four-variable flow in x, y, $v_x = n dx/ds$, and $v_y = n dy/ds$:

$$\frac{dx}{ds} = \frac{v_x}{n}$$
$$\frac{dy}{ds} = \frac{v_y}{n}$$
$$\frac{dv_x}{ds} = -n^2 \frac{2GM}{c^2} \frac{x}{r^3}$$
$$\frac{dv_y}{ds} = -n^2 \frac{2GM}{c^2} \frac{y}{r^3}.$$

(11.62)

Trajectories of light from the ray equation for several collimated rays are shown in Fig. 11.3 passing by a black hole. For rays with a large impact parameter, the rays are deflected towards the "optic axis." Closer rays are increasingly deflected. Rays with an impact factor less than a critical value cannot escape and are captured by the black hole.[9] An unstable circular light orbit exists at a radius of $1.5R_S$.

Rays with a large impact factor (larger than shown in Fig. 11.3) are deflected through small angles by the local index gradient

$$d\phi \approx \frac{\partial n}{\partial y} dx.$$

(11.63)

[9] See Misner, p. 673. The use of the ray equation for strong gravity is only approximate, and the light capture behavior in Fig. 11.3 is qualitative.

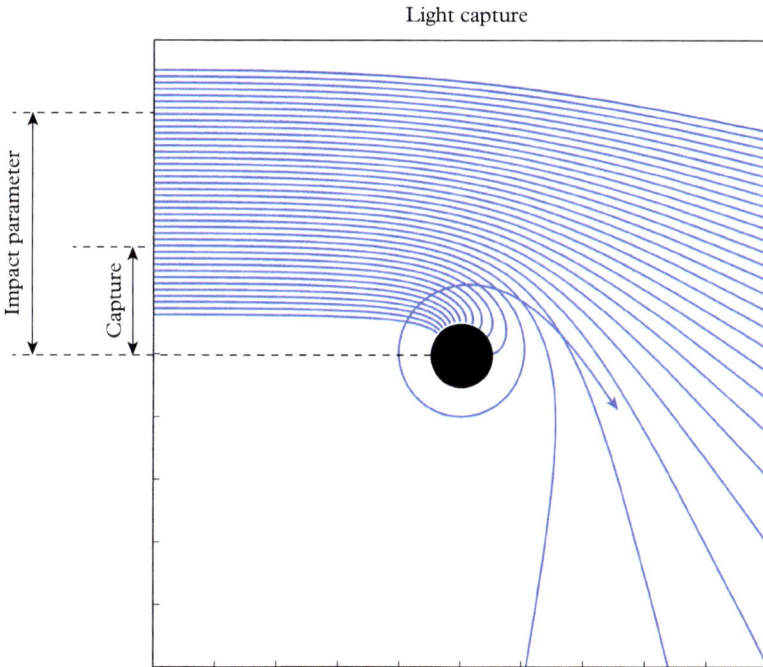

Light capture

Figure 11.3 *Light trajectories near a black hole. Rays with a small impact factor are captured by the black hole. One ray near the critical impact parameter circles the black hole once and escapes. There is an unstable circular light orbit at a radius of $1.5R_S$.*

The total deflection is obtained by integrating over the "interaction length" of the light with the gravitating body

$$\Delta\phi = \int d\phi = n^2 \frac{2GM}{c^2} \int_{-\infty}^{\infty} \frac{y}{r^3} dx.$$ (11.64)

For large impact parameters, the y position of the ray changes negligibly during the interaction, the index of refraction is nearly unity, and the integral simplifies to

$$\Delta\phi = \frac{2GM}{c^2} \int_{-\infty}^{\infty} \frac{b}{\left(x^2 + b^2\right)^{3/2}} dx,$$ (11.65)

leading to the equation for the deflection of light:

Deflection of light: $\Delta\phi = \dfrac{4GM}{c^2 b}.$ (11.66)

For light passing by the surface of the Sun, $b = R_{Sun}$, and this angle is 1.7 arcseconds, as predicted by Einstein and measured by Eddington in his 1919 expedition.

11.6.2   Gravitational lensing

At first glance, the deflection of light by a spherically gravitating object would appear to make a gravitational lens that could focus light. However, the angular deflection by gravity is inversely proportional to the impact parameter, while for a thin glass lens it is linearly proportional to the impact parameter. Therefore, a gravitational "lens" behaves very differently than an optical lens, and it has very high distortion. For instance, if a black hole is along the line of sight to a distant point source (let's say a quasar), then the quasar would appear as a bright ring around the black hole, known as an *Einstein ring*. The subtended angle of the Einstein ring is

$$\theta_E = \sqrt{\frac{L_{SL}}{L_s L_L} \frac{4GM}{c^2}},$$ (11.67)

where L_S is the distance to the source, L_L is the distance to the lensing body, and L_{SL} is the distance of the source behind the lens. Depending on the orientation of a gravitational lens relative to a distant source, and the mass distribution in the lens, the distant object may appear as a double image, or a smeared out arc. The principle of gravitational lensing is shown in Fig. 11.4. A massive object along the line of sight to a distant quasar bends the light rays to intersect with the Earth. Such "lensing" is not image-forming because is does not "focus" the rays, but

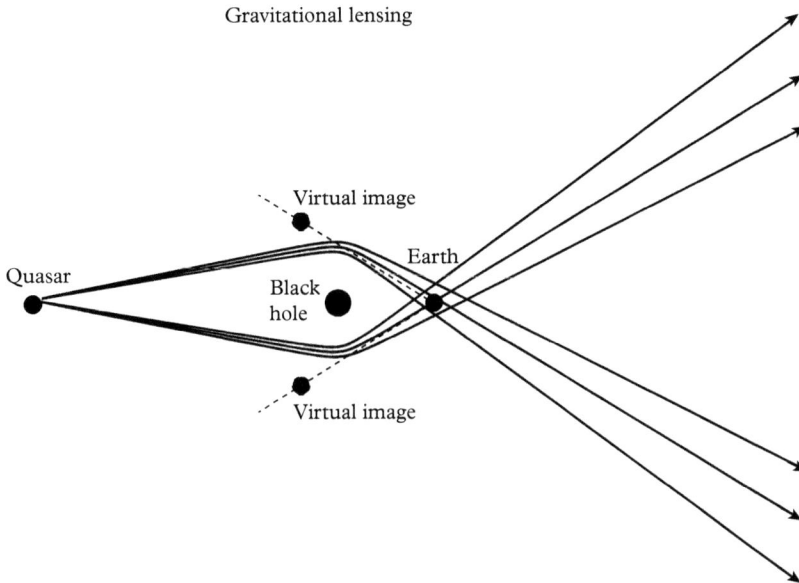

Figure 11.4 *Principle of gravitational lensing. A strongly gravitating body on the line of sight to a distant object (quasar) bends the light rays to intersect with the Earth. The figure is not to scale; the angular deflection is typically a microradian.*

it does produce arcs. Note that the image formed by the gravitational lens is a virtual image, or arc, that is closer than the original object.

There are many astronomical examples of double images of galaxies seen through an intermediate galaxy. One famous example is shown in Fig. 11.5 with numerous Einstein arcs. Gravitational lensing has become a valuable tool in the exploration of dark energy and dark matter. One piece of evidence in favor of dark matter comes from the bullet cluster which is the collision of two galaxy clusters. The luminous matter of each cluster collided with the other causing a viscous damping of the matter velocities. The gravitational lensing by all matter has been mapped out on both sides of the collision, but the X-ray sources from luminous matter are separated from the bulk of the gravitationally lensing matter. This separation of lensing matter vs. emitting matter provides evidence for non-interacting dark matter within large-scale structures such as galaxies and galaxy clusters. What dark matter is, and what physical laws it obeys, is one of the greatest unsolved problems in physics today!

11.7 Planetary orbits

The second notable prediction that Einstein made concerning his new theory of GR was an explanation of the precession of the perihelion of the planet Mercury. In a perfect $1/r$ potential, the major axis of the elliptical orbit is constant and does

Figure 11.5 *Gravitational lensing. The "nearby" galaxy is causing gravitation lensing of objects behind it. Distant galaxies are smeared into arcs by the strong aberrations of the gravitational lens.*
Image credit: NASA, ESA, the Hubble SM4 ERO team, and ST-ECF.

not precess. However, if there are radius-dependent corrections to the potential, then the axis of the ellipse precesses, which is observed for the orbit of Mercury.

The equation of a planetary orbit in the Schwarzschild geometry begins with the invariant four-velocity

$$g_{ab}\dot{x}^a\dot{x}^b = -c^2,\tag{11.68}$$

where the dot is with respect to the proper time $\dot{x}^a = dx^a/d\tau$. In the Schwarzschild geometry this is

$$-\left(1-\frac{R_S}{r}\right)c^2\dot{t}^2 + \left(1-\frac{R_S}{r}\right)^{-1}\dot{r}^2 + r^2\left(\dot{\theta}^2 + \sin^2\theta\dot{\phi}^2\right) = -c^2.\tag{11.69}$$

Planetary motion in this metric is planar, just as in the case of Newtonian gravity, and it is convenient to choose $\theta = \pi/2$ to give

$$-\left(1-\frac{R_S}{r}\right)c^2\dot{t}^2 + \left(1-\frac{R_S}{r}\right)^{-1}\dot{r}^2 + r^2\dot{\phi}^2 = -c^2.\tag{11.70}$$

The Lagrangian $L(x^a,\dot{x}^a;\tau) = \frac{1}{2}g_{ab}\dot{x}^a\dot{x}^b$ has no explicit time dependence or ϕ-dependence, leading to two conserved quantities through

$$\begin{aligned}\frac{\partial L}{\partial\dot{x}^0} &= g_{ab}\dot{x}^a\delta_0^b \\ \frac{\partial L}{\partial\dot{x}^3} &= g_{ab}\dot{x}^a\delta_3^b\end{aligned}\tag{11.71}$$

that are related to the constants E and l (energy and angular momentum). In the presence of a metric that differs from the Minkowski metric, relativistic energy and angular momentum are expressed as

$$E/mc = -g_{00}c\dot{t}$$
$$l = mg_{\phi\phi}\dot{\phi} = mr^2\dot{\phi}. \tag{11.72}$$

Both of these quantities are conserved. Using these constants in Eq. (11.70) gives

$$-\left(1 - \frac{R_S}{r}\right)^{-1}\frac{E^2}{m^2c^2} + \left(1 - \frac{R_S}{r}\right)^{-1}\dot{r}^2 + \frac{l^2}{m^2r^2} = -c^2, \tag{11.73}$$

and after some rearrangement

$$\frac{1}{2}m\dot{r}^2 + \frac{1}{2}\frac{l^2}{mr^2}\left(1 - \frac{R_S}{r}\right) - \frac{GmM}{r} = \frac{1}{2}\frac{E^2}{mc^2} - \frac{1}{2}mc^2. \tag{11.74}$$

The right-hand side is a constant of the motion with units of energy,[10] which will be denoted by the expression T_∞, and the equation is finally

$$\frac{1}{2}m\dot{r}^2 + \frac{1}{2}\frac{l^2}{mr^2}\left(1 - \frac{R_S}{r}\right) - \frac{GmM}{r} = T_\infty. \tag{11.75}$$

This is recognized as the equation for the non-relativistic central force problem, but with an extra factor of $\left(1 - \frac{R_S}{r}\right)$ in the angular momentum term. This extra factor is the general relativistic correction that leads to a deviation from the perfect $1/r$ potential, and hence leads to precession of the orbit.

The differential equation in Eq. (11.75) is commonly expressed in terms of derivatives of the angle ϕ by using Eq. (11.72) to yield

$$\left(\frac{1}{r^2}\frac{dr}{d\phi}\right)^2 + \frac{1}{r^2}\left(1 - \frac{R_S}{r}\right) - \frac{2GMm^2}{rl^2} = \frac{2m}{l^2}T_\infty. \tag{11.76}$$

The substitution $u = 1/r$ makes this

$$\left(\frac{du}{d\phi}\right)^2 + u^2 = \frac{2m}{l^2}T_\infty + \frac{2GMm^2}{l^2}u + R_S u^3. \tag{11.77}$$

Differentiating with respect to u leads to a simple form

$$\frac{d^2u}{d\phi^2} + u = \frac{GMm^2}{l^2} + \frac{3GM}{c^2}u^2, \tag{11.78}$$

which is valid for particle (or planet) speeds much less than c.

[10] By using the result from special relativity, $E^2 = p^2c^2 + m^2c^4$, the constant can be expressed as $T_\infty = \frac{1}{2}\left(\frac{E^2}{mc^2} - mc^2\right) = \frac{p^2}{2m}$, which is the non-relativistic particle kinetic energy far from the gravitating body.

In the absence of the last term of Eq. (11.78), this is the classical orbital result for an inverse square law

$$\frac{d^2u}{d\phi^2} + u = \frac{GMm^2}{l^2},$$ (11.79)

which has the elliptical solution

$$\frac{1}{r} = \frac{GMm^2}{l^2}(1 + \varepsilon \cos \phi) = u_0,$$ (11.80)

where ε is the ellipticity. When this ideal solution is substituted into Eq. (11.78), the result is

$$\frac{d^2u_0}{d\phi^2} + u_0 = \frac{GMm^2}{l^2} + \frac{3G^3M^3m^4}{l^4c^2}\left[1 + 2\varepsilon\cos\phi + \frac{\varepsilon^2}{2}(1 + \cos 2\phi)\right].$$ (11.81)

Only the second term in the brackets leads to first-order effects, giving the approximation to the solution as:[11]

$$u_1 = u_0 + \frac{3G^3M^3m^4}{l^4c^2}\varepsilon\phi\sin\phi,$$ (11.82)

which can be rewritten

$$u_1 = \frac{GMm^2}{l^2}\left[1 + \varepsilon\cos\left(\phi\left[1 - 3\left(\frac{GMm}{\ell c}\right)^2\right]\right)\right].$$ (11.83)

When ϕ equals 2π then the angle at the maximum radius has shifted (precessed) by the angle

GR precession angle: $\qquad \Delta\phi = 6\pi\left(\frac{GMm}{lc}\right)^2.$ (11.84)

In the case of the orbit of Mercury, this is 43 seconds of arc per century. This was what was derived by Einstein in 1915 during a particularly productive two weeks that also included his prediction of the deflection of light by the Sun.

The result in Eq. (11.78) describes a two-variable flow equation in the variables u and $v = \dot{u}$:

$$\dot{v} = -u + \frac{GMm^2}{\ell^2} + \frac{3}{2}R_S u^2$$
$$\dot{u} = v.$$ (11.85)

[11] Details of the secular solution can be found in Section 8.9 of Thornton and Marion, *Classical Dynamics of Particles and Systems*, 5th ed. (Thomson, 2004).

A single trajectory is shown in Fig. 11.6 for normalized parameters with $GMm^2/l^2 = 1$ and $R_S = 0.013$. The ellipticity and precession angle are large in this example.

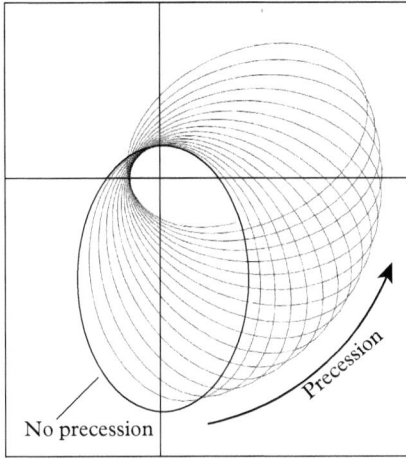

Figure 11.6 *Example of a highly elliptical orbit subject to gravitation precession with normalized units with* $GMm^2/l^2 = 1$ *and* $R_S = 0.013$.

11.8   Orbits near a black hole

Stars during a supernova can undergo gravitational collapse when the gravitation pressure exceeds the light pressure. The collapse can stop when electrons combine with protons to create a neutron star for which the nuclear pressure exceeds the gravitational pressure. However, if the original star mass was sufficiently large and the residual mass of the neutron star is 1.5–3 times a solar mass, then the nuclear pressure is insufficient to support the gravitational pressure, and the neutron star will continue gravitational collapse until its physical size decreases below the Schwarzschild radius. A black hole remains, and all further information from the star is removed from our sensible universe. Black holes can have masses ranging from 1.5 times a solar mass to super massive black holes that can contain the mass of over a billion suns. The Schwarzschild radius is

$$R_S = \frac{2GM}{c^2} = 2.95\text{km} \left(\frac{M}{M_\odot} \right) \tag{11.86}$$

in units of kilometers for a mass M relative to a solar mass $M_\odot$. A super massive black hole with the mass of a billion suns has a radius as large as our solar system.

Orbits around black holes are strongly perturbed compared to the small precession corrections of Mercury orbiting the Sun. To obtain an understanding of black hole orbital behavior, it is helpful to look at Eq. (11.75) as a central force problem with an effective potential

$$\frac{1}{2}m\dot{r}^2 + \Phi_{eff}(r) = T_\infty, \tag{11.87}$$

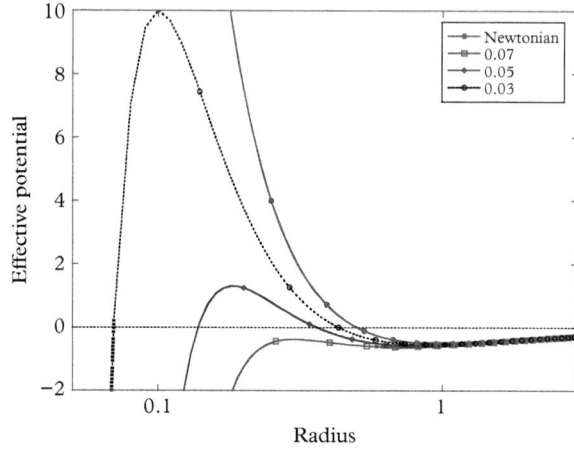

Figure 11.7 *The GR effective potential has an attractive short-range contribution that dominates the centrifugal repulsion.*

where the effective potential is

$$\Phi_{eff}(r) = -\frac{GmM}{r} + \frac{1}{2}\frac{l^2}{mr^2} - \frac{R_S}{2}\frac{l^2}{mr^3}.$$ (11.88)

The first two terms are the usual Newtonian terms that include the gravitational potential and the repulsive contribution from the angular momentum that prevents the mass from approaching the origin. However, the third term is the GR term that is attractive and overcomes the centrifugal barrier at small values of r, allowing the orbit to collapse to the center. Therefore, not all orbits around a black hole are stable, and even circular orbits will decay if too close to the black hole. A graph of the effective potential is shown in Fig. 11.7. In the classical Newtonian case shown in the figure, the orbiting mass can never approach the center because of the centrifugal divergence. The radius units in the figure are in terms of the circular orbit radius. Therefore, the Newtonian minimum is at $r = 1$. But in the GR case, there is the additional attractive term that dominates at small radius values. This causes the effective potential to have a centrifugal barrier, but the effective potential then becomes negative close to the origin.

To find the conditions for circular orbits, it is sufficient to differentiate Eq. (11.88) with respect to the radius to find the minimum that gives the equation

$$GmMr^2 - \frac{l^2}{m}r + 3\frac{R_S}{2}\frac{l^2}{m} = 0.$$ (11.89)

Solving for r yields

$$r = \frac{\frac{l^2}{mc^2} + \frac{l^2}{mc^2}\sqrt{1 - 6GMR_S\frac{m^2}{l^2}}}{mR_S},$$ (11.90)

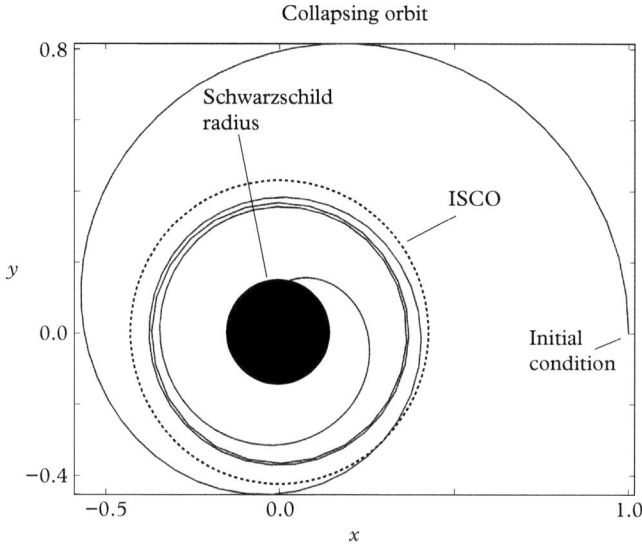

Figure 11.8 *Orbit simulation with conditions near the homoclinic orbit. R_S = 0.15. ISCO = 0.44. A particle that begins with an ellipticity settles into a nearly circular orbit near the homoclinic saddle point, after which it spirals into the black hole.*

where the positive root is the solution for the stable circular orbit. There is an innermost stable circular orbit (ISCO) that is obtained when the term in the square root vanishes for

$$l^2 = 3R_S^2 m^2 c^2, \tag{11.91}$$

which gives the simple result

$$r_{ISCO} = 3R_S. \tag{11.92}$$

Therefore, no particle can sustain a circular orbit with a radius closer than three times the Schwarzschild radius, and it will spiral into the black hole.

A single trajectory solution to the GR flow[12] in Eq. (11.85) is shown in Fig. 11.8. The particle begins in an elliptical orbit outside the innermost circular orbit and is captured into nearly circular orbit inside the ISCO. This orbit eventually decays and spirals with increasing speed into the black hole. The origin of these nearly stable circular orbits can be seen in the state space of Eq. (11.85) shown in Fig. 11.9. A circular orbit is surrounded by strongly precessing orbits that are contained within a homoclinic orbit that terminates at the homoclinic saddle point (similar to the homoclinic orbit in Fig. 3.13). The nearly stable circular orbits in Fig. 11.8 occur at the homoclinic saddle, and the accretion discs around black holes occupy these orbits before collisions cause them to lose angular momentum, and they spiral into the black hole.

[12] Equation (11.92) applies for particles that have low velocity. However, special relativity effects become important when the orbital radius of the particle approaches the Schwarzschild radius.

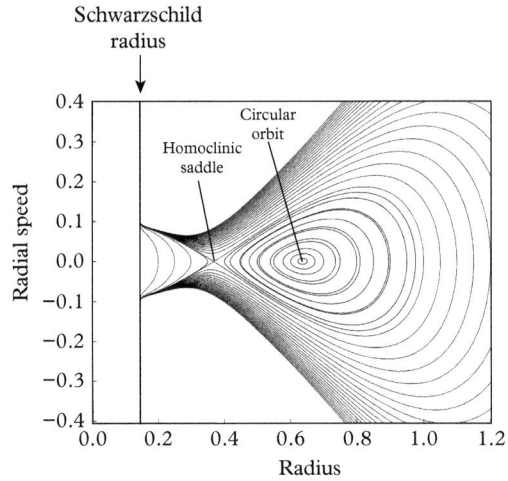

Figure 11.9 *State space diagram with flow lines for one value of (conserved) angular momentum. The ISCO is at 0.44, although for the angular momentum in this simulation, a stable circular orbit occurs at 0.65. There is a homoclinic saddle point between stable orbits, and orbits that are captured by the black hole. The Schwarzschild radius is at 0.15.*

11.9 Summary

The intrinsic curvature of a metric space is captured by the *Riemann curvature tensor* which also can be contracted to the *Ricci tensor* and the *Ricci scalar*. Einstein took these curvature quantities and constructed the *Einstein field equations* that relate the curvature of space–time to energy and mass density. For an isotropic density, the solution to the field equations is the *Schwarzschild metric* that contains mass terms that modify both the temporal and the spatial components of the invariant element. Consequences of the Schwarzschild metric include *gravitational time dilation*, *length contraction*, and *redshifts*. Trajectories in curved space–time are expressed as geodesics through the warped metric space. Solutions to the geodesic equation explain the *precession of the perihelion of Mercury* and the *deflection of light by the Sun*.

Riemann curvature tensor and Ricci tensor and scalar

The curvature of a metric space is captured by the Riemann curvature tensor and its contractions into the Ricci tensor and scalar:

$$R^a_{bcd} = \partial_c \Gamma^a_{bd} - \partial_d \Gamma^a_{bc} + \Gamma^e_{bd} \Gamma^a_{ec} - \Gamma^e_{bc} \Gamma^a_{ed}, \tag{11.5}$$

$$R_{ab} = R^c_{acb} = g^{cd} R_{dacb}, \tag{11.7}$$

$$R = g^{ab} R_{ab}. \tag{11.8}$$

Einstein field equations

The field equations of general relativity relate the curvature of space–time, captured in the Einstein tensor

$$G^{ab} = R^{ab} - \frac{1}{2}g^{ab}R, \tag{11.24}$$

to the mass–energy density T^{ab} through

$$G^{ab} + \Lambda g^{ab} = \frac{8\pi G}{c^4}T^{ab}, \tag{11.25}$$

where Λ is the "cosmological constant," which is an ongoing topic of astrophysical research through its connection to *dark energy* and the expansion of the Universe.

Spherically-symmetric metrics

For spherical symmetry, the weak-field metric in Cartesian coordinates is

$$ds^2 = -\left(1 - \frac{2GM}{c^2r}\right)c^2dt^2 + \left(1 + \frac{2GM}{c^2r}\right)\left(dx^2 + dy^2 + dz^2\right). \tag{11.45}$$

The Schwarzschild metric (for arbitrary field strength) in spherical polar coordinates is

$$ds^2 = -\left(1 - \frac{2GM}{c^2r}\right)c^2dt^2 + \frac{dr^2}{\left(1 - \frac{2GM}{c^2r}\right)} + r^2\left(d\theta^2 + \sin^2\theta\, d\phi^2\right). \tag{11.46}$$

Gravitational time dilation and length contraction

Clocks higher in a gravitational potential tick faster than those lower in the potential. (Clocks in satellites tick faster than clocks on Earth.) Clocks at the Schwarzschild radius of a black hole stop ticking altogether (when considered from a position far away). Yardsticks parallel to the radius vector of a gravitating body compress as they approach the body, and compress to zero length at the Schwarzschild radius.

Angular deflection of light by the Sun

Light passing by a spherically gravitating body with impact parameter b is deflected by an angle

$$\Delta\phi = \frac{4GM}{c^2b}, \tag{11.66}$$

which, for the Sun, is 1.7 arcseconds.

Orbital precession angle

The general relativistic contribution to orbital mechanics produces a deviation from the perfect inverse square law of Newtonian orbits. Elliptical orbits with an orbital angular momentum l precess by an angle

$$\Delta\phi = 6\pi \left(\frac{GMm}{lc} \right)^2, \tag{11.84}$$

which, in the case of the orbit of Mercury, is 43 arcseconds per century.

11.10 Bibliography

J. J. Callahan, *The Geometry of Space–Time* (Springer, 2001).

S. Carroll, *Space–time and Geometry: An Introduction to General Relativity* (Benjamin Cummings, 2003).

Ta-Pei Cheng, *Relativity, Gravitation, and Cosmology: A Basic Introduction* (Oxford University Press, 2010).
One of the most accessible textbooks on GR. It includes an ingenious approach to light deflection that treats warped space–time as a refractive index and the geodesic equation as a ray equation from optics.

R. D'Inverno, *Introducing Einstein's Relativity* (Oxford University Press, 1992).
At a little higher mathematical level than Mould or Schutz. Good derivation of the deflection of light by the Sun.

D. F. Lawden, *Introduction to Tensor Calculus, Relativity and Cosmology* (Dover, 2002).
A very direct and easy-to-follow description of the full topic.

C. Misner, K. Thorne and J. Wheeler, *Gravitation* (Freeman, 1973).
Charles Misner, Kip Thorne, and John Wheeler have written the most complete textbook on gravitation. This is the bible of general relativity.

Richard A. Mould, *Basic Relativity* (Springer, 1994).
The chapter on accelerating frames is done in great depth.

B. F. Schutz, *A First Course in General Relativity* (Cambridge University Press, 1985).
Excellent descriptions of vectors, covectors, and basis vectors. Derivations of the Newtonian correspondence for the laws of motion and of gravity waves are particularly clear.

Steven Weinberg, *Gravitation and Cosmology: Principles and Applications of the General Theory of Relativity* (Wiley, 1972).

11.11   Homework exercises

1. **Geodesic equation:** Find the geodesic equations (but do not solve them) for the metric $ds^2 = x^2 dx^2 \pm y^2 dy^2$. Are they straight lines?

2. **Ricci tensor:** Derive the Ricci tensor for a saddle (positive and negative curvature).

3. **Einstein tensor:** Derive Eq. (11.29) for the Einstein tensor in terms of the D'Alembertian operator.

4. **Null geodesic:** For a photon, show that conservation of four-momentum in curved space–time is equivalent to

$$\nabla_{\vec{p}} \, \vec{p} = 0$$

which defines a null geodesic.

5. **Newtonian gravity:** In the Newtonian limit, the equation of motion for a particle is

$$\frac{d^2 x^\mu}{ds^2} + \Gamma^\mu_{00} \left(\frac{c \, dt}{ds} \right)^2 = 0, \qquad \text{(a geodesic curve)}$$

where $\Gamma^\mu_{00} = -\frac{1}{2} g^{\mu\upsilon} \frac{\partial g_{00}}{\partial x^\upsilon}$. By expanding the metric in terms of flat Minkowski space, $g_{\alpha\beta} = \eta_{\alpha\beta} + h_{\alpha\beta}$, derive the expression for g_{00} in terms of G, M, c, and r. Use the basic Newtonian differential equation for a particle falling in a gravitational potential to make the connection between g_{ab} and M.

6. **Gravitational potential:** From $g_{00} = -(1 + 2\phi)$ derive the Newtonian correspondence relations

$$\Gamma^\mu_{00} = \frac{\partial \phi}{\partial x^\mu} \qquad R^\mu_{0\upsilon 0} = \frac{\partial^2 \phi}{\partial x^\mu \partial x^\upsilon}.$$

7. **Schwarzschild metric:** The Schwarzschild metric for space–time containing a weak spherically symmetric gravitational field is

$$ds^2 = -(1 + 2\phi) \, c^2 dt^2 + \frac{dr^2}{(1 + 2\phi)} + r^2 \left(d\theta^2 + \sin^2\theta \, d\phi^2 \right),$$

where $\phi(r) = -GM/c^2 r$ is the gravitational potential.

(a) To first order in ϕ, state the g^{ab} for this metric.

(b) Calculate all the nonzero Christoffel symbols of the second kind in terms of ϕ.

(c) Calculate the only nonzero components of the Riemann curvature tensor, which are (in isotropic coordinates):

$$R_{trtr} = 2\phi/r^2 \qquad R_{t\theta t\theta} = R_{t\phi t\phi} = -\phi/r^2$$
$$R_{\theta\phi\theta\phi} = -2\phi/r^2 \qquad R_{r\theta r\theta} = R_{r\phi r\phi} = \phi/r^2$$

(d) Is Schwarzschild space curved?

8. **Equivalence principle:** A photon enters an elevator at right angles to its acceleration vector **g**. Use the geodesic equation and the weak-field metric to show the trajectory is parabolic.

9. **Tidal force:** A photon enters an elevator at a right angle. The space is subject to a tidal-force metric

$$ds^2 = -c^2 dt^2 + (1 - 2\phi)\left(dx^2 + dy^2 + dz^2\right).$$

What is the photon trajectory to lowest order in ϕ?

10. **Photon orbit:** A photon can orbit a black hole in an unstable circular orbit. Derive the radius of this orbit.

11. **Schwarzschild metric:** There are a total of 13 Christoffel symbols that are nonzero in the Schwarzschild metric

$$ds^2 = -(1 + 2\phi)\,c^2 dt^2 + \frac{dr^2}{(1 + 2\phi)} + r^2\left(d\theta^2 + \sin^2\theta\,d\phi^2\right).$$

Four of them are

$$\Gamma^1_{11} = -\frac{1}{(1 + 2\phi)}\frac{GM}{c^2 r^2}, \qquad \Gamma^1_{00} = (1 + 2\phi)\frac{GM}{c^2 r^2},$$

$$\Gamma^0_{10} = \Gamma^0_{01} = \frac{1}{(1 + 2\phi)}\frac{GM}{c^2 r^2}.$$

What are the other nine?

12. **Christoffel:** Consider a spherically symmetric line element

$$ds^2 = -e^{2\Phi}c^2 dt^2 + e^{2\Lambda}dr^2 + r^2\left(d\theta^2 + \sin^2\theta\,d\phi^2\right).$$

Derive the nonzero Christoffel symbols:

$$\Gamma^t_{tr} = -\partial_r\Phi \qquad \Gamma^r_{tt} = -(\partial_r\Phi)\,e^{2\Phi - 2\Lambda} \quad \Gamma^r_{rr} = \partial_r\Lambda \qquad \Gamma^r_{\theta\theta} = -re^{-2\Lambda}$$
$$\Gamma^r_{\phi\phi} = -r\sin^2\phi\,e^{-2\Lambda} \quad \Gamma^\theta_{r\theta} = \Gamma^\phi_{r\phi} = r^{-1} \qquad \Gamma^\theta_{\phi\phi} = -\sin\theta\cos\theta \quad \Gamma^\phi_{\theta\phi} = \cot\theta$$

13. **Gravitational lensing:** Derive the expression for the angle subtended by the Einstein ring from gravitational lensing.

14. **Precession of the perihelion of Mercury:** Derive the secular solution Eq. (11.82) for the precession of the perihelion of Mercury.

15. **Black hole orbital mechanics:** Explore the fixed point classifications for the flow in Eq. (11.87) as a function of angular momentum and particle kinetic energy.

Index